ELECTRONIC GREYHOUNDS

Electronic Greyhounds

The *Spruance*-Class Destroyers

Capt. Michael C. Potter

Supply Corps, U.S. Naval Reserve

NAVAL INSTITUTE PRESS   ANNAPOLIS, MARYLAND

Library of Congress Cataloging-in-Publication Data

Potter, Michael C., 1950–
    Electronic greyhounds: the Spruance-class destroyers / Michael C.
Potter.
        p.   cm.
    Includes bibliographical references (p.   ) and index.
    ISBN 1-55750-682-5 (alk. paper)
    1. Destroyers (Warships)—United States.   I. Title.   II. Title:
Spruance-class destroyers.
V825.3.P68   1995
623.8′254′0973—dc20   94-22912

Printed in the United States of America on acid-free
paper ⊗

02 01 00 99 98 97 96 95   9 8 7 6 5 4 3 2
First printing

For my two favorite kids, Alice and Joan

Contents

Preface

This book is the history of the *Spruance* (DD 963)-class destroyers and their sisters in the *Kidd* (DDG 993) and *Ticonderoga* (CG 47) classes. The first of these warships entered service in 1975. All of them remain on duty in the new post-Cold War military.

My objective is be complete and accurate in an unclassified work. Technical information provided here gives general readers insight into these ships' past and future operations. Technical details were current as of 1994. My primary sources are official records, professional literature, and interviews with participants. The draft manuscript was cleared for open publication under CNO case numbers 93-221, 94-285, and 94-293. All statements, opinions, and conclusions are mine and do not represent views of the U.S. government or its contractors.

The *Spruance*-class destroyers changed the U.S. military. Despite their operational significance in recent history and in current events, no previous work has accurately covered their history. Many articles about these ships are polemical or flat-out wrong, and major parts of the story have never appeared publicly at all. I wrote this book because I knew that there was a story here, and that it had never been told. I came to appreciate these ships more while writing about them.

Samuel Eliot Morison quoted Polybius that "Historians should be men of action; for without a personal knowledge of how things happen, a writer will inevitably distort the true relations and importance of events." I am unsure whether I qualify as a man of action, but experience in both the military and industry has, I think, given me insight into this subject. During 1978–80 I was supply officer and a repair party officer aboard USS *Elliot* (DD 967). *Elliot* was the fifth destroyer of the *Spruance* class, and while I was abroad she became the first to operate in the Indian Ocean and the Arabian Sea. I have performed military duties on other ships, at the Naval War College, in the Pentagon, and as a defense contracting officer. As a systems analyst and project manager in industry, I lead and have won competitive proposal efforts of the sort in which the *Spruance* class was designed.

Terminology

The U.S. Navy and the Office of the Secretary of Defense referred to the new destroyer class as "DX" in 1966–68 during planning ("Concept Formulation"). DX became DD 963 in 1968 for the design ("Contract Definition") phase. DD 963 became the *Spruance* (DD 963) class in 1970. "*Spruance*" and "DD 963" are identical in meaning. "USS *Spruance*" refers specifically to the first ship of this class.

Every American warship has a unique, usually sequential, hull number within her designation. Thus USS *Bainbridge* (DD 1), authorized by Congress in 1898, was the first U.S. Navy destroyer (DD) and *Spruance* (DD 963) is the 963rd. In this text any ship with a hull number of DD 963 or higher is a *Spruance*-class destroyer. A ship with a hull number of DDG 993 or higher is a *Kidd*-class guided missile destroyer. A ship with a hull number of CG 47 or higher is a *Ticonderoga*-class guided missile cruiser, also called an Aegis cruiser. New warships, of course, often carry forward esteemed names. Different type designations and hull numbers usually designate different ships, even if the name is the same. Table 2.1 identifies modern destroyer classes by hull number. True to nautical lore that renaming a ship invites bad luck, the Navy rarely changes warships' names during their careers. However, ships may get new designations and hull pennant numbers. In particular, USS *Ticonderoga* (CG 47) was ordered as DDG 47 but became CG 47 when the Navy redesignated her as a cruiser.

During the period covered by this book, the term *frigate* changed in meaning in the U.S. Navy. To avoid ambiguity, I use present-day cruiser and destroyer designations when discussing the warships that until 1975 were known as guided missile frigates (DLG-series hull sequence numbers). Before 1975 the ships we now call frigates were designated as destroyer escorts, ocean escorts, or patrol frigates. Again to avoid ambiguity, I refer to these ships as destroyer escorts.

For ease of reading I have omitted the unvarying *AN/* prefix before electronics systems; for example, the AN/UYK-43 computer appears here as simply the UYK-43 computer.

Acknowledgments

I owe foremost thanks to Rear Admiral Nathan Sonenshein, USN (Ret.), and Vice Admiral Thomas Weschler, USN (Ret.), for hours of interviews, notes on key events, and documents on the origin of the *Spruance* class. Rear Admiral Donald A. Dyer, USN, provided detailed accounts of action off Grenada and Lebanon and in the Persian Gulf. Dr. Reuven Leopold and his staff at JJH, Inc., provided thorough insight into the *Spruance*, *Kidd*, and *Ticonderoga* designs. Richard F. Cross III shared his files on the DD 963 procurement history and on the General Dynamics design. H. E. Buttelmann and Thomas Buermann of Gibbs & Cox provided details of their DD 963 design for Bath Iron Works. Commander David Gillingham, USNR, and Commander Gregory Dies, USNR, educated me on combat systems. Dr. Norman Friedman, Mr. Chistopher Wright, and Commander Ken Smith, USN, made important corrections and suggestions about the outline and draft text. The men named in this paragraph are the experts. Their help and encouragement made this book a reality. Any errors are mine.

I am grateful to many more for recollections and assistance: General Al Gray, USMC (Ret.); Vice Admiral Raymond Peet, USN (Ret.); Rear Admiral Roger Horne, USN; Rear Admiral Mark Woods, USN (Ret.); Rear Admiral Richard Henning, USN (Ret.); Rear Admiral George Wagner, USN; Captain Gary Bier, USN; Captain J. R. Baylis, USN (Ret.); Captain Zeke Foreman, USN (Ret.); Captain Richard Cuciti, USNR; Commander Peter Ligman, USNR; Commander Douglas Leathem, USNR (Ret.); Lieutenant (j.g.) Michael Mitchell; Commander Naval Surface Force, U.S. Pacific Fleet public affairs; Commander James Schonberger, USN; Lieutenant Tom Manning, USN; Lieutenant Eric Clayton, USN; Lieutenant Chandler Swallow, USN; Sonar Technician First Class Norman Parker, USNR; Radioman Third Class Pickens, USN; Lee Jackson, one of the first women to serve aboard a DD 963 at sea; Al Lee; Dr. Michael Palmer; Leonard Schwartz; Russell Murray II; David Krepchin; and James Willis.

For the ships' operations I drew primarily on the command history files at the Naval Historical Center and on the archives at the U.S. Naval Institute. The Army Times Publishing Company library supplied texts of *Navy Times* articles on the ships' operations. The libraries at Naval Air Station North Island, Naval Air Station Miramar, and Naval Station San Diego assisted me. I am particularly grateful to the San Diego Public Library for locating obscure works, for access to newspaper clipping files, and for keeping useful books in the stacks. Notes cite specific sources and references. My editors, Mary Yates and J. Randall Baldini, corrected and improved every page of the text. Barbara Boyd and Beth Fitzgibbons at OpNav managed the security review.

I owe special thanks to Paul Wilderson of the Naval Institute Press, who backed this project from its inception. Finally, five close personal friends provided, in different ways, loyal and essential support: Glenn Anderson; Joan and Alice, who endured endless weekends and evenings of Daddy's writing; and above all my father, Dr. Brian Potter, and my wife, Jan. Both read and reread the developing text, shared a reader's insight into it, and continuously encouraged me to pursue this book.

ELECTRONIC GREYHOUNDS

Introduction

Except for six nuclear-powered cruisers, every major American surface combat warship designed since 1964 is either a destroyer of the *Spruance* (DD 963) class or a descendant of it. Before this decade is out the entire American force of surface combat warships could consist of *Spruance*-class destroyers or closely related designs. *Spruance*-class destroyers have made a revolutionary impact on the modern U.S. Navy, principally by introducing gas turbine propulsion, all-digital weapons systems, advanced underwater silencing, and Tomahawk cruise missiles. LAMPS helicopters, systems engineering for warships, integrated logistics support planning, and modular shipbuilding all originated with the *Spruance* class. The *Ticonderoga*-class variant introduced the Aegis combat system and multiple-purpose vertical missile launchers. The *Spruance* class ranks with the battleship *Dreadnought*, the fast aircraft carriers *Furious* and *Lexington*, Germany's U-boats, and the nuclear submarines *Nautilus* and *George Washington* as among the most innovative and influential warship designs of the century. They are also the most controversial American warships built in the last 50 years, perhaps in the last 100 years.

These ships play central roles in present-day naval operations, and several have come to world attention. Since 1975 half of all American warships firing weapons in action have been *Spruance*-class destroyers or their sisters in the *Kidd* and *Ticonderoga* classes. They supported commandos in Grenada, bombarded Lebanon, captured drug smugglers, searched for the wreckage of KE-007, helped to intercept terrorists over the Mediterranean Sea, defeated Libyan and Soviet interference with freedom of navigation, escorted convoys to Kuwait through the Persian Gulf, and destroyed Iranian outposts. One mistakenly shot down an Iranian airliner. These ships fired the first shots of Operation Desert Storm against Iraq. They launched most of the Tomahawk missiles that shattered Iraqi air defenses, communications, and, in 1993, the Iraqi nuclear weapons laboratory and secret-police headquarters. Equipped as electronic surveillance posts, they perform intelligence missions on the high seas and along unfriendly coasts.

Warships are among the most complex and fascinating engineering systems in the modern world. Of course, they deploy guns and missiles. It should be equally obvious that they must support their weapons with utility services, spare parts, combat information, and a skilled and alert crew. They must be mobile over enormous distances and house their crews for prolonged and remote operations. They operate in a hostile environment that includes the sea, electromagnetic and acoustic energy spectrums, and increasingly lethal threats from enemy missiles, bombs, mines, torpedoes, gunfire, and chemical or even nuclear attack. They must be affordable to build and to sustain within the national budget. The *Spruance* class exploits this complex environment. While this book was being completed, the Department of Defense was planning to reduce military forces. Under these plans, almost all ships in the *Spruance*, *Kidd*, and *Ticonderoga* classes will remain in active service well into the 21st century. More heavily armed cruisers and newer, cheaper frigates are being decommissioned. The military clearly values the *Spruance* class. These ships are weapons that work.

Yet even within the military an air of mystery still surrounds these giant destroyers. Even their crews rarely know their ships' full history. In a mystery something is unexplained, in this case the

USS *John Young* (DD 973), seen here in 1993, is one of 31 *Spruance*-class destroyers delivered to the Navy in 1975–83. Subsequently armed with a vertical-launch system, she fired Tomahawk strikes from the Mediterranean Sea against Iraq in 1991. The vertical-launch system is behind the forward gun.

facts to reconcile the ships' outstanding technical and operational success with highly negative accounts of their origin. From the start, opponents condemned the ships, the shipbuilder, their armament (or perceived lack of it), and the policies that created them. Criticism of modern weapons is, of course, commonplace. What is unusual in this case is that critics of the *Spruance* class were mostly naval professionals whom one would have expected to champion advanced new systems. Instead a naval architect on the design team alleged that the design effort was "dishonest," and that the winning contractor had suppressed cheaper and less risky alternatives.[1] A serving senior officer charged in 1971 that "the U.S. Navy does not want the 963-class destroyer as presently conceived," and that the new design was "now, on paper . . . , inferior to competitive Soviet ships which are already at sea."[2] The respected naval annual *Jane's Fighting Ships* reprinted that assessment for years. Congressmen and journalists charged that there was a massive cost overrun. A scholarly study of defense procurement grouped "the C-5A, the Cheyenne helicopter, the DD-963 destroyers, the MBT-70 (Main Battle Tank), the Skybolt missile, [and] the F-111" together as defi-

ciency-ridden "failures."[3] With due respect to the commentators, all the preceding statements (and many more) are wrong. The present book seeks in part to correct the flawed historical record and to show how such errors came into it.

The *Spruance* class epitomizes the developments that altered the Navy, and indeed the entire federal government, after World War II: large standing forces to confront the Soviet Union, central control of these forces, and interdependence between the military and industry.[4] The *Spruance* class was planned, designed, and introduced into service during 1966–75. For the American military—misused in Vietnam, hated by draft-age youth, demoralized from domination by Defense Secretary Robert McNamara, opposed by escalating Soviet military power, and staggering from worn-out equipment, gutted budgets, cost overruns, inflation, drugs, racism, and an absence of clear strategic purpose—this period was the worst of the century.[5] Within the Navy, suspicion lingers that the innovations in the *Spruance* class were symptoms of a bad era. A Navy-sponsored history published in 1989 reflected this attitude, alleging that the *Spruance* procurement was "a disaster for the Navy."[6] Was it a disaster? Do the

The *Kidd* (DDG 993) class of guided-missile destroyers is a variant of the *Spruance* design with the Tartar-D medium-range missile system. Originally ordered as cruisers by the shah of Iran, four ships entered U.S. Navy service in 1981–82. (U.S. Navy photo)

USS *Ticonderoga* (CG 47) introduced the revolutionary Aegis long-range missile and battle-management systems, mounted on the *Spruance* design. Key elements of President Ronald Reagan's competitive strategy against the Soviet Union, 27 Aegis cruisers entered service during 1983–94. (U.S. Navy photo)

ships represent a time of military decline, or the time of military resurgence that followed it?

The *Spruance*-class destroyers were intended to be revolutionary in how they were designed and how they were built. The DD 963 project was in part a political attempt to reform defense contracting and to revolutionize the American shipbuilding industry through the initiative known as Total Package Procurement. DD 963 was the largest Total Package Procurement contract ever issued by the Department of Defense, and the last. Congress and the press later attacked this initiative as a failure and a scandal. The criticism of the *Spruance* class is significant because subsequent American surface warships have been designed to avoid provoking a repetition of the controversy. Valuable innovations of the *Spruance* design, such as reserved areas for future weapons, have actually been lost in newer warship designs as a result.

Finally, the story of the *Spruance*-class destroyers is a fascinating chapter of military history. The *Spruance, Kidd,* and *Ticonderoga* classes have exerted an unusually direct influence on world events. When they were joining the fleet in full force, the United States under the Reagan administration directed naval power against Soviet Russia and other oppressive regimes. President Ronald Reagan exploited American competitive advantages over the Soviet Union to demoralize Soviet leaders and to force them to choose between maintaining their empire or meeting exacerbated domestic needs.[7] It worked, liberating Russia and the Soviet satellites. Naval operations supporting an explicit maritime strategy were one such competitive advantage, exploiting Soviet Russia's geographical vulnerability to threats from the sea. Other naval actions have supported police-action campaigns against erstwhile Soviet clients and independent thug regimes. Qualities of surveillance, communications, armament, reliability, and endurance put the subjects of this book on the cutting edge of contemporary military operations. When the guns begin to shoot, it is no coincidence that *Spruance*-class destroyers are almost always present.

Destroyers during the Cold War, 1948–1962

Destroyers in Postwar Strategy

The history of the *Spruance*-class destroyers begins with national defense strategy and the changing requirements for the U.S. Navy within it. From the end of World War II until the collapse of the Soviet Union, the United States sought to deter encroachments on clients and allies and to contain the Soviet empire, through threat of war if necessary to suppress Soviet military moves. While those objectives stayed constant over decades, the national strategy to achieve them fluctuated frequently and sometimes radically. Warship construction was authorized when the political leadership defined a national military strategy that required a different configuration of the fleet. Changes in strategy provided the impetus to proceed with new warships and the urgency to act quickly.

National strategy was, of course, only one influence. Five policy considerations underlay the origins and development of all post–World War II American surface warships:

- The fluctuating role of convoys and aircraft carriers in national strategic planning, since they required defensive escorts
- The growth of Soviet Russia's military power at sea, including nuclear submarines and anti-ship cruise missiles
- The need to build, operate, and update (economically during peacetime) adequate types and numbers of warships to carry out national military strategy for war
- The costs and opportunities to implement new military technology, in particular to counter Soviet threats
- The effectiveness of warships on contemporary operations, principally off Cuba and Vietnam, and in the Mediterranean after the Soviets built up their fleet there in 1967

At all times officials planning and designing ships for future service had to judge the relevance of the past and the trend of the future in these areas. Additional transient influences arose (notably political interest in the DD 963 procurement as a tool to revolutionize warship design, defense contracting, and the American shipbuilding industry), but these five stand out because they existed before the *Spruance*-class destroyer program began, and they endured or reappeared throughout it. Later, critics of the *Spruance*-class destroyers invariably compared them unfavorably with earlier postwar destroyers. It is instructive to trace the evolution of the postwar destroyer against the background of the governing strategies.

The First Postwar Destroyers

At the end of World War II the United States demobilized but simultaneously devised strategies to wage a new war in case Soviet hostility turned into outright aggression. Anticipating that a Soviet invasion of western continental Europe could be successful but would overextend the Red Army, in 1946 Chief of Naval Operations Admiral Chester Nimitz and his staff planned a sea-based offensive to recover Europe. The Navy would exploit its strength in aircraft carrier–based air power and expertise in amphibious assaults to destabilize the Soviet position by attacking in force on unexpected fronts. Aircraft carrier task forces would attack Soviet ground forces within range in Europe and the Middle East, suppress air defenses obstructing land-based bomber flight paths into the Soviet Union itself, and establish air superiority for amphibious counterattacks in the Mediterranean and the Persian Gulf.[1] The strategy seems to have been to outlast the Soviet Union and to defeat it by a gigantic siege, which had succeeded against Germany in World War I and Japan in World War II.

This strategy engendered little interest outside the Navy, as the almost nonexistent early postwar shipbuilding budgets showed. Even the newest of the several hundred destroyers in the victorious wartime fleet were prewar designs that were already becoming obsolete in the face of technological advances in submarine performance and guided weapons.[2] To stay abreast of developing technology, the small shipbuilding budget paid for the first postwar destroyers: the specialized antisubmarine warfare (ASW) ship *Norfolk* (DL 1, ex–CLK 1) and four antiaircraft destroyers of the *Mitscher* (DL 2, ex–DD 927) class, all ordered in 1948. They all served principally as experimental platforms incorporating wartime research into seaworthiness, high-pressure steam propulsion, automatic-loading guns, long-range radar, and ASW improvements.[3] In addition, several war-construction destroyers were completed as specialized ASW hunter-killer destroyers (DDK and DDE designations).

After the Soviets blockaded Berlin in 1948, the Truman administration authorized the military to plan for war against the Soviet Union, to include use of nuclear weapons. The newly independent U.S. Air Force insisted that rapid advances in aviation, nuclear weapons, and intelligence about the Soviet Union would make a land-based bomber offensive a fast and efficient method of eradicating the Soviet Union as a military power during a war. Under strong congressional pressure, the Truman administration adopted the Air Force strategy, or at least the procurement plans for it, in preference to the Navy's.[4]

Cancellation of the aircraft carrier *United States* (CVA 58) and nine destroyers in favor of Air Force B-36 bombers precipitated the "Revolt of the Admirals" in 1949, when senior naval officers objected before Congress about cutbacks in naval forces.[5] This was one of the many sensational national political outbursts over naval aviation; the Billy Mitchell affair of 1921–25 had been the first. (Future disputes would include the F-111B fighter during 1962–68, advocacy of nuclear-powered aircraft carriers during 1962–63 and 1977–79, and lobbying against the B-2 stealth bomber after 1989.) *Spruance*-class destroyers *Spruance, Arthur*

W. Radford, Conolly, Stump, and *Kinkaid* commemorate officers involved in the Revolt of the Admirals.

The new strategic focus was geographical containment. The Soviets' detonation of a nuclear bomb in 1949 reduced interest in a mobile, sea-based strategy, since one or two nuclear bursts would devastate a beachhead even the size of the Normandy assault. (That argument worked both ways: The defender could not risk concentrating his armies either. Small, dispersed landings therefore might still open a strategic front in the face of nuclear defenses.) With the formation of the North Atlantic Treaty Organization (NATO) in 1949, the United States committed itself to garrison a large ground force in Europe. At least for the key European theater in the late 1940s, the Navy was considered to be primarily a convoy service to reinforce Army and Air Force operations during conventional phases of a war against the Soviet Union. A forward-deployed NATO garrison would hold ports and air bases to launch naval and air strikes. This would buy time for the United States to remobilize its industrial and manpower strength to provide reinforcements after war broke out or was politically deemed unavoidable.

During 1949–51 Chief of Naval Operations Admiral Forrest Sherman reasserted a maritime strategy within the new national strategy. In event of war, naval striking forces would guard Middle Eastern oil fields and confront the Soviet Union with a southern front from the Mediterranean Sea. New nuclear weapons were physically smaller and could be carried by fighter-size aircraft from existing carriers. By 1951 Sixth Fleet *Midway*-class carriers in the Mediterranean Sea carried nuclear bomb components for strikes on Soviet Black Sea submarine bases and other theater targets.[6] General Dwight Eisenhower, NATO supreme Allied commander, now planned to use carrier-based air power to strike Soviet ground forces within range of the Mediterranean and the North Sea.[7]

Mobilization convoys from the United States and Canada to Great Britain, the NATO front, and other bastions would sail under war conditions with naval escorts to defend them against Soviet attack. Since full-scale war production would need about two years to ramp up, transports and modern destroyers could themselves be built during national mobilization. The second series of post–World War II destroyer classes included design features for mass production by industry.

These were 17 destroyer escorts of the *Dealey* (DE 1006) and *Claud Jones* (DE 1033) classes, ordered during 1951–57, and 18 destroyers of the *Forrest Sherman* (DD 931) class, ordered during 1953–56. The destroyer escorts were slow but inexpensive and adequate for convoy-screening duties against submarines of the era. The new destroyer class commemorated Admiral Sherman, who had died in office in 1951. Plans for the *Forrest Sherman*–class destroyers emphasized high speed, over 34 knots, to support aircraft carrier flight operations and to thwart submarine torpedo attack. When actually designing the new destroyers, however, the Navy settled for barely 30 knots for a reason that the *Spruance* class would correct 20 years later:

> The conclusion was reached that the Navy did not want to pay for the increase in ship speed in terms of a ship's size . . . but the cheapest thing by far in a ship is the steel hull. . . . The Bureau [of Ships] could obtain the speed desired in this ship, provided the armament was left alone, *i.e.*, fixed, and the design then completed to give the speed and endurance required. Unfortunately the thinking seems to be that when a large ship is designed it must be loaded up with offensive weapons, with a resultant enormous increase in cost.[8]

Admiral Arleigh Burke, an influential naval planner, ordered that the *Forrest Sherman* class must be rated officially at 31 knots, because the nation knew him as "31-Knot Burke" from his destroyer exploits and thought 31 knots was fast. His policy and nickname were ironic. During World War II then-Captain Burke had refused to order speeds above 31 knots because boiler superheater watches and complicated split-plant steaming would exhaust his engineers right before battle. Admiral William Halsey's staff coined the nickname, which meant "slowpoke," to tease him, since his *Fletcher* (DD 445)–class destroyers were rated at 35 knots.[9]

Upon the outbreak of the Korean War in 1950, President Harry Truman ordered a naval blockade of North Korea. Under international law every section of blockaded coast was put under surveillance by warships, not aircraft, once every 24 hours.[10] Postwar cutbacks had so reduced the Pacific Fleet that at first most of the blockading ships were British. The sea-based amphibious assault at Inchon in 1950 was the strategic turning point of the war. The blockade diverted much

North Korean effort to moving supplies from China down long, vulnerable land routes. Carrier air strikes were essential for destroying enemy targets far behind the front lines and for providing air superiority. In an interesting theater initiative, carrier-based aircraft and Air Force B-29 bombers from Japan combined to raid heavily defended North Korean targets during 1951–52.[11] With no naval threat at sea off the Korean coast, most destroyer actions were shore-bombardment missions. Five destroyers were damaged by mines, and over 20 were damaged by enemy artillery.[12] The continuing blockade helped to coerce the enemy into signing an armistice in 1953.

Life aboard destroyers after the war, reported Lieutenant (junior grade) H. Ross Perot, chief engineer of USS *Sigourney* (DD 643) in 1954, was godless and debauched, a floating version of the cities of the plain.[13] Radar picket destroyers (DDR designation) patrolled to detect marauding Soviet aircraft well out to sea. Transports carrying nuclear warhead components to test sites in the Marshall Islands sailed in convoys under full wartime conditions, blacked out, radio-silent, and with destroyers guarding them.[14]

The Surface Missile Ships

The Korean War showed that the United States would need two years to build a force to challenge the Soviet Union, and concern grew that the Soviets might be able to start and end a war in less time than that. The success of aircraft carriers off Korea, their nuclear quick-strike capability, and a much larger defense budget led to construction of the supercarriers of the *Forrestal*, *Kitty Hawk*, and *Enterprise* classes, ordered during 1951–57, and modernization of older carriers.[15] President Eisenhower established a new national defense strategy in 1953, the "New Look," promising prompt nuclear retaliation in the event of war in Europe. The goal was to deter Soviet aggression and, if deterrence failed, to avert disaster at the front by blunting the Soviet ground drive and to cripple the Soviet Union by preemptive strikes while the United States mobilized. President Eisenhower thought that nuclear weapons were all that deterred the Soviets. The New Look emphasized force deployments to frustrate a Soviet knockout blow. Duplicate weapons and overlapping service missions were accepted as the costs necessary to accelerate deployment of new weapons to ward off a new world war, as were globe-polluting nuclear tests and appalling human experiments. Aircraft carriers were among the bases designated to launch initial nuclear retaliatory strikes.[16] Dedicated hunter-killer task groups centered on ASW aircraft carriers (CVSs) would advance ahead of the attack carrier groups into the eastern Mediterranean and the Norwegian Sea to clear them of Russian submarines.[17] Destroyers were essential to both types of carrier task group. They were to sink submarines and to patrol around, and especially ahead of, the attack aircraft carriers to prevent hitherto undiscovered submarines from gaining firing positions.

Every large U.S. Navy surface combatant built after 1948 has been designed as an aircraft carrier escort. Accompanying the revival of aircraft carrier task forces in national strategy, during 1956–62 the Navy ordered 54 large guided missile–armed cruisers, frigates, and destroyers as fast task force escorts, and converted 9 World War II cruisers to mount guided missile batteries. New ships were the *Farragut* (DLG 6), *Charles F. Adams* (DDG 2), *Leahy* (DLG 16), and *Belknap* (DLG 26) classes, and the nuclear-powered *Long Beach* (CGN 9), *Bainbridge* (DLGN 25), and *Truxtun* (DLGN 35). Originally the Navy typed the DLG/DLGN classes as cruisers, which traditionally bore the names of cities. USS *Bainbridge*, which predated the *Leahy* design despite her hull sequence number, was originally to have been designated a cruiser and named USS *New York City*. Concerned that Congress would think "cruiser" too expensive, the Navy switched the designation to "frigate," which it described as a type of destroyer and thus would bear a late naval hero's name. The first choice was a five-star fleet admiral, but those not already so commemorated were still alive. The Navy's first destroyer, DD 1, had been named *Bainbridge*, so this name was chosen to help characterize the new ship as a destroyer.[18]

The surface missile ships would provide air defense against land-based bombers, both by directing carrier-based interceptors onto distant approaching targets and by shooting down aircraft that might intrude as far as the task force screen. Their main batteries of surface-to-air missiles (Talos, Terrier, and Tartar) would be the first line of defense when weather was too severe for flight deck operations, or if many fighters were rearming or on missions elsewhere, or if the carrier was not on alert.

In fleet exercises the new weapons did not necessarily perform to designers' expectations. A ship with a Terrier guided missile battery needed to detect and classify targets at up to 150–250 miles' range, and to track them actively in from 100 miles to hit them by the time the targets closed to missile-intercept range of 10–20 miles. One ship could keep at most four targets engaged at once. These systems could be overwhelmed by aircraft attacking at low altitude, by missiles popping up from a submarine or from a ship emerging from a nearby coast, or simply by a mass air assault.[19]

The missiles and guidance systems were almost experimental. Old shipboard electrical and air-conditioning plants provided a lethal environment for delicate electronics. Constant engineering changes defeated attempts to provide the supporting inventory of spare parts essential for reliability of the new weapons.[20] Army Nike-Hercules and Nike-Ajax surface-to-air missile systems had the same problems and on the average were operational about one hour a day. The crew of the destroyer *O'Hare* (DDR 889), testing an early version of the Naval Tactical Data System (NTDS), found it spectacular to see half the North Atlantic displayed from the composite input of a half-dozen ships, when the system worked, which was about 10 percent of the time.[21]

By 1956 Soviet development of transportable thermonuclear bombs made massive-retaliation strike plans obsolescent, since Soviet long-range bombers with their weapons could scatter to hundreds of airfields before a U.S. attack arrived, and indeed they might even strike first. Soviet submarines equipped to fire nuclear land-attack missiles entered service. The need to escape a Soviet first strike prompted the development of solid-fueled fast-launch ballistic missiles, including Polaris, and of aggressive preventive-war plans. Reasoning that the Soviets would devastate the United States despite retaliatory nuclear strikes, Admiral Arleigh Burke, now Chief of Naval Operations, recommended building a Polaris submarine fleet as an indestructible strategic reserve weapon. President Eisenhower regarded Polaris as an alternative not to Air Force bombers for destroying Soviet cities and command centers, but to aircraft carrier task forces, in particular for suppressing Soviet air defenses at the outbreak.[22] After 1957 President Eisenhower sanctioned building just one aircraft carrier, the oil-fueled *America* (CVA 66).

The surface warship fleet lost its offensive role after the Navy canceled the Regulus ship-to-shore nuclear-warhead cruise missile project in 1958 in favor of Polaris. Destroyer sailors still looked to their tradition of torpedo boats, valuing speed, maneuverability, firepower, and compactness in destroyers, the greyhounds of the sea.[23] The effect of these decisions became apparent in the fleet configuration of the early 1960s.

FRAM and DASH

The urgency of need for antisubmarine escorts led to the Fleet Rehabilitation and Modernization (FRAM) programs for World War II destroyers. FRAM I began in 1958 after the successful modernization of six *Fletcher* (DD 445)–class destroyers for transfer to Germany.[24] Under FRAM I, 79 *Gearing*-class destroyers were rebuilt to mount a modern ASW armament: the SQS-23 medium-range attack sonar, ASW homing torpedoes, DASH drone helicopters, and an ASW rocket (ASROC) box launcher for eight rocket-thrown torpedoes and nuclear depth charges. The FRAMs, as the modernized destroyers became known, were successful and popular warships in service. A FRAM could make almost full speed on three of her 600 psi boilers, so that the engineers could work on the fourth at sea while the ship still met operational commitments. That allowed her engineers time off in port. Another 52 wartime destroyers underwent the less extensive FRAM II ASW conversion, receiving DASH and a towed variable-depth sonar rather than the new hull-mounted SQS-23 sonar. Early variable-depth sonars shared the shipboard electronics of the old hull-mounted attack sonar, a configuration that permitted use of only one sonar at a time. A contact held by variable-depth sonar could be lost when the ship switched to her hull-mounted sonar for attack. DASH was another story.

DASH was a small remotely controlled drone ASW helicopter. The DASH mission was to attack any submarine that a destroyer detected attempting to penetrate an ASW screen around a task group or other shipping formation, without alerting the submarine to take evasive action.[25] Hopes were high for DASH, and all the FRAMs and new destroyer escorts received hangars and flight decks for the drones. In service, correlation of sonar and radar plots aboard ship, necessary to guide the helicopter to the contact area, was shaky. DASH could not locate or classify contacts. It proved almost impossible to guide because it lacked a feedback loop to show the operator how its flight surfaces were set. The radio data link was vulnerable to environmental radio-frequency noise and

to electronic jamming. Reports told of drones flying upside down or crashing into ships. Over half of the 746 DASH drones were lost at sea.[26] Exercise kills of submarines were negligible. The failure of DASH left the destroyer force without helicopter capabilities, since the DASH hangar was too small and the flight deck too weak for a manned helicopter.

The FRAMs' strengths were gun firepower against land targets, a simple and robust propulsion plant, and short-range ASW effectiveness. Their limitations included hull strength, seakeeping, habitability, long-range ASW, and antiaircraft defense. They carried air search radars but had almost no capability to defeat jet aircraft.[27] Further modernization of the FRAM destroyers was not feasible because they were too noisy and had too little remaining internal space, stability margin, and electrical and cooling capacity to support new radar, sonar, and display systems. This created a significant block-obsolescence problem, since the majority of the Navy's surface ASW and gunfire support ships were FRAMs.[28] The need to build a large number of new destroyers would lead to the DX program and the *Spruance* class.

New ASW Ships and Task Force Escorts

In 1959 Navy planners decided that all new fleet escorts should mount the long-range SQS-26 sonar, which was too large to backfit to existing ships.[29] New fast submarines could approach targets quickly and attack them with new longer-range weapons. Destroyer-type ships had to be able to detect and to fight submarines at longer ranges. Longer-range sonar permitted greater separation between escort ships, so that fewer were necessary. All the new surface missile ships mounted ASW weapons in addition to their air defense armament. The growing Soviet strategic nuclear arsenal and submarine fleet made obsolete the earlier mobilization plan to build transports and convoy escorts after the outbreak of a war, hitherto expected to be long. This led to the construction of a large series of ASW destroyer escorts that would be ready at the outbreak. These were the *Garcia* (DE 1040), *Brooke* (DEG 1), and *Knox* (DE 1052) classes, designed during 1958–64 and ordered during 1960–67.

The primary war missions of the destroyer escorts were, officially, to operate with the ASW aircraft carriers (CVSs) in ASW hunter-killer groups, and to defend replenishment ships and

The Soviet Union built Echo-II class nuclear submarines to attack U.S. carrier task forces with salvos of high-flying supersonic cruise missiles. Soviet patrol aircraft would locate targets and in effect would command missile-launching bombers, submarines, and surface warships into intercept positions. (U.S. Naval Institute collection)

amphibious assault groups against submarine attacks. Hunter-killer groups would screen attack carriers during early-war nuclear strikes, and afterward would provide midocean convoy defense. *Essex*-class ASW carriers carried a four-aircraft fighter-bomber detachment (usually A-4 Skyhawks) for air defense and surface attack, indicating that midocean ASW task groups did not anticipate much air or surface action. The destroyer escorts similarly mounted minimal air defense and antiship armament. Their most significant feature was normally invisible: the long-range SQS-26 sonar mounted beneath the bow. They were slower, ASW-oriented counterparts of the surface missile ships, whose high costs meanwhile inevitably shrank the fleet as aging warships from World War II were retired with fewer replacements. To offset the reduction in fleet numbers, the successive new destroyer escort classes featured more powerful boilers and mounted heavier armament so that they could perform some destroyer missions requiring better speed and gun firepower.[30]

The *Leahy* and *Belknap* classes of guided missile cruisers and the *Knox*-class destroyer escorts comprised a new series of surface warships.[31]

They were much larger than their predecessors, to increase internal volume for improvements demanded by the fleet. They had greater fuel capacity, internal torpedo tubes, internal space for a variable-depth sonar, a magazine and loader for the ASROC launcher, and a rapid-firing semiautomatic 5-inch 54-caliber gun to supersede the old hand-loaded 5-inch 38-caliber guns. Other improvements included a larger and more efficient combat information center layout, better habitability, larger storerooms, and better access to equipment.[32]

ASW forces included the postwar destroyers, destroyer escorts, FRAMs, nuclear submarines, land-based maritime patrol aircraft, and helicopters and S-2 Tracker aircraft flying from CVSs. Nuclear submarines could set barriers between islands and across the approaches to the North Atlantic and other oceans to trap Soviet submarines attempting to penetrate south or to return to base. The combination of maritime patrol aircraft and the seabed-mounted Sound Surveillance System (SOSUS) reduced the need for midocean ASW patrols by CVS-centered hunter-killer groups. Enemy submarines faced destruction in offensive minefields, by strikes on their bases and anchorages, or during transit by attacks from American submarines and patrol aircraft.

The Rise of the Soviet Navy

In the Soviet Union Premier Nikita Khrushchev consolidated power after Joseph Stalin's death, and in 1956 he promoted Admiral Sergei Gorshkov to command the Soviet Navy. Khrushchev and Gorshkov canceled Stalin's program for a Nazi-style fleet of cruisers and submarines in favor of an innovative new fleet to deter the United States from honoring its commitments to NATO and other alliances. The U.S. Navy was built around aircraft carriers and the need to sail them to within aircraft striking range of their targets on land and at sea. Admiral Gorshkov built the Soviet Navy around missiles and the need to place missile-launching submarines, ships, and land-based naval bombers within missile range of their targets. The Soviets armed their new fleet for nuclear strikes against the United States. By 1962 nearly 50 Soviet submarines mounted nuclear land-attack missiles. All Soviet submarines could fire nuclear-warhead torpedoes into harbors, in particular military embarkation and debarkation ports.[33] Submarines for these missions had entered service in 1956, several years before Polaris.[34]

The American aircraft carrier task forces' nuclear strike threat against the Soviet Union forced a shift in Soviet national strategy for use of its fleet. The Soviet Navy's contribution to national defense would be to sink or disable the aircraft carriers before they approached within air strike range. Fast nuclear-powered submarines carried long-range nuclear torpedoes to fire into ports or into advancing task force formations from ahead of the destroyer screen. The Soviets maintained another 350 modern torpedo attack diesel submarines. Adapting cruise missile technology for antiship attack, during 1958–61 the Soviets introduced seven types of antiship cruise missiles. They built more than 50 submarines and surface warships mounting their first-generation long-range antiship cruise missiles (Soviet P-5, P-7, and P-25; all called SS-N-3 by NATO). It is likely that most, perhaps all, of the operational missiles carried nuclear warheads. Since the missiles had much greater range than the surface search radar of the launching ships and submarines, Soviet naval aviation received land-based bombers for long-range reconnaissance, for targeting, and for launching more antiship missiles.[35] American intelligence was aware of Soviet technological developments. Work began in 1955 on two new guided missile systems—the shipboard Typhon and airborne Phoenix (originally Eagle)—to defeat saturation attacks by Soviet bombers and cruise missiles.[36]

Despite the buildup, Soviet officers doubted that they could defeat alert Western naval forces in battle.[37] Admiral Gorshkov's strategy therefore was a Battle of the First Salvo. Soviet antiship attack forces would concentrate and open fire during the ambiguous peace of a crisis, to destroy American warships before they were released to attack Soviet forces. Soviet military articles on naval tactics after 1959 emphasized First-Salvo strikes with nuclear missiles, to destroy multiple ships in one long-range attack.[38] Omnipresent Soviet tattletale surveillance ships showed the U.S. Navy that such targeting could be feasible. The small number of deployed American aircraft carrier task forces could be targeted almost all the time, making imminent First-Salvo attack indistinguishable from normal surveillance and harassment. Details of Soviet destroyer designs hinted that the Soviets planned to wage chemical and biological warfare as well as nuclear.[39]

The Cuban Missile Crisis

In 1960 the first integrated nuclear target plan, SIOP-62, still tasked aircraft carrier task forces with strategic nuclear strike missions, until more Polaris submarines became operational. Shortly thereafter John F. Kennedy, a Navy war hero who wore a PT 109 tie clip until the day he died, became President. The early years of the 1960s were a high-water mark for the Navy as Admiral Burke's work came to fruition, although its underlying strategy was being phased out. *Enterprise*, the first nuclear-powered aircraft carrier, joined the fleet, as did the ill-fated first fast attack submarine of the *Thresher* class, the fossil-fueled aircraft carriers *Kitty Hawk* and *Constellation*, and the nuclear-powered guided missile cruisers *Bainbridge* and *Long Beach*. A new generation of aircraft—the F-4 Phantom II, A-5 Vigilante, A-6 Intruder, E-2 Hawkeye, P-3 Orion, H-2 Seasprite (then a single-engine utility helicopter), and SH-3 Sea King—entered service.

Soviet naval expansion was under way, though not far advanced, when the Soviet Union began shipping bombers and nuclear ballistic missiles to Cuba during 1962. Khrushchev was aware that American reconnaissance satellites were discovering that his nuclear menace to the United States was weak, and he acted to strengthen it quickly. President Kennedy ordered a naval quarantine of Cuba in October 1962, in preference to air strikes on the Soviet weapons. The original blockade line was 800 miles from Cuba, beyond the range of Soviet aircraft based there. Kennedy moved the line inward to 500 miles, giving Khrushchev more time to turn his transports around before they encountered American destroyers.[40] A fleet of 180 warships surrounded Cuba to search Soviet-bloc transports at sea for offensive weapons. If a transport tried to run the blockade, destroyers were to fire warning shots or, if necessary, to disable her rudder by gunfire. The Navy let ships not capable of carrying offensive weapons pass, but it challenged others. The sole Soviet freighter carrying nuclear warheads stopped far short of the quarantine line.

The new aircraft carrier *Enterprise* was assigned to guard Guantánamo Bay. She made long-range nighttime dashes, outrunning her escorting destroyers, to frustrate any Soviet-Cuban attempt to locate the already renowned warship by daylight searches of her last area of operations.[41] Air defense was a concern for aircraft carriers, in particular during the ambivalent peace of an international crisis, when an approaching military aircraft might be engaged in routine reconnaissance or in a preemptive attack. *Enterprise*'s SPS-33 height-finding radar, critical for vectoring fighters to targets at multiple altitudes, was probably inoperable during the missile crisis. This was a new electronically scanned phased-array radar, but it never achieved full operation and was discarded in 1979. Its capability and problems with its reliability influenced the later Aegis project.

The Soviets deployed five diesel submarines near Cuba, then a very rare out-of-area operation for the Soviet Navy.[42] Antisubmarine warfare denied the Soviets their only potential military use of the sea during the missile crisis. President Kennedy regarded the ASW operation as the most intense moment of the crisis.[43] He ordered that destroyers were to track the Soviet submarines and to force to the surface any that submerged. Destroyers hounded them with sonar and explosive signal charges until the submarines' batteries and air supplies gave out. The destroyer *O'Hare* quickly surfaced one submarine, but another tried high-speed underwater dashes, silent running, air-bubble decoys, and other maneuvers to evade sonar prosecution. *O'Hare* and other destroyers held contact for over 24 hours, but the submarine refused to surface. A P-3 maritime patrol aircraft sortied from Florida to the scene, and the destroyers moved off to break sonar contact. Deceived, the submarine surfaced under the lurking P-3's cameras within an hour.[44]

Khrushchev seemed as surprised as if the naval quarantine had intercepted rockets en route to the moon. To get the missiles out, President Kennedy conceded to withdraw obsolete nuclear strike missiles from Turkey and gave "assurances against an invasion" of Cuba that were so conditional as to be no assurance at all. The Soviets removed all the bombers entirely under naval pressure. Departure by sea of the bombers, over Fidel Castro's enraged protest, was the quid pro quo for lifting the quarantine.

It was an article of faith within the U.S. Navy that it had taught the Soviet Union a lesson. Admiral Isaac Kidd Jr., later a commander of the Sixth Fleet, looked back and wrote:

> The single most important event leading to the rapid buildup and change in character of the Soviet Navy was the Cuban missile crisis of 1962. . . . [It] taught them a lesson that they appear to have learned well. From that time, the Soviet Navy has been transformed from a primarily coastal defensive force to a blue-water power through construction of longer-legged combatants and support forces capable of sustaining them in open-ocean operations.[45]

In fact the Soviets' naval buildup had been under way for years to meet their own strategic needs. The quarantine showed the value of surface warships in controlling a crisis before an opponent could escalate to combat. It may have influenced plans for the *Knox*-class destroyer escorts, ideal ships for the quarantine missions of antisubmarine warfare and naval presence. Yet despite the spectacular display of sea power during the missile crisis, the stature and relative strength of the U.S. Navy were eroding. Concentration on fighting a nuclear war, the threat of massed cruise missile attacks on aircraft carrier task forces, patrols by Soviet missile submarines off American coasts, and a necessarily increased emphasis on defensive ASW all combined to throw the Navy on the strategic defensive.[46] In the strongest challenge to the vitality of the Navy since before the Korean War, the carrier-based combat fleet was losing its national strategic role.[47] A measure of this was that construction of destroyers, the mainstay of the fleet, came to a practical halt for a decade after 1962. Construction would resume with the *Spruance* class.

Destroyers on Trial: McNamara, Rickover, and Vietnam

The McNamara Revolution

The *Spruance* class introduced a technological revolution after 1970 in part because a 10-year interruption of surface warship development had delayed the adoption of weapons and systems that could have entered service before then. Since the interruption largely coincided with the 1961–68 term of Secretary of Defense Robert McNamara, naval lore blames him for it. That conveniently obscures the underlying problem of the Navy's own weak response to the need for better weapons quality. The problems with missile systems, and later with high-pressure steam boilers, eroded both confidence and interest in building additional high-complexity surface warships. One reason a radical approach to the design and construction of the *Spruance* class became attractive was that the Navy's surface warship plans of the 1960s were disappointing.

Advocates of greater service independence today project McNamara's strong but deeply flawed personality onto the Office of the Secretary of Defense (OSD), which they charge has always been anti-Navy. Critics associate McNamara with changes, in particular the decline of surface warships in national strategy in 1960, that were in fact well under way before he took office. The *Spruance* class's origin as a product of McNamara's policies would make the ships future targets for especially harsh criticism from Navy partisans.

In 1958 Congress had transferred all military procurement authority to OSD to strengthen central control over the military forces. This was essential to respond quickly to the threat of a surprise Soviet nuclear attack. Further, President Eisenhower was dissatisfied with the services' obedience to his policies and wanted OSD to control weapons programs. McNamara took office at age 44 in President Kennedy's New Frontier cabinet in 1961 and was committed by the President to strengthening the military without increasing the defense budget's share of national expenditures.[1] OSD soon dominated the other branches of the Department of Defense, namely, the three uniformed services and the Joint Staff.

The Kennedy administration rejected the Massive Retaliation strategy of initial nuclear strikes and deleted carrier-based nuclear strikes from revisions of the SIOP (single integrated operational plan). The new strategy was Flexible Response, in which the United States would respond with force at the same level initiated by an aggressor. A nuclear exchange might still occur but could be withheld while conventional forces attempted to contain a crisis. Whether or how Soviet naval power would threaten Western strategic interests during a crisis or war was a matter of dispute. Even 20 years later an authority wrote, "Most studies on the Soviet Navy have assumed that it has been built and operated under the influence of relatively short-term goals and objectives. In that context, tactical changes have been interpreted as major departures in Soviet planning."[2]

The result of American military uncertainty about Soviet naval capabilities and objectives was political uncertainty about the size and type of forces needed to control the sea and to exploit it against Soviet opposition. This uncertainty implicitly challenged the Navy to justify the expense of maintaining a powerful fleet, in particular aircraft carriers and their escorts. With the growth of the U.S. fleet ballistic missile submarine force, and given the exposure of ships to Soviet submarine and cruise missile attacks, OSD doubted that aircraft carrier task forces were cost-effective. They could not contribute enough in a nuclear war against the Soviet Union to be worth the investment in ships, aircraft, equipment, and manpower necessary for future success against the strengthening Soviet threat. OSD described its evaluation of carrier task force effectiveness to Congress early in 1962:

> The principal use of aircraft carriers in the years ahead will be in the limited war role. As we acquire larger forces of strategic missiles and Polaris submarines, the need for the attack aircraft carrier in the general war role will diminish. However, they will still maintain a significant nuclear strike capability which could augment our strategic retaliatory forces. But in the limited war and Cold War roles, the attack carrier force provides a most important and unique capability.[3]

Professional naval officers agreed with this definition, but it described what was a secondary strategic priority. After 1957 the Navy hoped to build aircraft carriers at a rate of one every three years. President Eisenhower had rejected nuclear propulsion for the CVA 66 aircraft carrier, ordered in 1960, considering such a ship to be useless in a large war and unneeded in a small one.[4] After operational experience with *Enterprise* during the Cuban missile crisis in 1962–63, the Navy requested additional funds to convert the projected CVA 67 aircraft carrier design to nuclear reactors. Advantages of nuclear propulsion for aircraft carriers included sustained high speed, freedom from stack-gas corrosion of aircraft, better antenna locations, reliability, and fewer diversions from operations to refuel the ship. The issue was whether the ship's mission in the new national military strategy was significant enough to justify the additional investment for nuclear propulsion.[5]

Political strength decided the issue. Chief of Naval Operations (CNO) Admiral George Anderson had impressed President Kennedy as unreliable during the missile crisis quarantine. Admiral Anderson had considered it to be a purely military operation directed against Soviet ships, rather than as coercive diplomacy directed against Nikita Khrushchev.[6] He was dismissed during the CVA 67 debate, as was Secretary of the Navy Fred Korth over other issues. (Three of the five preceding CNOs—Admirals Louis Denfield, William Fechteler, and Robert Carney—had also been prematurely dismissed as unsuitable.) Secretary McNamara ordered that CVA 67 would be fossil-fueled.[7]

Navy officers were offended that OSD held their experience of little account in the evaluation. As in 1947–49, they feared that loss of the Navy's institutional authority, such as authority to decide its own ships' characteristics, would relegate the service and themselves to a minor strategic role. Relations between the Navy and OSD threatened to become poisonous.[8] Officer resignations doubled between 1963 and 1964.[9] President Kennedy, however, had been disturbed by the enthusiasm of the Joint Chiefs of Staff for attacking Cuba even at the risk of nuclear war. Afterward he regarded Secretary McNamara, who had the services under control, as the most important member of his cabinet.[10] This reputation stayed with McNamara when Lyndon Johnson became President upon Kennedy's assassination in November 1963. The controversial CVA 67 became USS *John F. Kennedy*.

Today a common Navy view is that the McNamara era was a time of decline.[11] During his tenure budget cuts averaged nearly 22 percent of the services' budget estimates annually, demolishing plans and priorities developed within the services.[12] Shipbuilding cuts exceeded $30 billion during 1962–68, while block obsolescence was removing large numbers of World War II–built ships from duty.[13] Cynicism flowed down from senior officers that the Secretary of Defense was undermining the Navy.[14] Admiral Elmo Zumwalt later told Congress, "If you look at the years 1962–1972, in its shipbuilding appropriations the Navy was down to less than $1 billion per year at a time we should have been spending $3 billion a year on new ships. . . . The effect of the Vietnam War has been, in essence, to cost us the equivalent of about a generation of shipbuilding."[15] With the decline of aircraft carrier battle groups in national strategic plans, naval construction during the 1960s shifted sharply toward defensive ASW and protection of shipping. Rear Admiral Wayne Meyer wrote, "Surface warships were seen merely as defensive forces: their future role blurred and ill-defined. This era saw not a single new missile, launcher, radar, or secondary defense landed in the surface combatant."[16]

Table 2.1 summarizes warship construction during three successive seven-year periods. It shows ships built under the programs of the dominant officials of each period: CNO Arleigh Burke, Secretary McNamara, and CNOs Thomas

Table 2.1 Navy shipbuilding programs, 1956–76

The relevant decision period was about two years before the fiscal year; for example, FY 1969 was planned during calendar year 1967. FY 1961S–62S were supplemental appropriations to accelerate Polaris submarine construction.

New Ship Construction	FY 1956–62 (Burke)	FY 1961S–62S and 1963–69 (McNamara)	FY 1970–76 (Moorer, Zumwalt)
Attack submarines	25	36	27
Ballistic missile submarines	19	22	4
Aircraft carriers	4	2	2
Cruisers and destroyers	61	2	34
Destroyer escorts (frigates)	12	54	10
Amphibious assault ships	10	44	4
Combat auxiliary ships (tenders, logistics, salvage)	8	33	5
Totals	149	183	86

Moorer and Elmo Zumwalt. The Navy's mission certainly fluctuated, but the shipbuilding record of the McNamara era does not show overall decline. Concern about decline in the 1960s centered on the striking forces: carriers, cruisers, and destroyers. The decline in their construction is explained by ineffective weapons, Navy infighting over nuclear propulsion, and deletion of the aircraft carriers' nuclear strike mission from the SIOP in April 1964.

The Tonkin Gulf Incident

For most Americans the Vietnam War began with the destroyer action known as the Tonkin Gulf incident. The Kennedy administration had initiated covert actions against North Vietnam to induce it to abandon its assault on South Vietnam, without provoking Chinese intervention as had happened in Korea. Destroyers were to obtain communications intelligence of North Vietnamese infiltration by sea and to locate coastal defenses and radars to be attacked, or avoided, during small-boat raids or air attacks.

On July 30, 1964, the *Allen M. Sumner*–class destroyer *Maddox* (DD 731) recorded North Vietnamese radar and radio activity stirred up by a U.S.-sponsored small-boat raid. During the night of August 1–2 *Maddox* intercepted radio traffic revealing that the North Vietnamese were planning hostile retaliation against her. That afternoon three radar-equipped Soviet-built P-4 PT boats approached *Maddox* at over 40 knots. *Maddox* turned away at 27 knots, went to general quarters,

and, having no doubt that she was under attack, opened fire at 9,800 yards. The PT boats closed and fired three torpedoes from 3,000–5,000 yards, which *Maddox* evaded. Boat T-339 closed to 1,700 yards and sprayed machine-gun fire at the destroyer. *Maddox*'s shellfire severely damaged her, and the other PT boats turned away. As they retreated, four F-8 Crusader fighters from USS *Ticonderoga* (CVA 14) attacked them, sinking the crippled T-339. One F-8 was damaged. *Maddox* fired 283 shells during the action and was hit by one penetrating bullet. North Vietnamese public radio broadcasts quickly reported this incident and praised the two surviving boat crews for supposedly driving the destroyer away.

Maddox remained on patrol, reinforced by USS *Turner Joy* (DD 951), the newest and last destroyer of the *Forrest Sherman* class. After dark on August 4, both destroyers classified a pattern of radar echoes as a surface contact approaching from the north at 38 knots. *Turner Joy* opened fire when the range closed to 10,000 yards. The contact emitted no radar or radio signals and appeared to turn away at 6,000 yards. *Turner Joy* continued to shoot at radar echoes, but *Maddox*, whose radar operators had fresh experience in tracking PT boats, evaluated nothing as a contact, nor did aircraft that had arrived in support. *Maddox* fired about 60 rounds at echoes reported by *Turner Joy*. The destroyers steered ramming courses across their wakes and even dropped shallow-fuzed depth charges in case any PT boats were trailing. *Turner Joy* claimed gunnery hits, but at dawn she found no debris or oil slicks. A captured North Vietnam-

An officer aboard USS *Maddox* painted the old destroyer's close-quarters fight against North Vietnamese PT boats in the Tonkin Gulf in August 1964. The action gave President Johnson the pretext he sought for escalating the Vietnam War. (courtesy Cdr. Robert J. Ference, USN [Ret.])

ese naval officer in a position to know later denied that any boats had sortied that night. The radar echoes were evidently normal sea returns from waves.[17]

Telegraphic messages from the ships during this action arrived in Washington during the workday, and eager OSD staffers grabbed them.[18] Acting on the initial contact reports without permitting any analysis by the Navy, President Johnson ordered carrier aircraft to raid several North Vietnamese naval bases and a fuel depot on August 5. Secretary McNamara, again without review by the Navy, gave Congress an inaccurate and misleading account of the Tonkin Gulf incident. Congress passed the Tonkin Gulf Resolution on August 7 to authorize President Johnson to use military force in Southeast Asia. White House staffers had written it months earlier for use at a suitable opportunity.

The Tonkin Gulf incident provoked the most divisive series of constitutional crises since the Civil War. These included the Vietnam War, the Pentagon Papers case, part of the Watergate scandal, and the War Powers Resolution.[19] In publishing the leaked Pentagon Papers in 1971, the *New York Times* chose to start with the Tonkin Gulf incident as the most sensational episode in the whole 23-year chronicle. It is speculation whether an independent Navy analysis really would have dared to disagree with OSD and the White House. The Navy's public account of the *Vincennes* incident in 1988 has a similarity to the Tonkin Gulf incident: silence.

Destroyer Operations during the Vietnam War

During the Vietnam War, destroyers were stationed along the enemy-held coast for surveillance, antiair warfare, anti–PT boat operations, helicopter refueling, aircraft navigation reference, and combat search and rescue (SAR) of downed pilots under fire.[20] Vietnam showed that warships required the Naval Tactical Data System (NTDS) to track jet aircraft. The most capable destroyers and cruisers operated close to the enemy coast on Positive Identification Radar Advisory Zone

(PIRAZ) station. A ship on PIRAZ required NTDS, an aircraft navigation beacon (TACAN), three-dimensional air search radar (bearing, range, altitude), and multiple radio channels for air intercept control of fighters.[21] Ships operating off North Vietnam on PIRAZ and other missions had to be prepared to defeat simultaneous attacks by aircraft and PT boats. Often it was necessary for two ships to cover each other on a single mission. A guided missile cruiser might embark a combat search-and-rescue helicopter (usually an armored, machine gun–armed HH-2 Seasprite) and provide air intercept control and antiaircraft defense, with a gun-armed destroyer in company for anti–PT boat defense and shore bombardment.[22]

Destroyers operated off the coast of South Vietnam on the Operation Market Time blockade and provided gunfire support for troops inland. Rear Admiral Thomas Weschler, commanding the naval support base at Da Nang, learned from experience there that destroyers needed versatile, piloted helicopters for reconnaissance, search and rescue, cargo and passenger transfers, and other missions. Later, this thought would be the stimulus for the LAMPS program and the large helicopter deck on the *Spruance* class, as will be seen. The semiautomatic rapid-fire Mark 42 5-inch/54-caliber gun mount disappointed destroyer crews:

"Experience in Vietnam has demonstrated time and again that it takes more than one gun in a ship to provide effective fire support. Indeed, when it comes to the [Mark 42] 5″/54 caliber gun, experience has shown that it frequently takes two guns just to keep one firing," wrote Captain L. D. Caney, commander of Destroyer Squadron 5, in a letter to the U.S. Naval Institute *Proceedings*.[23] (The Naval Institute is the naval officers' professional organization, and the letters section of its monthly journal *Proceedings,* "Comments and Discussion," is the Navy's unofficial public-access bulletin board.) It was often prohibitively difficult for a Mark 42 mount crew to fire mixed salvos of star shell and high-explosive shell for naval gunfire support (NGFS) of ground forces at night.[24] Nonetheless, the amphibious forces appreciated a 5-inch/54-armed destroyer:

> The main advantage of the 5″/54 over its older sister, the 5″/38, in shore bombardment is range (about 13 miles, rather than 8.5), not rate of fire. There are few situations in South Vietnam, particularly in the IV Corps area [the Mekong Delta], where rapid, sustained fire would make the difference between success and failure. But there are many places in the delta where shallows extend far out to sea,

The handsome *Belknap*-class cruisers supported air strike operations against North Vietnam. Although 40 percent larger than the *Farragut* class, their installed armament was similar. The extra size went to more fuel, NTDS, larger magazines, longer-range sonar, and a facility for a small utility helicopter. These improvements made them much more useful warships. (U.S. Naval Institute collection)

making it impossible for the shorter-range 5″/38 to hit targets that are not close to the shoreline.[25]

Objections to the New Destroyer Escorts

Sailors believed that the *Brooke*-class destroyer escorts (designed by the Navy during Admiral Burke's tenure as CNO) epitomized Secretary Mc-Namara's view of warships. The class had one system each of many complex types: one SQS-26 sonar, one 3-D air search radar, one missile guidance system, one gun, one high-pressure steam propulsion plant, one propeller. The slow warships became known to their crews as "lemons," the "McNamara class," although in fact he too considered them poor designs.[26] Officers looked on their counterparts aboard the *Brooke* class with actual pity, explaining that without redundancy almost any equipment casualty crippled the ships for a mission.[27] The true difficulty was not inherent unreliability but that new maintenance skills, such as high-pressure boiler welding, had to be learned. To a busy, FRAM-experienced crew the *Brooke* class lacked ease of use, to use a later phrase. During the 1960s destroyer missions included prolonged gunfire support off Vietnam, surveillance, and PIRAZ. Destroyer escorts like the *Brooke* class were about all that was being built. "The resultant reduction in firepower, speed, staying power, versatility, and flexibility [of the future destroyer force] is obvious," Captain Caney wrote in *Proceedings.*[28]

Following the *Brooke* class, 46 even cheaper, single-gun, single-screw *Knox* (DE 1052)–class ASW destroyer escorts were ordered during FY 1964–67. The first was not commissioned until 1969, but all their primary systems had been in service since 1963. Secretary of the Navy Paul Nitze, a McNamara ally, liked the design. The *Knox* class would be good at the missions of the 1962 Cuban missile crisis: ASW, unopposed naval presence, and many ships. Secretary Nitze made ASW a personal priority.

In deep water *Knox* and the other destroyer escorts could deal with short-endurance conventional submarines or with submarines that might approach to attack guarded shipping or naval task groups. However, the DASH drone helicopter, planned as their longest-range ASW weapon, proved useless and was abandoned in 1967.[29] In active mode their long-range SQS-26 sonar was inferior in shallow water to the smaller SQS-23 on

the FRAMs,[30] and it was not officially accepted as fully operational until late 1968.[31] Again the true issue was ease of use: exploiting the capabilities of the long-range SQS-26 sonar and its new displays required careful operational planning. Judged against the tangible demands of the Vietnam War, and against the threat that Soviet warships would operate to counter U.S. ASW forces, the lone 5-inch Mark 42 gun was inadequate. As designed, the ships mounted no point-defense antimissile weapons. Their seakeeping and 27-knot speed were inadequate for operating with aircraft carriers or for open-ocean pursuit of nuclear submarines.[32] The ships relied upon a complex, acoustically noisy steam propulsion plant. Secretary Nitze forbade use of high-pressure 1,200 psi boilers, which had powered all postwar oil-fueled destroyers thus far, in specifications for new ships. This ban was due to maintenance requirements and pushed the Navy toward gas turbines.[33]

The Failure of the Typhon Air Defense System

Ships without capable antiaircraft, antisubmarine, and antishipping weapons were at risk from the growing Soviet fleets of submarines, land-based bombers, and surface ships, all mounting long-range remotely targeted antiship cruise missiles (ASCMs). The Soviets could attack with ASCMs and nuclear warheads in weather conditions in which a carrier could not operate her aircraft. The Typhon missile system was the first attempt at a shipboard air defense system to counter the increasing danger of a massed air raid or ASCM attack against a task group. The Typhon program included a medium-range missile and a ramjet-propelled long-range missile. A massive radar combined air search and missile guidance functions. The new missiles worked well enough, but the radar system did not.

Navy officers sought to divert Typhon funds toward upgrading the Talos, Terrier, and Tartar missile systems already in the fleet.[34] OSD reasoned that problems with Typhon and the existing missile systems made investment in additional surface missile ships futile until their weapons' defects were fixed. The axe fell:

> The readiness, reliability, and effectiveness of all three "T" missiles are being substantially enhanced through a comprehensive program to bring these weapon systems to their full potential. Tartar guided missile ships already

with the fleet are being modified on a regular schedule to fire an improved, long-range version of the missile. The Typhon weapon system, which was intended to provide a more advanced air defense for the fleet, turned out to be far too large, complex, and costly for deployment, and the termination of the project was announced on January 7, 1964. Work is continuing on the development of a new standardized missile to replace Tartar and Terrier in the near future, and for long-range improvement a completely new surface-to-air fleet missile system is under development.[35]

That was the origin of the Standard missile family and the Advanced Surface Missile System, later renamed Aegis. The surface missile "get-well" program modernized the Terrier missile ships with NTDS. This was an illuminating use of new electronics to improve the firepower of an outwardly unchanged missile battery. Reliable all-digital systems were developed, including medium- and long-range Standard missiles; the SPG-51D Tartar-D missile fire control system; the Mark 86 gunfire control system; and the Mark 26 fast-loading missile launcher, able to fire multiple weapons (thus exploiting commonality) and to interface with the digital fire control systems. Another project begun about the same time was to develop a small point-defense antimissile system.

Building ships to mount these weapons was another matter. High new-construction costs had engendered plans to convert the early postwar all-gun *Norfolk*, *Mitscher*, and *Forrest Sherman* classes to guided missile destroyers. OSD canceled all but six of these DDG conversions and canceled a Typhon-armed cruiser that Congress had authorized without OSD's request.

Gas Turbines and Project Seahawk

Caught between OSD's demands for cost-effectiveness and the low tolerance at OpNav (the Office of the Chief of Naval Operations) for innovation and risk, the Naval Material Command's administration of shipbuilding was unsuccessful in advancing naval technology.[36] The major case in point was the adoption of gas turbines for warship propulsion. Steam boilers were the largest single maintenance burden on the fleet. The Navy had investigated the possibilities of gas turbines for ship propulsion as early as 1938 and began a close study of the Royal Navy's work with

gas turbines in 1946, even acquiring British engines for experiments. British designers, enthusiastic about the advantages of gas turbines, put combined steam-and-gas turbine (COSAG) propulsion plants aboard the Royal Navy's new Tribal-class frigates and County-class destroyers in the late 1950s. They justified gas turbines to politicians as a feature that would enable the ships to break out of port before a nuclear attack.

The low-powered (7,500 hp) industrial gas turbines in the Royal Navy ships proved extremely successful. Gas turbines shifted the traditional naval architectural concern of matching hull design with full-speed propulsive power to the concern of matching hull design with cruising speed. With the gas turbine plant available for high speed, the ships' hull design and steam plant were optimized for good cruising endurance. Required to cover 5,000 miles at 12 knots, the British Tribal-class frigates could do it at 18 knots with fuel to spare. Under gas turbine boost power, their full speed of 26 knots was also better than planned. The gas turbine plant consumed fuel heavily at high speed, but the ships remained fuel-economical over their operating profile because they often steamed at cruising speeds with the gas turbines off line and consuming nothing.[37]

The Royal Navy soon switched to lighter aircraft-derivative turbines that could be removed and overhauled ashore. British "stokers" preferred gas turbine plants over boilers because they made for less watch standing, faster light-off, and less maintenance of the steam plant.[38] The Soviet Kashin-class destroyers began entering service during 1963 and were the world's first warships to be powered entirely by gas turbine engines. In 1967 the Royal Navy decided that all its new warship designs would use gas turbine propulsion exclusively.

During the 1950s the long-range SQS-26 sonar seemed too large for a destroyer, so a "sonar scout" was conceived to mount the SQS-26 sonar and a variable-depth sonar. This ship was to be unarmed, with the sole mission of active-sonar detection of underwater contacts for accompanying destroyers to prosecute using smaller, short-range attack sonars. The sonar scout required silencing and seakeeping improvements so that her own machinery's operation and her hull's passage through the water would not deafen the sonar.[39] By the late 1950s the increasing speed and endurance of Soviet submarines and the longer range of their torpedoes made it essential to equip all new

Table 2.2   Projected ASW destroyers after the *Knox* (DE 1052) class, 1962–67

Internal Navy disputes blocked construction of all these ships.

Project Seahawk ASW Destroyer (July 1963 Design)

Displacement	3,524 tons (light); 5,829 tons (full load)
Dimensions	Length 450 feet (waterline); beam 49.9 feet; draft (keel) 16.8 feet
Propulsion	CODOG 4 gas turbines, 4 diesels; 56,000 maximum sustained shp, 30 knots; range 4,580 miles at 25 knots
ASW weapons	SQS-26 sonar; new keel-mounted passive sonar; variable-depth sonar (VDS); DASH; ASROC (standard or extended-range version); tubes for Mk 48 21-inch wire-guided torpedoes
Other armament	1 5-inch gun, Mk 56 director; 2 3-inch/50 guns

FY 1969 DE Steam-Powered Design

Displacement	4,880 tons (light); full load over 6,000 tons
Dimensions	Length (waterline) 507 feet; beam 54.8 feet
Propulsion	3 600 psi boilers, 2 shafts 40,000 shp
ASW armament	SQS-26 sonar; SQS-35 independent VDS; Mk 116 integrated digital underwater fire control; 12.75- and 21-inch torpedo tubes; ASROC; DASH
Other armament	2 5-inch/54 guns, gun director unknown

FY 1969 DEG Gas Turbine–Powered Design

Displacement	3,535 tons (light); 5,185 tons (full load)
Dimensions	Length (waterline) 440 feet; beam 47.5 feet
Propulsion	Gas turbines (COGAG), 1 shaft 28,000 shp, 27 knots; range 6,000 nm at 20 knots
ASW armament	SQS-26 sonar; Mk 116 integrated digital underwater fire control system; 12.75-inch torpedo tubes; ASROC
Other armament	1 5-inch/54 Mk 45 gun, gun director unknown; 1 Tartar battery

destroyers with the long-range SQS-26 sonar. Research showed that destroyers could mount the large SQS-26, but no corresponding long-range destroyer-launched ASW weapon existed. Project Seahawk was a design for a prototype destroyer to test new ASW systems and to have improved propulsion (see table 2.2). Seahawk would mount almost exclusively ASW weapons. Its mission was similar to that of the destroyer escorts in that it was to operate with CVS hunter-killer groups in midocean or in company with attack carriers.

During the early 1960s the Bureau of Ships (BuShips) and Pratt & Whitney developed the FT4A marine gas turbine, a 14,000 hp engine derived from the Air Force J-75 turbojet and the JT4 turbojet of the early Boeing 707 commercial transport. Seahawk was designed in several variants, some of which were to use four FT4As in a lightweight, quiet, highly automated propulsion plant delivering about 56,000 shp to two shafts. Since Seahawk would be a major ship, BuShips began a program to ensure that both the gas turbine and the ship installation were the best available. The Naval Boiler and Turbine Laboratory in Philadelphia obtained five J-75 turbojet engines from the Air Force and converted them to FT4As. The testing of these gas turbines for Project Seahawk was more thorough and arduous than anything ever attempted on any steam or diesel plant of similar power.

A series of progressively more severe experiments sought out the gas turbine's limits as a main propulsion engine for a destroyer. The laboratory staff repeatedly stopped, started, and accelerated the engines using a variety of air inlet designs and fuel types, in particular cheap, dirty, sulfurous, seawater-contaminated fuels. Gas turbines mounted on barges were subjected to explosions to test their survivability against shock during combat.[40] The gas turbines met all naval requirements for maintainability and damage resistance. The tests showed that large marine gas turbines were feasible for large warships and offered significant advantages for naval warfare: rapid starting, acoustic silencing, and adaptability to automated controls.[41]

High fuel consumption and the design of reversible-pitch propellers or reduction gears for large aircraft-derivative gas turbines remained problems. Nearly all contemporary gas turbine–powered warship designs included a fuel-saving second power source to drive the propeller economically at cruising speed. Since gas turbines start quickly, the high-power gas turbines were used for high speed or for sudden response, such as moving the ship quickly out of port. Warship engineering plants combined steam and gas turbines (COSAG), diesels and gas turbines (CODAG or CODOG), or a smaller cruise gas turbine with a larger boost gas turbine (COGAG or COGOG).[42] The dilemma was that a separate engineering plant for fuel economy required more crew skills, more spare parts, and even additional fuel types, when a primary objective for the U.S. Navy was to reduce such logistical considerations. American warships typically operated at transoceanic distances from their bases, so gas turbines' high fuel consumption and short life between major overhauls were problems. Their acquisition cost was higher than that for a steam propulsion plant or diesel engines.[43]

Rear Admiral Hyman Rickover, the nuclear power lobby's single-minded leader, whom Congress pressed the Navy to keep on active duty far past retirement age, opposed gas turbines lest they compete with nuclear propulsion for surface warships. Circumstantial evidence suggests that nuclear-specialty officers, who needed Rickover's approval for career advancement, sabotaged Project Seahawk. The project became sidetracked over a demand for 40-knot speed. Ostensibly the destroyer now was to detect a submarine at long range, close on the target at 40 knots, and make a close-in attack with ASROC, DASH, or homing torpedoes. This speed was unachievable, so opponents insisted on returning to lower speed and nuclear-compatible steam propulsion.[44]

Officially Seahawk died because OSD accepted the nuclear propulsion officers' evaluation that other ASW forces such as nuclear submarines and the SOSUS/P-3 combination were more cost-effective for open-ocean ASW. The cancellation of Seahawk may have led to the expansion of the cheap *Knox* class.[45] None of these alternatives supported aircraft carrier strike operations. The high gas turbine intakes on the *Spruance* class, located so as to reduce salt contamination, are a visible legacy of Seahawk. The first warships to use the U.S. Navy's gas turbine research were Denmark's *Peder Skram*–class frigates, designed for CODOG plants with NavSEC assistance.

The Navy still had no seagoing gas turbine–powered ships. Assistant Secretary of Defense for Installations and Logistics Paul Ignatius stepped in to help. Exploiting his control of the Military Sealift Command, Ignatius instigated construction of a large gas turbine–powered fast vehicle transport.[46] While designed specifically for charter to the Navy, this ship, *Admiral Wm. M. Callaghan*, was commercially owned, so was beyond Admiral Rickover's paralyzing reach. As built she was powered by two FT4A gas turbines connected through a reversing reduction gear to a single standard fixed-pitch propeller. Engineers from the Navy's Philadelphia propulsion laboratory frequently tested maintenance and operating procedures for gas turbine systems aboard her.[47] She would have a major role in the creation of the *Spruance* class.

Cancellation of New Guided Missile Destroyer Designs

After the failure of Typhon and completion of the design for the *Knox* class in 1964, BuShips began design of a Tartar-armed guided missile destroyer (DDG) for construction beginning during 1966 (see table 2.3). These ships were planned as steam-powered follow-ons to the *Belknap*-class cruisers ordered during 1961–62. The series of weakly armed destroyer escorts did not eliminate Navy interest in building higher-capability ships to support aircraft carrier strike operations, and if anything stimulated it.[48] The Navy wanted 23 DDGs, an average ratio of 1.5:1 for 15 aircraft carriers. The Tartar guided missile system was a characteristic of all new guided missile ship designs. The short Tartar missile did not need large control fins to be attached manually before launching, so it could be loaded and fired more quickly than the larger Talos and Terrier. This speed suited it better to countering saturation attacks. Tartar missiles were stowed vertically in a magazine below the launcher, a compact installation. It had less impact on ship design, in particular on its requirement for centerline length. OSD objected to building missile ships until the missile problems were solved, and it canceled the FY 1966 DDG program.[49]

Following the rejection of the FY 1966 DDG class design, BuShips designed a class of four DDGs for the FY 1967 construction budget. The FY 1967 DDG design featured the improved Tartar-D digital missile fire control system integrated with NTDS. Commonality may have attracted OSD to the FY 1967 DDG because the promising Advanced Surface Missile System project would also use Tartar-D missiles and launchers. The design emphasized reliability and initially featured 600 psi boilers, similar to those (dating from World

Table 2.3   Guided missile destroyer design evolution, 1964–67

Only the FY 1967 DLGN was built, as the *California* (CGN 36) class cruisers.

FY 1966 DDG	
Displacement	4,994 tons (light); 6,666 tons (full load)
Dimensions	Length (waterline) 500 feet; beam 54 feet; draft (keel) 17.8 feet
Propulsion	1,200 psi steam 85,000 shp; range 5,000 nm at 20 knots
ASW weapons	SQS-26 sonar; ASROC; 12.75-inch torpedo tubes; combination of long-range ASW systems selected from VDS, DASH, ASROC (standard or extended-range version), 21-inch Mk 37 wire-guided torpedoes
AAW weapons	SPS-48 3-D radar; 1 Tartar battery, 2 SPG-51 directors; Naval Tactical Data System
Guns	2 5-inch/54 Mk 45 guns, 1 Mk 68 gun director
FY 1967 DDG Steam-Powered Design	
Displacement	5,748 tons (light); 7,463 tons (full load)
Dimensions	Length (waterline) 515 feet; beam 57 feet; draft (keel) 18.5 feet
Propulsion	600 psi steam 75,000 shp; range 5,000 nm at 20 knots
ASW weapons	SQS-26 sonar; VDS (?); ASROC; 12.75-inch torpedo tubes
AAW weapons	SPS-48 3-D radar; 1 Tartar battery, 2 SPG-51 directors; Naval Tactical Data System
Guns	2 5-inch/54 Mk 45 guns, 1 Mk 68 gun director
FY 1967 DDG Gas Turbine–Powered Design	
Displacement	6,173 tons (light); 8,450 tons (full load)
Dimensions	Length (overall) 552.5 feet; 525 feet (waterline); beam 60 feet; draft (keel) 19.2 feet
Propulsion	Gas turbines 56,000 shp; range 7,000 nm at 20 knots
Armament	Same as steam-powered design
FY 1967 DLGN USS California *(CGN 36)—As Built*	
Displacement	11,100 tons (full load)
Dimensions	Length (waterline) 560 feet; beam 63 feet; draft (keel) 21 feet
Propulsion	2 nuclear reactors 70,000 shp
ASW weapons	SQS-26 sonar; ASROC; 12.75-inch torpedo tubes; Mk 114 underwater fire control
AAW weapons	SPS-48 3-D radar; SPS-40 2-D radar; 2 Tartar batteries, 4 SPG-51 directors; Naval Tactical Data System
Guns	2 5-inch/54 Mk 45 guns, 1 Mk 86 gunfire control system

War II) in the popular FRAMs. Navy surface warfare officers were still skeptical about gas turbine propulsion, since it was unproven in large American warships.[50] However, Secretary Nitze objected to such a conservative design. Early in 1966 he ordered BuShips to redesign the DDG to use gas turbines for propulsion.[51]

The FY 1967 DDG design grew over 1,000 tons for conversion to gas turbines, as shown in table 2.3. Propulsion and auxiliary machinery weight decreased 300 tons, offset by a fuel increase of over 500 tons to increase range and to feed the high fuel consumption of the inefficient first-generation marine gas turbines. In this design, the main propulsion gas turbines drove electric generators, and reversible electric motors drove the shafts, since mechanical reversing couplings needed for the propulsive power of the FY 1967 DDG had never been built.

As with almost all modern warship designs, the FY 1967 DDG was volume-critical. Adding any

system required enlarging the ship structure to provide the necessary additional space. The effect is cascading, because the ship design must then be further enlarged for steel to maintain hull strength, and then enlarged still further to maintain stability and reserve buoyancy. In the FY 1967 DDG redesign, enlarging the intake and uptake ducting amidships needed for the gas turbines caused the design to grow by 600 tons. However, steel was the cheapest component of the ship.

Admiral Rickover insisted that any warship displacing over 8,000 tons, as the gas turbine FY 1967 DDG design would now do, had to be nuclear-powered. A destroyer hull of less than 8,000 tons had too little beam for the reactor box. Secretary of the Navy Fred Korth had agreed to this threshold. Secretary Nitze chose not to provoke Rickover's congressional sponsors by challenging it after he took over from Korth, whom McNamara had fired.[52] Congress rejected OSD's budget request for $153 million for detail design

Table 2.4 Summary of postwar destroyer construction programs and their basis in evolving national strategy

Strategic Era	Fast Battle Force Ships (High-end ships)			Protection of Shipping Escorts (Low-end ships)		
	Hull numbers	Built	Class	Hull numbers	Built	Class
1946–49: Early Cold War	DD 927–930 DL 2–5	 4	Became DL 2–5 *Mitscher* class	DL 1 (ex-CLK 1) DDE/DDK types	1	*Norfolk* Destroyer conversions
1950–54: Mobilization Preparation	DD 931–951 DDG 1	18 +3 1	*Forrest Sherman* class FMS *Gyatt* (conversion)	DE 1006–1032 DE 1033–1036	13 4	*Dealey* class + 14 FMS *Claud Jones* class
1955–61: Massive Retaliation/ Alliance Support	DD 952–962 DDG 2–30 DDG 31–36 DLG 6–15 CAG 1–CG 12 CG 13–15	 23 6 10 12 	First 8 became DDG 2–9; last 3 = FMS *Charles F. Adams* class + 6 FMS DD 931/DL 2 conversions *Farragut* class; redesignated DDG 37–46 Early guided missile cruisers Hull numbers not used	WW II DDs Sonar scout DE 1037–1039 DEG/FFG 1–6 DE/FF 1040–1051	131 2 6 10	FRAM conversions Canceled *Bronstein* class + 1 FMS *Brooke* class *Garcia* class + 2 FMS
1958–61: Improved Task Force Escorts (no strategic change)	DLG/CG 16–24 DLGN/CGN 25 DLG/CG 26–34 DLGN/CGN 35	9 1 9 1	*Leahy* class *Bainbridge* *Belknap* class *Truxtun*			
1962–66: Sea Control and Counterinsurgency	Typhon DLGN FY 1966 DDG FY 1967 DDG DLGN/CGN 36–37	 2	Canceled Canceled Canceled *California* class	Seahawk DE 1052–1107 DEG *Knox* FY 1969 DE	 46	Canceled *Knox* class; 10 canceled Canceled Canceled
1967–70: Flexible Response	DD 963–992, 997 DXG DLGN/CGN 38–42	31 4	*Spruance* class (DX) Canceled *Virginia* class (DXGN); CGN 42 canceled			
1971–77: Sea Control	DG/Aegis CSGN 1 F-DD 993–998		Canceled Strike cruiser; canceled FMS; 2 canceled, 4 became DDG 993–996	PF 109/FFG 7–61 PHM 1–6 DH 1	51 6	*Oliver Hazard Perry* class + 4 FMS *Hercules* class Sea Control Ship; canceled
1978–89: Maritime Strike	DDG 993–996 DDH 997 CG/CGN 43–46 DDG/CG 47–CG 73 DDG 48–50 DDG 51–78	4 27 28	*Kidd* class (ex-Iranian FMS) Canceled Not used *Ticonderoga* class Hull numbers not used or became CG 48–50 *Arleigh Burke* class Flights 1–2	DD 998–1003 FFG 62–70	6	*Spruance* class for convoy command; canceled Canceled
1990–present: Crisis Response	DDG 51 Flight 3 DDG 79	 6+	Canceled *Arleigh Burke* class Flight 2A	NFR-90		NATO frigate replacement; canceled

Note: FMS = Ships procured for foreign military sale (previously called offshore procurement). FMS designs varied and were not always the same class as the hull number might suggest.

and long-lead components for two fossil-fueled FY 1967 DDGs.[53] Instead Congress authorized funds for a Rickover-supported cruiser mounting the weapons planned for the defunct FY 1966 DDG, despite rapid advances in digital combat systems that made that armament obsolescent. Secretary McNamara opposed all nuclear surface ship construction,[54] and the contract to build the cruiser USS *California* (CGN 36) and a sister that Congress authorized the following year was issued only after he left the Pentagon in 1968.[55] Difficulty in integrating her mixed analog and digital combat systems kept her virtually inoperable for a year after her commissioning in 1974.

Later, Admiral Rickover would try to kill the DD 963 project when it became clear that the new destroyers would feature gas turbines. He opposed gas turbines from parochialism, but perhaps with foresight: no further nuclear-powered surface combat warships were authorized after the *Spruance* class showed the superiority of gas turbines. Despite his opposition, Admiral Rickover had three important and positive influences on the DD 963 project. First, a key reason for its success was its use of land-based testing and training facilities, which gas turbine proponents copied from the nuclear propulsion program. Second, in the 1960s Admiral Rickover stood almost alone in arguing the case for a powerful, strike-oriented surface force. Third, advocates of nuclear propulsion correctly pointed out that higher-quality ships, which in the 1960s meant nuclear-powered ships, improved crew motivation and performance.[56]

Cancellation of the FY 1969 Destroyer Escort

During 1966 the Naval Ship Systems Command and the ASW project office began feasibility studies for a new destroyer escort for the FY 1969 budget (see table 2.2). The FY 1969 DE design incorporated the all-digital integrated ASW combat system originally planned for Project Seahawk. One plan was to build these ships in a pair of specialized destroyer escort classes, both sharing a common hull design and propulsion plant, either steam or gas turbines in various ship design versions. One class would be armed for ASW and another as an air defense guided missile ship. However, after the *Brooke* class, OpNav ruled out long-range air defense armament for a destroyer escort. The ASW project office required the FY 1969 DE to have a 5,000-mile range at 20 knots and a maximum speed of 33 knots.[57] Planning for the FY 1969 DE was absorbed into the DX/DXG project.

Meanwhile, Congress and the Johnson administration had found that the costs of Vietnam and the Great Society were hugely greater than planned, and cut military construction to pay for them. All 10 planned FY 1968 *Knox*-class escorts were canceled, including one prototype ship (DE 1101) to be powered by FT4A gas turbines.[58] The Navy again asked for two DDGs in the FY 1968 budget, but now OSD rejected them in favor of designing a new guided missile destroyer tentatively called DXG. The final ship construction authorization for FY 1968 included a single surface combat warship (a second *California*-class cruiser, again added by Congress and again not ordered until McNamara left OSD) and for FY 1969 none at all. FY 1969 was the last year when the federal government balanced its budget.

Thus, as shown in table 2.4, during the mid-1960s every planned destroyer class was unacceptable to OSD or to Congress or to factions within the Navy. During a time of rapid technological advance, block obsolescence of the World War II fleet, and a rapidly growing Soviet military challenge, not one significant surface warship was under construction. It was against this background that the concept for the *Spruance*-class destroyers formed.

The Navy Adopts Total Package Procurement

The Cost-Effectiveness Challenge

Despite large budgets, Secretary McNamara's dictatorial management style provoked resentment in the services. McNamara made pervasive changes that showed the individual services that he had pulled decision-making authority into the Office of the Secretary of Defense (OSD). Commonality, often associated with McNamara, applied the economic principle that mass production reduces unit cost. Combining mission requirements created a single "common" weapon system for multiple missions that could be produced in a larger and more efficient production run. Commonality and cost-effectiveness were not necessarily compatible. A multimission weapon system had to be designed and equipped to be effective at all those missions. Weapons designed during the McNamara era tended to be large, so required large power plants, and thus were expensive.

The F-111 aircraft was the best-known example. It began as an Air Force deep-strike bomber that McNamara ordered the Navy to use as an interceptor, based on a sketchy analysis that this could reduce total cost. To get rid of it, the Navy increased the performance demands to impossible levels, the same tactic later copied by opponents of the Seahawk destroyer. Navy aircraft must touch down on a carrier with their payload and accelerate to flying speed in about 300 feet along the angled deck, in case the tailhook misses the arresting wires. With a payload of six heavy Phoenix missiles, the Navy F-111B variant could not do that. (Nor could its successor, the F-14A Tomcat, but the F-14 remains a far better fighter than the F-111B ever could have become.) Arming fighters only with long-range missiles proved worse than useless in Vietnam. Costs soared for the Air Force F-111 versions. Pilloried for years about the F-111, McNamara let the services lead subsequent projects.[1]

It was longstanding practice to evaluate the total cost of a proposed ship class (or any other weapon system) with design, construction, operation, crew size, and logistical support requirements. Procurement decisions for new weapons emphasized logistics and budget requirements as much as anticipated mission performance. What was new was that Secretary McNamara publicly rejected experience-based military judgments and predictions about the mission requirements and effectiveness of alternative weapons. In private he kept a closet staff of three officers, Navy Captain Raymond Peet, Army Colonel Alexander Haig, and Air Force Colonel Alexander Butterfield, to give him military advice on defense, such as evaluating his lists of targets to bomb in North Vietnam.[2] McNamara created the Office of the Assistant Secretary of Defense for Systems Analysis, or OASD(SA), to challenge the services' assumptions and evaluations.

Although derided as "whiz kids" whose naivete could make military knuckles turn white, the OASD(SA) staff included capable military officers and defense industry executives.[3] Lieutenant Commander Charles DiBona, an OASD(SA) staff officer educated at Oxford as a Rhodes scholar, wrote:

> One of the most fundamental changes wrought by Secretary Robert S. McNamara is that the defense budget is no longer planned in terms of Army, Navy, or Air Force. Funds are allocated to military missions, not services. . . . [For example,] attack carrier forces compete with Air Force tactical aircraft and Army tactical missiles.

As strategic, technological, and political changes take place, new missions for the Armed Forces will develop, some old ones will take on new importance, some will decline. The size and importance of the Navy of the future will depend on how imaginative and convincing the Navy is in demonstrating the superiority of sea-based systems for both new and old missions. We must continue to evolve and develop new, ingenious, and increasingly effective ways of using ships.[4]

DiBona had married a Swede, and while on leave in her homeland he visited the large Götaverken shipyard at Arendal. He observed how the Swedes had increased productivity through automation and modular construction techniques for building large oil tankers. DiBona identified four ways to gain more for the money spent on shipbuilding:

- Ensure during preliminary design that each specified feature is justifiable on the basis of total added life-cycle cost as well as of added mission effectiveness
- Standardize designs, build ships in large blocks, and award the contract for all ships of one design to one contractor
- Evaluate the costs of ship design and construction features that could reduce operational costs
- Modernize American shipyards[5]

The OSD staff had its own view of why new destroyers were not being built. OASD(SA) was skeptical about conservatism in shipbuilding and the unreliability of new naval weapons. Prevailing ship procurement practice was for new class feasibility studies to be prepared and a preliminary design to be drafted by the Naval Ship Systems Command (NavShips; called BuShips, the Bureau of Ships, until 1966). Cost-effectiveness trade-offs, systems analysis, and commonality were not new to the military. Ship design feasibility studies were straightforward cost-effectiveness trade-off evaluations of the sort that had been in use for decades. Battleship design in the 1930s had used a trade-off process wherein "cost" was the ships' treaty-limited "standard" displacement. If the Office of the Chief of Naval Operations (OpNav) approved the preliminary design, then NavShips prepared a contract design and sent it out for construction bids. Shipbuilders had little opportunity to optimize the design for more efficient production. This procedure was unchanged since the 1920s.[6]

Independent research (normal in the aerospace industry) in basic ship features occurred at some foreign shipyards but was rare in American shipbuilding. OSD publicly contrasted the shipbuilding industry with foreign competition: "It has become increasingly apparent in recent years that our shipbuilding industry, both public and private, has fallen far behind its competitors in other countries. Not only does it cost twice as much to build a ship in this country, it also takes twice as long. . . . The American shipbuilding industry is generally technically obsolescent compared to those of Northern Europe and Japan."[7]

OASD(SA) suspected that conservatism in warship design, whereby each successive class was only slightly improved over its predecessor, contributed to enlarged ships, higher cost, and longer building times.[8] Only Navy-tested technology went into warship designs, favoring caution and incremental design evolution over innovation and risk. American warships still relied exclusively on boilers for propulsion, while the world's other navies were adopting gas turbines for boost power and quick breakouts from port. Even small foreign warships mounted antiaircraft missile batteries, yet only the largest U.S. Navy warships did. OASD(SA) wondered whether the Navy, anticipating that any war would be nuclear, had forgotten lessons from World War II about the need for ship defenses and survivability.[9] Development of small shipboard point-defense missile systems commenced in 1962, but years later Israeli and North Korean attackers found the intelligence ships *Liberty* and *Pueblo* unprotected.

Standardization and configuration control became a focus for cost-effectiveness. The Navy's contract design for each class specified only a warship's layout, dimensions, weapons, propulsion, and other major features. Each shipbuilder prepared its own detail design for each contract. The detail design specified the construction plans and the individual hull, mechanical, and electrical (HM&E) equipment to be bought or built for the ships. The shipbuilder's detail design thus determined the range of HM&E spare parts to support the ship during her operational career. During 1948–65 the Navy built 351 major warships under 209 contracts, with no contract for more than five ships and 203 for three or fewer.[10] The *Wall Street Journal* described the resulting problem: "One Navy officer notes that ships of the same class often differ greatly because each contractor buys equipment such as pumps, boilers, and winches on the open market; this equipment varies depending on where and when it was purchased. . . . Of the 157,000 spare parts the Navy stocks for its ships, 33% can be used on only one ship."[11]

A case in point was the DDG 2–class destroyers, in which only 29 percent of the HM&E components were common to all 23 ships of the class.[12] Installing identical equipment aboard all ships of a class would minimize the range of spare parts and maintenance skills needed to support the class in service. Concentrating spare-parts investment on the common parts would reduce costs and would improve the likelihood that a given part would be in the supply system when needed. Similarly, it would be easier to train sailors to operate and maintain the ships. Savings from standardization would be greater for a larger class of ships. Building many ships to a standard design would let shipyard staffs develop skills and efficiency from experience, so that, for example, ships 11–20 would be cheaper than ships 1–10. This learning curve was essential to realizing economies of scale.[13]

Economies of scale supported the idea of using a single contractor to build an entire class of ships. All the later DE 1052–class ships were built at Avondale in Louisiana in an attempt to realize economies of scale. Instead other problems occurred. Construction delays and cost overruns beset the DE 1052 program. At the same time, in the same shipyard the Coast Guard's *Hamilton*-class high-endurance cutters of about the same size and technological sophistication (and featuring CODOG propulsion and far superior appearance) were on budget and on time. The Navy ordered design changes periodically for what it regarded as logical and routine updates.[14] Avondale sued the Navy for more money, alleging that design changes disrupted the yard's workflow. Navy officials suspected, but could not prove in court, that Avondale was concentrating on the Coast Guard cutters because the profit was better. Many other shipbuilders sued and won damages on similar grounds.

OSD beat NavShips bloody over costs. Design change orders were a major problem. The seriatim cancellations of destroyer construction programs prevented the Navy from incorporating new technology other than by modifying existing designs. An underlying problem was that OpNav often did not look at ship designs very closely until construction began, and then suddenly decided that changes were essential. This made it impossible to estimate ship construction costs accurately, a problem further compounded by court-ordered damages to shipbuilders.

The block-obsolescence problem was becoming more urgent. Most of the destroyer force and all of the antisubmarine aircraft carriers and gunfire support ships had been built for short service in World War II. Spare parts were scarce and overhauls long. The old ships required large, costly crews, but poor habitability discouraged reenlistments by skilled sailors. Submerged Soviet submarines could outrun many destroyers in heavy seas, and often they were undetectable by older destroyer sonars. U.S. Navy destroyers carried no helicopters for long-range submarine detection and attack. Regional powers found ex–U.S. Navy FRAMs adequate for local service. During 1971–74 the Philadelphia Naval Shipyard modernized two destroyers for Iran, with new sonar, air conditioning, a small manned ASW helicopter with a hangar, and eight Standard-MR missiles for air defense and short-range antiship attack. The gunfire control radar illuminated targets in both cases. Taiwan and South Korea also used extensively modernized FRAMs. Regional powers faced weaker opponents, and the ships were below U.S. Navy damage control standards. Requirements for silencing, habitability, ship stability, internal volume, topside space, cruising endurance, electrical power, and cooling made it impossible to modernize the old ships to meet future American demands.[15]

A variety of studies recommended solutions. In 1965 Russell Murray of OASD(SA) contended that the Navy's shipbuilding program could survive the attacks of the cost-effectiveness analysts only through the use of the techniques of mass-production industry to bring costs down.[16] Captain Elmo Zumwalt prepared a Major Fleet Escort Study. Captain Ray Peet headed a Ship Acquisition Study that identified the change order problem. It pointed out that the Navy had little useful policy for sorting out mission requirements, technological opportunities, and numbers needed of new warships. Captain Peet suggested a Force Level Advisory Group within OpNav to focus operational insight on ship characteristics and force levels. OASD(SA) and other Pentagon offices saw that suggestion as a threat to their prerogatives and rejected it.[17] But OSD liked a new concept called Total Package Procurement.[18]

The Navy Adopts Total Package Procurement

In 1964 an Assistant Secretary of the Air Force with an aerospace industry background charged that defense contractors offered unrealistically low prices to win the competitive initial contract to develop a weapon. Afterward, when other competitors had been eliminated, they made inordinate profits from sole-source production, spare parts, and technical support. The Assistant Secretary outlined Total Package Procurement (TPP) as a major reform of defense contracting. Under TPP each competitor for a contract was required to give firm prices at the outset for development, all production, spare parts, and other costs associated with introducing a weapon system. Stiff contract terms bound the manufacturer to the prices, performance levels, and delivery schedules in future years.[19] Secretary of Defense McNamara and Secretary of the Navy Paul Nitze were zealous proponents of TPP for ship acquisition.[20] On July 30, 1965, Nitze signed a directive that "suggested . . . new methods for designing, contracting for, and constructing ships."[21]

The key to TPP was for the service to specify what it required a new weapon to do in terms of performance. Each competing contractor would then design a candidate weapon and commit to a price and method for building it.[22] An aerospace manufacturer designed aircraft to exploit the capabilities of his production line and support organization. Giving shipbuilders similar design freedom might enable each to design ships that would be cheapest for that yard to build. The Navy could then select the design and yard offering the lowest cost, or even better, the lowest life-cycle cost, including both construction and operation. Spares would be delivered with the ships. Competitively set prices would reduce the need to go back to the contractor to negotiate costly sole-source contracts for follow-on production or spare parts. With a multiship, multiyear TPP contract, a shipbuilder could make long-term plans and cost-saving investments to modernize the shipyard. Applying TPP to warship procurement might induce modernization of the American shipbuilding industry (assuming that more than one shipyard would win TPP contracts) and create a modern fleet that would be logistically simpler and cheaper to support.

The greatest attraction of TPP was the opportunity it offered the Navy to pursue innovation by going outside its own in-house warship design staff at BuShips. There had been breathtaking, literally space-age advances in digital electronics, computer software, aviation, communications, materials, and large-scale mass production. American industry certainly appeared able to design innovative warships. As an example, General Dynamics built submarines (at Electric Boat) and destroyer escorts (at the famous Bethlehem Fore River shipyard in Quincy). It had a staff of naval architects who did the detail designs for these ships. The same firm built the Navy's Standard missiles and had corporate expertise in advanced naval combat systems. It was designing an integrated ASW combat system at the Naval Electronics Laboratory in San Diego. (This evolved into the Canadian CCS-280 and U.S. Navy Mark 116 underwater fire control systems for the *Iroquois* and *Spruance* classes, respectively.) The firm's Convair Division had designed and built seaplanes and was eager to apply automated seaplane flight control technology to warships to reduce crew size and to improve combat response time.[23] Such an industrial titan appeared able to design advanced and original warships from scratch, and to build them cheaply using an infusion of aerospace mass-production methodology.

The comprehensive theory of Total Package Procurement was compelling. The Navy first attempted to use it for a class of fast deployment logistic (FDL) transports. The FDL project manager was Rear Admiral Nathan Sonenshein, a naval architect and marine engineer educated at the Naval Academy and the Massachusetts Institute of Technology. During the Korean War he had been chief engineer of USS *Philippine Sea* (CV 47), one of the few aircraft carriers still in service following World War II and one whose success off Korea helped to justify the *Forrestal* class of supercarriers. As Pacific Fleet maintenance officer in 1960–62 he discovered that the biggest item in the budget was boiler repair. This experience later led him to seek a new naval fuel type that would reduce boiler maintenance and, subsequently, to get rid of boilers altogether.[24] After attending the Harvard Business School, in 1965 Sonenshein became Assistant Chief of BuShips for Design, Shipbuilding, and Fleet Maintenance. The project manager concept was a Navy initiative originated to develop the completely successful Polaris fleet ballistic missile submarine system. Its extension to FDL was an OSD initiative. As FDL project manager in the Naval Material Command (NavMat), Admiral Sonenshein had authority independent of NavShips and the Navy's other institutions. He held positions in NavMat and in the surface warfare directorate at OpNav to coordinate both OpNav's "requirements side" and NavMat's "producer side" of the FDL project.

Admiral Sonenshein and others were willing to exploit TPP as a method to build large numbers of new warships while reducing the problem of contractor claims. The advantages of TPP for the Navy included project stability ensured by using a multiyear contract, and the ability to concentrate control over budgets and schedules with the project manager to enforce the TPP contract. The rapid increase in intricate electronics for each ship class saturated the NavShips design staff.[25] Under TPP, industry designers would supplement NavShips, which was at its authorized hiring ceiling. The Navy would design nuclear-powered ships or other specialized ships but would employ TPP for large classes where economies of scale seemed possible.[26] Above all, making the shipbuilder unambiguously responsible for the ship contract design would eliminate the two major premises of shipbuilders' financial claims: that Navy-ordered design revisions disrupted the shipbuilding process, and that Navy-provided ship designs were sometimes not feasible to build at all.

Admiral Sonenshein met with the Air Force to learn more about TPP. While the Air Force depended entirely on industry designers, commercial shipbuilders had no preliminary-design experience and relied on the Navy to supply that. Perversely, this reliance let delinquent contractors sue the Navy for issuing them allegedly infeasible designs. Aircraft procurement provided for prototypes, whereas ship designs went straight into construction. Thus Sonenshein had to adapt TPP to the Navy.[27]

"Too Many Innovations": The FDL Project

While freighters may seem a long way removed from destroyers, the FDL project profoundly influenced the DD 963 project. The FDL concept was for a fleet of mobile floating arsenals for Army and Marine Corps heavy equipment and stores, with a corresponding fleet of Air Force transports for the troops and their immediate-use equipment. (The highly successful Boeing 747 airliner was designed under an FDL CD contract. Lockheed won

the FDL aircraft production contract but nearly went bankrupt building its design, the C-5A.)

In 1965 BuShips had prepared a preliminary design, later designated FDL(X), for the FDL ship. The FDL(X) design was weight-critical, so for buoyancy the hull had excess volume. That made it easy to incorporate gas turbines, and the FDL(X) design featured gas turbine propulsion. After OSD canceled Seahawk and all other new warship projects, the FDL transport was the only remaining shipbuilding project that might have had enough ships to make TPP viable. OSD ordered the Navy to restart the project as a Total Package Procurement from industry. Secretary Nitze described the FDL ship project to Congress:

> The Fast Deployment Logistic (FDL) ship project . . . is being used as a trial application of a new approach to ship design and construction. Under this approach the Navy is, for the first time, applying the contract definition process to ship design and the "total package" approach to ship procurement. The Navy hopes to realize the following advantages from these methods:
>
> 1. Added impetus to the modernization of shipbuilding facilities.
> 2. Lower average costs of ships.
> 3. Increased standardization of ships.
> 4. Increased industry input into naval ship design and construction.[28]

Internal Navy opposition developed against the radical and untested TPP idea when FDL(X) was abandoned in favor of the new methods. The commander and deputy commander of NavShips both resigned in protest late in 1965 over the new FDL plan, shipyard closings, reductions in their authority, and the implication that Total Package contractors would replace NavShips as the primary source for warship designs.[29]

In 1966 the OpNav Ships Characteristics Board was abolished, confirming that the project manager rather than the Chief of Naval Operations staff would decide such matters.[30] NavShips was reorganized, with the ship design offices moved to a new subordinate division called the Naval Ship Engineering Center (NavSEC, not to be confused with the Secretary of the Navy, SecNav). NavSEC was removed to a remote suburb of Washington, D.C., away from the NavShips headquarters. This geographical split reduced the Navy designers'

Litton Industries won the Total Package Procurement competition for Fast Deployment Logistic (FDL) ships with this design. Congress refused to appropriate funds to build them. The Navy used the Total Package Procurement procedure developed for the FDL project in the DD 963 class competition. (U.S. Naval Institute collection)

access to the commander of NavShips and increased the influence of the project managers.[31] At NavSEC many naval architects were soon doing administrative work instead of design work.[32]

TPP (officially Contract Definition, or CD) began with Concept Formulation (CF, or CONFORM), when the Navy specified a ship class's required mission capabilities and performance.[33] The Navy issued the FDL performance specification to industry in a formal request for proposals. Interested firms responded with descriptions of their capability to design and build the specified ships. In 1966 the Navy awarded funded CD contracts to three firms to develop a new preliminary FDL ship design, production plans, the associated support plans such as training and spares, and the construction price.[34] These firms (Litton Industries, Lockheed, and General Dynamics) all were principally aerospace manufacturers. Tex Thornton, chairman of Litton Industries, had been Robert McNamara's mentor in the World War II Army Air Force and postwar at Ford.

The contractors' preliminary designs and construction plans were impressive in their originality, although OSD had hoped for more radical approaches, such as detachable cargo hulls to be connected to seagoing tugs as on Mississippi River barge-trains. Litton showed that its FDL design minimized the cost per potential unloading port, based on the number of the world's ports where

ships of that size could dock. The commercial designs specified smaller crews for the fleet of fewer, larger, more highly automated FDL ships than had NavShips' FDL(X). Reducing crew size by one man would save $1 million over the life cycle of each ship.[35] Litton's FDL design was for 30 ships, each displacing 40,500 tons and carrying 10,850 tons of combat cargo. The FDL(X) plan had been for 54 ships, having larger crews but displacing only 28,000 tons and carrying only 6,200 tons.[36] The Integrated Logistics Support plan for the FDL was the first Navy use of this procedure. It defined crew size, training, spare parts, and maintenance requirements, and showed the costs associated with them. The Navy selected Litton's proposal for award and forwarded this recommendation to OSD.

OSD requested funds from Congress early in 1967 to start building the ships at Litton's Ingalls Shipbuilding yard. The two-year delay for Contract Definition resulted in a more cost-effective FDL ship design and production plan, but the times had changed. After his F-111 and CVA 67 decisions and the Tonkin Gulf Resolution, McNamara had antagonists in Congress who were willing to scuttle FDL to oppose him and OASD(SA), which had selected FDL as the test case for Total Package Procurement. Many in the Navy remained hostile. OpNav told Congress that the service did not want funds diverted from any of the Navy's own ship construction projects (which then included a request for the first five DD 963 destroyers) in order to build transports for the Army: "The Navy's position has been and still is that, while concurring with the concept, funding of the FDL should be in addition to, not at the expense of the Navy's SCN [Shipbuilding and Conversion, Navy] budget."[37]

Congress, under pressure from opponents of further global-policeman roles in Vietnam-type conflicts and from shipping lobbyists who feared Navy competition for hauling military cargo, rejected funds for any FDL ships. Senator Richard Russell, chairman of the Senate Armed Services Committee, was particularly concerned that FDL ships would allow escapades in Africa. OSD failed to win funding for three straight years and abandoned FDL after 1969. "It was a great artistic success that failed at the box office," concluded the FDL technical director, Captain Richard Henning.[38] Another officer wrote, "The FDL project failed because it contained too many innovations. Contract Definition, total package procurement, and shipyard modernization: each aroused opposition. . . . Moreover, the strategic purpose which the FDL was intended to serve lacked congressional approval."[39]

Admiral Sonenshein's systematic approach toward TPP nevertheless earned him OSD's confidence.[40] Innovation, improved standardization, modernization of the shipbuilding industry, and lower average ship costs remained attractive benefits. OSD was eager to use Contract Definition and TPP in two new shipbuilding programs, the *Tarawa* (LHA 1)–class amphibious assault ships and a new class of destroyers.

The need for FDL became real in 1979, when post-shah Iran turned violently anti-Western. In an effort that commanded White House attention, a small rapid-deployment force of seven transports was loaded with Marine Corps equipment. They were stationed at Diego Garcia, specifically to support a prompt if small military response on land in Southwest Asia. The force grew to 37 ships. During 1990–91 they carried out the bulk of the Operation Desert Shield/Desert Storm buildup.[41]

Soviet Naval Expansion Accelerates

Soviet naval expansion accelerated as part of the enormous military buildup following Nikita Khrushchev's overthrow in 1964 by Leonid Brezhnev. Soviet naval expansion had already started in 1956, so construction and deployment of new ships and submarines would probably have increased after 1962 even without the Cuban missile crisis.[42] By 1967 the Soviet fleet of nuclear tactical submarines outnumbered their Western counterparts by 50 percent, 45 to 30.

With both the American and Soviet ballistic missile submarine construction programs growing, the Soviets had shifted their naval strategy from anticarrier tactics to anti-Polaris and counter-ASW. Changing their fleet to match, the Soviets began building a series of ASW submarines and large ASW surface warships. These had both an ASW mission against the Polaris submarines, whose early ballistic missiles could reach their targets only if launched from waters close to the Soviet Union, and a counter-ASW mission against Western naval forces menacing the Soviets' own offensive submarine operations.

A much larger construction program was just beginning. Naval analyst Michael MccGwire explains:

> In 1966, the Soviet political leadership ruled that a world war would not necessarily be nuclear or involve massive strikes on their homeland. It thus became logically possible, and therefore necessary, for the Soviets to adopt the wartime objective of avoiding the nuclear devastation of the Soviet Union. This meant they had to forgo nuclear strikes on the United States, because that would result in retaliation on the Soviet Union. The U.S. military-industrial base would therefore remain intact, which made it essential for the Soviets to deny the United States a bridgehead in Europe from which to mount an offensive once it had built up its forces, as it had in World War II.[43]

Large-scale production of the Yankee-class second-generation nuclear ballistic missile submarines and associated Victor-class guardship/anti-Polaris submarines absorbed almost all of the Soviet Navy's allocation of nuclear reactors, so that the new Charlie I–class antiship attack submarines featured just one reactor, while the other submarine classes had two.[44]

The Soviets concentrated on surveillance and electronic warfare, emphasizing decoys and jamming to turn the radioelectronic environment to their advantage. Central command posts, at first exclusively ashore but later sometimes aboard cruisers, were built to correlate data from all sources and to pass targeting assignments to submarines, warships, and land-based naval aircraft. All these were platforms to launch conventional or nuclear missile strikes against Western task forces, a mission the West assigned to aircraft carriers.[45] The Soviets planned to achieve local superiority at sea in selected theaters during a longer war.[46]

Still unable to build as many nuclear submarines as their huge military expansion demanded, the Soviets built surface ships for missions that the West assigned to submarines: surveillance, barrier ASW, and antishipping warfare to defend against Western ASW forces and to block oil shipments and military reinforcement convoys. Apparently the Soviets anticipated that their submarines and land-based bombers carried adequate numbers of anticarrier nuclear weapons and would be less exposed to task force surveillance. The Soviets changed their plans to build two new classes of anticarrier cruisers. One class would have mounted long-range nuclear antiship missiles (SS-N-12, Soviet P-35) to break up carrier battle group defenses. Another, more numerous, class would have mounted more numerous shorter-range missiles (SS-N-9, Soviet P-50) to hunt down ships damaged but not sunk by the long-range attack. Poorly modified instead for ASW, these two transitional designs emerged as the Kresta I and II classes.[47] Like the Kresta II, the next generation of Soviet surface warships, the Kara and Krivak classes, did not mount dedicated antiship cruise missiles.[48] Misinformed comparison of these ships with the *Spruance* class was to have an enormous impact on the U.S. Navy, as will be seen in chapter 7.

The Soviet Union's naval missions included support of worldwide Soviet diplomacy, which had elements of psychological warfare against the United States: harassing American ships and aircraft on the high seas, attempting to cause collisions or other accidents, and displaying that Soviet forces could congregate to threaten massed initial strikes on American naval forces during a crisis.[49] The Soviets acted like gamblers during diplomatic confrontations, frequently risking military action to advance their goals.[50] Their objective may have been to discover which nations weakened in the face of threats. While the first-salvo threat reflected the fact that Soviet warships, lacking magazine capacity, reliable equipment, a combat information center, and effective damage control practices, were not survivable, it high-lighted the need for U.S. naval forces to be able to defeat such an attack.

A Limited Revival of the Maritime Strategy

OSD support for destroyers developed during a resurgence of interest in naval operations for the Flexible Response national strategy. Curiously, the Navy had little to do with stimulating OSD's interest. In 1967 NATO formally adopted Flexible Response to replace the Massive Retaliation plans last updated in 1956. Under Flexible Response the NATO allies would initially defend against Soviet aggression at the level, conventional or nuclear, chosen by the attacker. Flexible Response was a defensive strategy that relied on holding Western Europe long enough for nuclear threats to induce the Soviet Union to back down after an invasion. This was a shaky premise for a strategy, but the alternative of intercontinental nuclear war looked worse. Offensive strategies such as overrunning the Russian heartland or siege must have seemed impossible.

Flexible Response depended significantly on use of the seas. It assumed that war with the Soviet Union would come only after a time of rising tensions, during which the United States would mobilize its reserves and deploy them to the Central Front by sea (REFORGER: return of forces to Germany). ASW forces and fast convoys were central to this strategy. U.S. investment in naval forces might restore NATO solidarity, shaken by growing American reluctance to defend Europe at the risk of Soviet attacks on American cities.

As the United States and its allies looked to their navies to widen their strategic options, the Soviet Union prepared to close off those options.[51] NATO recognized that the Soviets could make strategically destabilizing incursions along NATO's flanks, particularly in Norway and by attacks on oil shipping routes. Aircraft carrier task forces would provide air support for defense of the flanks in Norway and the Mediterranean. Outside NATO, defense of Japan similarly relied on naval support. Flexible Response required prevention of Soviet ASW actions against transiting American and British fleet ballistic missile submarines, whose weapons would be held in reserve.[52]

Secretary Nitze, an influential strategist, suggested responding to Soviet threats on land by taking naval action against Soviet interests at sea. A prompt U.S. naval blockade would warn the Soviets to control the outbreak of conflict. This was similar to operations planned for the first phase of the Maritime Strategy devised around 1980. Further, the Soviets might use their huge and growing submarine force to blockade Europe to deter U.S. reinforcement of NATO.[53] Warfare at sea might even see use of nuclear weapons such as ASROC depth charges, which would signal U.S. fortitude but carry little risk of escalating to nuclear attacks against either homeland. It was consistent with contemporary strategic nuclear targeting plans that attempted to provide follow-up targeting options, controlled escalation, flexibility, and sequential attack:[54]

> The United States, perhaps without allied support, [might elect] deliberately to escalate with nuclear weapons at sea in response to Soviet actions or threats elsewhere. In the early 1960s, the Navy began to consider the possibility of a limited tactical nuclear war, at least initially restricted to the seas, as a plausible and safer alternative to escalation on land. The *War at Sea* studies initiated by then Secretary of the Navy Paul H. Nitze examined the implications of a solely maritime war "where the objective is to coerce the antagonist solely by maritime pressures. . . . Land, overland airspace and inland waters are, in general, taken as sanctuaries." In such a war, without concurrent land combat, "The political objectives of the contestants are to be reasonably well defined and do not encroach upon the vital interests of any nuclear power. This is recognized by all of the major contestants."

> By the late 1960s, various *War at Sea* Operations Plans (OPLANs) were being drawn up by the Navy. The Atlantic Fleet's OPLAN 2300 [1966–67] provided for fleet operations in contingencies involving the Soviet Union in which, among other cases, the United States wanted to carry out "a controlled response to Soviet military aggression short of strategic nuclear attack." The plan called for a gradual escalation of naval actions toward the "highest intensity of operation" which would entail "the entire Atlantic Fleet in a complete confrontation with Soviet forces at sea in warfare short of the strategic use of nuclear weapons."[55]

Sequential attacks would gradually increase in intensity to coerce the enemy by threatening worse to follow. Contemporary views differed violently over whether (1) conventional bombing under this theory was failing to coerce North Vietnam, but President Johnson was too vain to admit it; or alternatively, (2) bombing had succeeded in bringing North Vietnam to the point of collapse, but the Johnson administration was too deluded to exploit it.

Another reason for the new interest in war at sea was that the Soviets were developing the naval capability to coerce a diplomatic adversary by declaring certain sea areas closed to navigation or by establishing a Cuban-style quarantine themselves.[56] Egypt, then a Soviet client, declared it was closing the Gulf of Akaba to Israeli shipping in 1967, triggering the Six-Day War. Israel received 90 percent of its oil through the Gulf of Akaba, an obvious parallel with Western dependence on the Persian Gulf and other straits.

The idea of prosecuting a war only at sea attracted OSD for crisis control and helped to justify new investments in naval surface forces.[57] This objective first appeared in a 1968 defense budget request to Congress:

> In addition to providing naval support for the regional contingencies, we also want to have a capability for successfully concluding a War at Sea. . . . Soviet (and to a lesser degree, Red Chinese) attack and cruise missile submarines are the main threat to our ability to win a War at Sea. . . . Our War at Sea strategy is based essentially upon the rapid emplacement of ASW forces, comprised of submarines and land- and sea-based ASW aircraft, between the enemy submarines and their potential targets.[58]

Aircraft carrier task groups continued to be highly useful even after OSD, evaluating the value and vulnerability of aircraft carriers as weapons for use against the Soviet Union, removed aircraft carriers from the SIOP (single integrated operational plan) because they offered too little assurance that targets assigned to them would be destroyed promptly.[59] Withdrawn from national strategic targeting, the carrier force became available for nuclear and conventional strike missions planned by theater commanders, particularly from the eastern Mediterranean against southern Soviet targets. Potential targets along the Russian littoral included submarine tenders and air defenses that would obstruct Air Force bomber routes. Submarine tenders and bases had always been significant targets for the Navy, to sink submarines there and to lay minefields to block fleeing submarines.

National strike plans since 1962 had listed Soviet ballistic missile submarine bases among the highest-urgency targets for nuclear attack.[60]

Of course, all actual combat had been conventional without Soviet participation. Contemporary American defense analysis called for military capability to wage two major 90-day campaigns and one lesser campaign simultaneously, although in practice, Congress and the administration supported at most a "one-and-a-half-wars" force level. Under Flexible Response even this commitment required many ships. The configuration and technological sophistication of the American armed forces showed the NATO allies what the United States considered necessary to carry out the new coalition strategy. A new national strategic requirement developed for naval operations, although the U.S. Navy as an institution, suppressed and overridden by Secretary McNamara, contributed little to it. Neither OSD nor the Navy addressed the use of surface ships to fight the Soviet Union after an outbreak, or to force an end to a war by siege or even invasion.

Aircraft carrier task groups had been in action off Vietnam since the first raids on North Vietnamese targets. Congress strongly supported nuclear-powered aircraft carriers. In 1966 President Johnson ordered OSD to drop its opposition to nuclear power and to proceed with the *Nimitz* (CVAN 68) class.[61] Plans for a new class of destroyers to escort the carriers and for other missions quickly followed. OSD followed this circuitous analytical path to arrive at the longstanding naval requirement for quiet, fast destroyers.

The DX/DXG Project

The Origin of DX and DXG

Given the reduced role of aircraft carriers, Captain Elmo (Bud) Zumwalt of the staff of the Office of the Secretary of Defense wrote in the Naval Institute *Proceedings* in 1962, the Navy had enough "complex" ships, the heavily armed classes for strike operations. He recommended building a "simplified mainstream" of ASW ships that the Navy could build cheaply and in quantity for convoy escort, hunter-killer ASW, and the peacetime naval presence mission. He advocated a gun-armed, minimally manned, long-range, 23-knot destroyer escort for the simplified mainstream.[1] The *Knox* class fulfilled this requirement. However, Soviet naval technology was advancing more quickly than expected. The sustained high speed of Soviet nuclear submarines and the increased range of their weapons increased the underwater threat to the attack carrier task forces. Aircraft carriers needed sonar-equipped escorts to screen them and to disrupt submarine attacks. This created a need for fast ASW ships to support attack carrier task force operations forward of the SOSUS barriers and closer to the Soviet Union.

By 1966 OASD(SA) (the Office of the Secretary of Defense for Systems Analysis) could see the scale of the problem of replacing large blocks of obsolete warships.[2] With the DE 1052–class production scheduled to end, a new continuing-construction program was needed to sustain a shipbuilding base as a mobilization contingency. The FDL project's success in generating creative designs for series production suggested that Total Package Procurement was a solution. TPP offered both a fleet of new ships and a modernized shipbuilding industry using faster, reduced-cost shipbuilding techniques such as modular construc-tion as used at Götaverken. The new destroyer project would be the confluence of strategic requirements, advanced but hitherto unused naval technology, and hopes for innovation and cost reduction from Total Package Procurement.

The *Spruance* class was not the Navy's idea. Perhaps aware of the common-hull, common-propulsion concept for the FY 1969 destroyer escorts, in 1966 OASD(SA) suggested building two notional destroyer classes. They would have as much commonality as possible, perhaps incorporating identical hulls and identical propulsion. Weapons would vary. DX would be a destroyer escort (DE) variant with a lighter ASW-only armament. DXG would be a cruiser (DLG) type mounting both ASW weapons and an air defense guided missile system.[3] *X* indicated different specialized missions, AAW (antiaircraft warfare) or ASW, for a basic common ship. It suggested that the traditional distinctions among ship types had lost significance. The key elements of the OASD(SA) approach were as follows:

- Economies of scale from building identical ships at a single yard
- Standardization to reduce training requirements and to provide life-cycle supply management for spare parts
- Modernization of the shipbuilding industry, in particular to use modular construction for faster shipbuilding and future modernization[4]
- Large numbers of ships for convoy escort duty[5]

In September 1966 OASD(SA) estimated that 18 DXGs built to the FY 1967 DDG design would cost $57.2 million each, and 85 DXs would cost $40.8 million each. DX would be a DXG "deconverted" to mount DE 1052 weapons, despite open Navy concern that the DE 1052 class was inadequately armed.[6] With 18 DXGs the Navy could retire all the converted guided missile cruisers and relegate the weak *Brooke* class from carrier duty to ASW service. OASD(SA) reduced the number of DXs to 75 to replace 250 FRAMs and other old destroyers.

The Office of the Secretary of Defense wanted DX to cost under $20 million each. An OSD analyst, or possibly Secretary McNamara himself, prepared a table showing that if initial DX construction costs were about $22 million, then progressive learning-curve savings as production built up should reduce the price to $19.7 million after two years. Secretary McNamara was prepared to support a destroyer costing less than $20 million and laid out his objectives for DX/DXG in a Draft Memorandum for the President in November 1966. This document provides an interesting evaluation of the surface fleet in 1966 and is reproduced as appendix A.[7]

It is not clear how OSD arrived at the $22 million cost.[8] It may have been a design-to-cost goal for the initial ships, to force the price below $20 million for later ships from learning-curve savings. As a cost estimate, it was dangerously misleading because $22 million covered roughly just the hull and machinery for a repeat DE 1052. Weapons and electronics that did not show on ship construction contracts (but were part of the standard ship construction budget) would add $10 million per DX even with no improvement at all over the DE 1052 class. Since DXG would be an aircraft carrier task force escort, commonality between DX and DXG would make DX a larger, faster, and more expensive warship than a DE 1052. Finally, a TPP contract was expected to cover a new mass-production shipyard and crew training facilities. OSD did not include any of these costs in the DX price estimate. Secretary McNamara warned that "all of these figures must be considered highly tentative."[9] Members of Congress remembered only the implausible estimate of $19.7 million; thought that it would be the total cost for the ship, weapons, and electronics; and were enraged when years later a fully equipped *Spruance*-class destroyer cost several times as much.[10] Thus the *Spruance* class and controversy about it both rose from the same source at the same time.

Secretary McNamara's November 1966 Draft Memorandum for the President approved adding $30 million to the FY 1968 budget for Contract Definition contracts for DX and DXG.[11] DX/DXG

"represents the most ambitious peacetime program ever conceived for surface warships."[12] It ended work in progress to resubmit the rejected FY 1967 DDG design, in favor of DXG and its virtue of commonality with DX. Norman Friedman later wrote, "That no DXG would be built at least for a decade was hardly obvious at the time. Whether the demise of DDG FY 67/68 is to be blamed on Admiral Rickover's success in achieving Congressional authorization for two nuclear rather than two conventional [guided missile ships] or whether it was much more a matter of the best (DXG) being the mortal enemy of perfectly good enough is by no means clear."[13] Would the FY 1967 DDG have been good enough? When the *California*-class cruisers joined the fleet in 1974 mounting almost identical weapons, Vice Admiral Eli Reich on the OSD staff described them as "loaded from stem to stern with technically achievable, but not very practical, systems and subsystems. . . . The Navy has done an inadequate job of specifying overall ship system integration design."[14]

The signal to the Navy was that if the Navy followed the TPP rules, then OSD would approve new destroyers. The last destroyer-type warships built to a completely new design were the *Leahy* (DLG 16) and *Bronstein* (DE 1037) classes, designed almost a decade earlier. This interval raised both cost and risk, since more components of the DX/DXG design would be new technology. The fundamental influences on warship development had all changed drastically since then: strategy, the Soviet threat, costs, technology, and operations. In addition there were the new considerations of OSD's domination and of political interest in the DX project as a policy measure to revolutionize warship design and the shipbuilding industry. A primary virtue of TPP was that it committed OSD, the White House Office of Management and the Budget, and Congress very early to a large production run on a definite schedule.[15]

The original DX TPP plan was to buy the entire ship commercially, with no government-furnished equipment (GFE). GFE exposed the Navy to risk of contractor claims for damages if the equipment was defective or was delivered late. For DX the Navy would specify weapons and weapons-related electronics, but the shipbuilder would buy them and coordinate deliveries and installation.[16] New weapons would be developed in parallel with the destroyer shipbuilding project and installed in a shipyard overhaul.

The Navy opened a DX/DXG program coordinator office late in 1966. DX/DXG needed wide support within the Navy and Congress lest it suffer the fate of the FDL transports, the FY 1967 DDG, and the F-111B interceptor, all casualties of naval and congressional antipathy. To sell DX/DXG as another destroyer escort class to the Navy, OSD brought Captain Ray Peet in as DX/DXG program coordinator. He was the first commanding officer of the nuclear-powered cruiser *Bainbridge* and had been a destroyer veteran since the 1943 Battle of Cape St. George, then-Captain Arleigh Burke's tactical masterpiece. As program coordinator, he would set the military requirements for DX/DXG. The Chief of Naval Operations, Admiral David McDonald, and the Chief of Naval Material, Admiral Ignatius Gallantin, personally gave Peet their guideline for DX/DXG: build the best ship. Competition like that between the staffs of the secretary of the Navy and the CNO was common.[17]

The first controversy over the *Spruance* class did not take long to occur. In preparing the Concept Formulation study defining the missions for DXG, Captain Peet emphasized the opportunity for a powerful surface ship. He outlined the operational mission requirements that would make DXGs more powerful than guided missile cruisers:

1. operate offensively, either independently or with strike, amphibious, or ASW forces; and to shield them, replenishment groups, and military and mercantile convoys against air, surface, and submarine threats;
2. detect and destroy missiles, aircraft, and submarines, alone or as part of a coordinated system;
3. destroy shore targets at close range (shore bombardment);
4. provide naval gunfire support for ground forces;
5. surveillance and trailing of hostile ships and submarines;
6. blockade;
7. air control for ASW, search and rescue, and patrol;
8. destroy surface targets at close range;
9. electronic intelligence collection and tactical deception;
10. provide a limited task unit flag plot and accommodations for a task unit commander and staff.[18]

With these missions Captain Peet clearly defined DXG as a high-end warship, not a destroyer escort.

DX would be a reduced DXG. On *Bainbridge* he had seen that with unlimited range at high speed and independence from oilers, one nuclear ship could maintain the same presence on patrol as two conventional ships from the same era. Nuclear power also supported large electronics loads. He knew that gas turbine technology was advancing, and he planned to evaluate costs and engineering developments in both conventional and nuclear propulsion. He preferred to see whether the results of the FDL or LHA shipbuilding projects validated Total Package Procurement before committing the destroyer program to it. Under TPP, minimal Navy input and prohibition of design changes risked locking the Navy into a potentially poor design, which would be fatal for a combat ship.

All this clashed with Secretary of the Navy Nitze's restrictions: use non-nuclear propulsion, follow TPP procedures, and build large numbers of low-cost ships for convoy duty. Secretary Nitze refused, however, to take public responsibility for the construction program or for the ships that might result from it.[19] Peet described his confrontation with the Secretary:

> I would appear before Nitze and give him briefings. Nitze began telling me not to consider nuclear power as well as saying that the ship shouldn't be this or shouldn't be that. Finally I said to Secretary Nitze, "Look, if you want to tell me this in writing, fine. Otherwise, I think that it's not proper. I think that I should get my military guidance from the Chief of Naval Operations." . . . I intended to lay the case out, look at the alternatives and decide what is in the best interest of the Navy, what would be the best escort for us to build. Nitze didn't like that. Nitze . . . went to [the CNO] and said that he wanted me relieved as program manager. He didn't feel that I could be objective as far as he was concerned. He was right, because I sure wasn't going to take the type of instruction that he was trying to give me to direct it.[20]

Getting fired for standing up to McNamara's men did not hurt Captain Peet's naval career. He was promoted, went on to a series of major operational commands, and retired as a vice admiral, declining an offered fourth star.

Published accounts of this incident have been polemical. In the following, the commentators refer to Captain Peet as an admiral, although that was not his rank yet. Secretary Nitze's protégé and assistant, Admiral Zumwalt, remembered it this way:

> The admiral [whom Nitze] put in charge of the project, together with the engineering duty officers on the development staff, found a host of technical reasons to recommend a larger, more expensive ship. [Nitze] fired that admiral and got a new one to manage the project. The new admiral came up with the same findings. My guess is that both of them thought they were speaking for the CNO. Anyway, three project managers later, when the new escort finally got designed, it was the far too expensive DD-963.[21]

Vice Admiral John Hayward, an officer with a career as distinguished as Admiral Zumwalt's, dissented:

> The original admiral in charge of the DX/DXG project, as it was known in those days, was Admiral Ray Peet. He had the courage to point out to Secretary Nitze the lack of wisdom associated with the total package concept for the procurement of ships. The secretary as well as the then-Captain Zumwalt were pushing for warships and merchantmen to be built in accordance with their utopian concept. Ray Peet had the courage to tell the secretary he had better get another man if he was to come up with a directed answer, which the secretary did. He did not come up with a host of technical reasons to come up with a larger ship as Zumwalt states. It is pure fiction! Peet was fired because he pointed out to Nitze the lack of wisdom of the total package concept.[22]

Afterward, Paul Nitze was promoted to Deputy Secretary of Defense. The reasons why he had removed Captain Peet were forgotten. The ironic consequence was that work continued toward getting exactly the same ship that Nitze had opposed, although under Total Package Procurement. The new Secretary of the Navy was Paul Ignatius, a creative and supportive official who had already helped the move to gas turbine propulsion through the *Admiral Wm. M. Callaghan* project.

DX/DXG Program Office Objectives

The DX/DXG program coordinator from February 1967 through source selection in May 1970

was Rear Admiral Thomas Weschler, an expert in ordnance, destroyers, and amphibious warfare. The night before he was to take a new command in San Diego, he was suddenly ordered instead to Washington as DX/DXG program coordinator. Captain Peet had just been dismissed. Admiral Weschler had commanded forces in Vietnam for the preceding 18 months and, no surprise, had never heard of DX/DXG.[23] Combat experience, initiative, intelligence, a diverse career, and knowledge of the Navy made him a perfect leader for the DX/DXG program.

Captain L. D. Caney, just arrived in Washington after commanding Destroyer Squadron 5, was another officer to receive sudden orders to join the DX/DXG project:

> [Captain Peet] had disagreed with the approach planned for the acquisition of this new class so I was told to get with it and make something happen. Soon thereafter I was directed to brief the CNO (Admiral David McDonald) in order to get a formal approval to set the project in motion. I gave the briefing to the CNO and a roomful of admirals. When I finished the CNO turned to Rickover and asked what he thought about the project. His response is unprintable but in essence he said, "I think it's a bunch of crap." Fortunately Vice Admiral [J. B.] Colwell [Deputy CNO for Fleet Operations and Readiness] commented favorably so we went ahead on his recommendation.[24]

The Chief of Naval Material was Admiral Ignatius Gallantin, a World War II submarine hero. Technically knowledgeable and politically astute about working inside OSD and Navy channels, he had supported Total Package Procurement for the FDL and LHA ship acquisition projects. Admiral Weschler had worked for him in the Polaris project office in the early 1960s and had impressed him. It was said of the Polaris project that it was the most successful thing the Navy had ever done, and would ever be allowed to do. Gallantin and Weschler now set up the DX/DXG program along the lines of Polaris.

As program coordinator, Admiral Weschler had five roles in turning DX/DXG from a budget request into a destroyer construction project. First, deciding the mission roles for DX/DXG was paramount. The Flexible Response strategy and the Soviet Union's malignant military growth precluded another destroyer escort class for the obsolete midocean ASW escort mission. An ironic result of Admiral Peet's departure was that his Concept Formulation plan for DXG became the mission basis for both DX and DXG. Captain Peet had expected to prepare a separate plan for DX but had been dismissed before starting it. During a presentation to OASD(SA) Captain Caney discovered that OSD still expected DX to be a destroyer escort and to be even cheaper than the *Knox* class. Captain Caney opposed such ships. The DX/DXG program office made many presentations to persuade OSD that the bare-bones-minimum ship needed was a far more powerful destroyer for forward operations. OASD(SA) required specific studies to justify every major ship characteristic. The program office prepared them quickly.[25] Later, OSD established the Defense Systems Acquisition Review Council (DSARC), which defined a new set of procedures to authorize weapons procurement. After still further justification studies, DX (by then redesignated DD 963) became the first defense program to pass a DSARC review.[26]

On duty in Vietnam in 1966, Rear Admiral Thomas Weschler commanded the U.S. Naval Support Activity at Da Nang before becoming DD 963 program manager. Bringing lessons about destroyer and helicopter operations along the Vietnamese littoral with him to the DX/DXG program, he turned the plan for DX from a cheap destroyer escort into a highly capable modern destroyer. (U.S. Naval Institute collection)

The program office had to judge which of many technological advances were feasible. Unlike aircraft procurement, DX would proceed straight to design and production without an R&D-funded prototype phase. The decision point for each particular possible technological advance was whether to commit the ships' price, effectiveness, and construction schedule to it. The program office required that any principal element not already fleet-proven must have a viable backup that could be installed without major impact on the ship construction project. The ship would be sized to take either alternative.

Second, the program coordinator supported Total Package Procurement. This was the only method under which OSD would approve ship construction, a fact that gave officers who understood TPP significant power within the Navy.[27] The TPP proponents would, along the way, specify the military characteristics of the destroyers. Admiral Nathan Sonenshein from the FDL project was a vital supporter of DX. He had long been a proponent of gas turbines. His ability to get ship construction approved enabled him to bypass centers of resistance to both TPP and gas turbines.[28] For reasons to be explained shortly, use of TPP eliminated nuclear power as a candidate for DX/DXG propulsion.

Third, the program coordinator briefed other Navy officials on the evolving plan for DX. Secretary Nitze settled for use of TPP and did not oppose the shift toward a more powerful destroyer.

DX would clearly be the largest destroyer ever built. Admiral Weschler expected that the ship would displace 6,000–8,000 tons. It had to come in below 8,000 tons to avoid provoking a fight with Admiral Rickover and the nuclear power lobby.[29] Fleet officers objected that 6,000–8,000 tons was too large for a destroyer. Admiral Weschler pointed out that a repeat of the popular but cramped 4,525-ton DDG 2 class with adequate accommodations and air conditioning would displace 5,500 tons for the same armament. The program office persuaded Navy and congressional audiences that DX would be an acceptable destroyer without having "gold-plated" leading-edge technology throughout. It was hard to sell, but it was the only way to keep technology from pushing up costs. The program office quoted, and possibly introduced into American usage, the slogan attributed to the Soviets, "Comrade, *better* is the enemy of *good enough!*" For carrier operations 30 knots in sea state 4 was good enough; 34 or 40 knots, which some wanted, was unnecessary.[30]

Fourth, congressional interest in DX/DXG was high, especially about technology, and required many briefings.

Fifth, the project manager organization had direct access to OASD(SA), the Office of the Chief of Naval Operations (OpNav), and other offices, and had to work beyond the confines of the Naval Ship Systems Command. The DX shipbuilding project would have financial and planning impact far beyond the ship construction budget, such as on helicopters and personnel rate structure requirements. Admiral Weschler coordinated these broader areas.

The Major Fleet Escort Study

With the costs of the Vietnam War and the Great Society straining the national economy, Secretary McNamara evidently began vacillating about whether to continue with DX/DXG. OSD and the Navy still did not share an understanding of the strategic uses of sea power.[31] The head of OASD(SA), Alain Enthoven, sensed that DX/DXG was in jeopardy. He urged Secretary Nitze and Admiral Zumwalt, director of the OpNav Systems Analysis Division, to justify the Navy's need for new destroyers in a presentation that Secretary McNamara would accept. Accordingly, Admiral Zumwalt prepared a new Major Fleet Escort Study in parallel with the DX/DXG program office's developing ship specification. Enthoven also recruited Admiral Sonenshein, the Navy's expert on TPP, to help with the DX project.[32]

The Major Fleet Escort Study's primary objective was to define the numbers of new-design destroyers needed. For costing and armament, the ship design foreseen in the Major Fleet Escort Study was consistent with the DX/DXG characteristics planned as of early 1967. This was for a destroyer with a light-ship displacement of about 4,000 tons, which suggests a full-load displacement of about 5,600 tons and a length of about 485 feet overall. Cost was estimated as $31.5 million. Propulsion was left to the DX/DXG design team. Armament would be two lightweight 5-inch/54 guns mounted fore and aft; a point-defense missile defense system; a small manned helicopter; ASROC; and a hull-mounted medium-frequency sonar (SQQ-23 PAIR). The study group doubted that longer-range ship-mounted active sonars

(SQS-26 and SQS-35 IVDS) could exploit both convergence-zone and bottom-bounce phenomena and thought that with the *Knox* class the fleet would have enough of those sonars anyway.[33] (Chapter 11 will discuss these sonar terms.)

The Major Fleet Escort Study pointed out that the inherently shorter range of sonar relative to air search radar meant that more ASW-capable ships were needed in a screen than AAW ships. Most fast guided missile escorts also mounted sonars and ASW weapons. Reducing the carrier screen from four guided missile ships to three, and adding three DXs for ASW, "while increasing the overall cost of carrier forces by 3%, increases carrier output by 15% by reducing losses to enemy attack." Once this density of screen was achieved, additional force "output" would be better increased by adding more carriers and strike aircraft rather than more escorts. To appease Admiral Rickover and his supporters in Congress, the Major Fleet Escort Study suggested a few nuclear-powered DXGNs (which became the *Virginia* [CGN 38] class) mounting DXG armament.[34] OSD could not complain about statistical imprecision in the study, although others thought that the numbers were examples at best.

On the basis of Admiral Zumwalt's Major Fleet Escort Study, Secretary McNamara recommended funding to design both DX and DXG.[35] The Navy requested 49 DXs and 33 DXGs, of which OSD accepted 40 and 24, respectively, for the five-year defense budget horizon.[36] Congress authorized funds for the first year of the five-year plan; the other four years forecast future requirements. Secretary McNamara reported on the DX/DXG program to Congress:

During the last year [1967] we have intensively studied the entire fleet escort force requirement. We now have a much better understanding of the numbers and types of escorts the fleet will need in the mid-1970s for anti-submarine warfare (ASW) and anti-air warfare (AAW). One of the major conclusions we have drawn from this study is that the ASW requirement should be the determining factor in computing the size of the escort force. . . .

The proposed new escort ship program, entailing an investment of about $3.0 billion, presents us with a unique and most important opportunity to effect a major advance in the management of the Navy's shipbuilding and operating programs, ranging over the entire life cycle of the ships—from design and development to construction, supply, maintenance, and operation. All three classes of ship involved [DX, DXG, DXGN] will have essentially the same operating profile and many of the same characteristics. . . .

Certainly, within each of the three classes, we can build identical ships. While each class of ship will differ somewhat in overall length and displacement, we can expect them to have essentially the same internal and external arrangement and outfitting, the same navigation and communications systems, and virtually the same ASW and gun systems. Propulsion and machinery systems could be common to the conventionally-powered destroyers and guided missile ships, and the missile systems could be common to both the conventionally-powered and nuclear-powered missile ships. By achieving this standardization, we would not only be able to reduce the development and construction costs through multi-year, total package procurements, but the lifetime operating costs of these ships as well, and we would also provide an additional strong incentive for our private shipbuilders to modernize their yards.

Whereas the DX had originally been envisioned as merely a more economical replacement for our DE construction program, it now appears that this ship should be a larger, faster destroyer type. The DX now envisaged would be heavier than our present DEs and be fast enough to escort our attack carriers. It would also have guns for gunfire support missions, and a Basic Point Defense (Sea Sparrow) missile system for close-in air defense as well as the latest ASW equipment.

The DXG would be somewhat larger with the same speed and endurance, and for air defense it would have the new, more capable Tartar-D system, which employs new fire control and search radars and the Standard missile. However, because it is the ASW rather than the AAW requirement which is controlling, we plan to install only one Tartar-D [fire control] system on each DXG. We believe it more advisable to have a greater number of ships with an AAW capability, thus permitting wider area coverage while reducing susceptibility to electronic countermeasures or loss,

than to concentrate the same missile capability on fewer ships. In addition, the DXG will have the latest ASW equipment and will mount one improved 5″ gun.

The DXGN would simply be a nuclear-powered version of the DXG and would be somewhat heavier. . . . Funds are [requested] in the FY 1969 budget for five DXs, for advance procurement of long lead-time items for the two DXGNs to be started in FY 1970, and for contract definition of the DXGN and the DXG. (Contract definition of the DX was funded in FY 1968.)[37]

For FY 1969 Congress approved $25 million for long-lead-time GFE for DX, but no funds for the ships themselves because the TPP construction contract would not be ready to award until FY 1970. Congress denied FY 1969 funding for Contract Definition of DXG because the Navy would not be ready with a DXG specification before FY 1970.[38] The DX program office had succeeded in defining DX as a highly capable warship. This increased the complexity of the ship that the Navy would require industry to design, and it raised the stake that the ship must be a success to justify its price.

Influence of Canada's *Iroquois*-Class Destroyers

The disruption of naval shipbuilding since 1962 had deprived the Navy of applied familiarity with new-technology systems such as gas turbines. DX program office personnel visited the two closest sister navies, the U.S. Coast Guard and the Royal Canadian Navy. The Coast Guard was consulted about gas turbine propulsion and about messing and berthing for an all-volunteer crew. Improvements in shipboard habitability were needed for the all-volunteer military, which, it was clear, would soon result from national resentment of the Vietnam draft.[39]

A team of 20 engineers and naval architects visited Ottawa to study the detail design for the Royal Canadian Navy *Iroquois* (DDH 280)–ciass destroyers.[40] These were the West's first new-construction warships with all-gas-turbine propulsion. Members included Captain J. R. Baylis, soon to be technical director of the DD 963 ship acquisition project, and Captain Richard Henning from the FDL office, soon to become the DD 963 ship acquisition project manager. The Canadian destroyers, also known as the Canadian Tribal class,

Although the relation is not externally apparent, Canada's *Iroquois*-class destroyers (the Canadian Tribal class) are the closest ancestors of the *Spruance* class. Throughout the DX/DD 963 program, U.S. Navy officers from the DX Program Office studied the *Iroquois* design intensively, especially its use of helicopters, gas turbine propulsion, and acoustic silencing. (Canadian Forces photo by WO Al Clarke)

were being ordered for open-ocean ASW. They featured an advanced all-digital combat system descended from the ASW combat system designed at the U.S. Naval Electronics Laboratory.[41] The Canadian Navy had been the world's first to operate helicopters as destroyer weapons. The *Iroquois* class was the Canadians' third generation of helicopter-armed destroyers (DDHs). Sikorsky CHSS-2 (SH-3) Sea King helicopters were their second generation of ASW helicopters to operate from destroyers, as distinct from ASW carriers. The U.S. Navy destroyer force had barely experimented with its first generation. Captain Henning studied Sea King flight operations aboard a Canadian frigate.

The *Iroquois*-class destroyer hull derived from a series of designs that originated with a postwar Royal Navy requirement for good seakeeping for operations against fast submarines in the North Atlantic. (The Royal Navy Type 12/*Leander*/Type 22 frigate classes were another adoption of the same specification.) Canadian destroyers were designed for ASW operations in icy seas, with less emphasis on range and endurance than the U.S. Navy.[42] With U.S. Navy assistance from the Seahawk experiments, the Canadians designed the *Iroquois* class for a COGOG propulsion plant of two 3,700 shp cruise gas turbines and two 25,000 shp FT4A boost gas turbines. The design had two propeller shafts and controllable/reversible-pitch (CRP) propellers. The U.S. Navy team was particularly interested in the shock mountings for the gas turbines. Armament was an Italian OTO Melara 5-inch/54-caliber gun, the Sea Sparrow point-defense antiaircraft missile system, and a flight deck and hangar for two large SH-3 ASW helicopters. All these engines and weapons except for the gun were American, but none was aboard any U.S. Navy warship in service or under construction.

The relation is not externally apparent, but the closest ancestor of the *Spruance* class is the *Iroquois* class. The DX/DD 963 specification showed many similarities to the *Iroquois* design: seakeeping, interest in innovative propulsion and auxiliary plants, Sea Sparrow, acoustic silencing, multiple sonars and ASW weapons, an all-digital ASW combat system, and a large helicopter facility. More subtly and more significantly, the Canadian design seems to be the baseline configuration of technology for DX.

The *Iroquois* class was about two years ahead of DX. The American team considered how they might improve DX in comparison. They noted, for example, that *Iroquois*'s outward-facing stacks for gas turbine exhaust would present an infrared target and decided that DX needed infrared suppression.[43] Later, DD 963 program officers visited the building yards in Quebec. Accepting that the *Iroquois* design was practical and would result in successful destroyers, the U.S. Navy DX project managers evaluated equivalent innovations as risk-free in the evolving DX/DD 963 design, even when the U.S. Navy itself had little or no operational experience with the technology. This helps to explain their confidence in insisting on advanced

all-gas-turbine propulsion and CRP propellers over doubt and opposition from OSD, powerful members of Congress, and other Navy authorities.

DX: Advancing Technology

The DX project office could in theory start with a clean sheet of paper and its knowledge of naval missions and technology.[44] In practice, it authorized only systems that the Navy had already tested satisfactorily. This reduced the risk in an

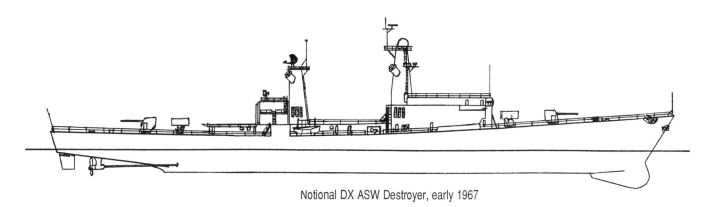

Notional DX ASW Destroyer, early 1967

Search radar:	SPS-49 2-D air search
Guns:	2 5" Mk 45 guns, Mk 86 fire control with SPQ-9 surface-search radar, 1 remote optic sight
ASW:	SQS-26 or SQS-53 sonar, ASW helicopter, ASROC
AAW:	Mk 115 Sea Sparrow with manual director
Propulsion:	Steam

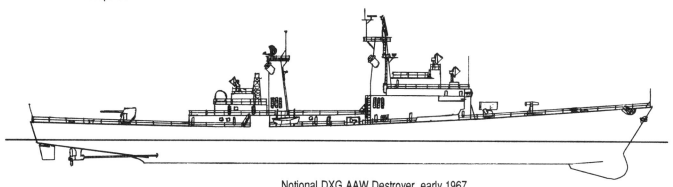

Notional DXG AAW Destroyer, early 1967

Search radar:	SPS-48 3-D air search, SPS-49 2-D air search
Guns:	1 5" Mk 45 gun, Mk 86 fire control with SPQ-9 surface-search radar, SPG-60 illuminator, 2 remote optic sights
ASW:	SQS-26 or SQS-53 sonar, ASROC
AAW:	1 Tartar-D launcher with Mk 74 fire control system, 2 SPG-51D illuminators, Shortstop ECM
Propulsion:	Steam

These drawings, based on several sources, depict typical Naval Ship Engineering Center (NavSEC) studies for DX (*top*) and DXG with a common hull and oil-fired steam boilers for propulsion. Hull length is about 485 feet overall; displacement for the ASW-oriented DX would be about 5,600 tons full load. In 1968 NavSEC prepared an unofficial DD 963-class design, probably very similar to this DX design. In particular, it still used steam propulsion, not gas turbines.

ambitious design project. It would speed delivery of the new destroyers, since it would take longer to develop more advanced combat systems than to build ships to mount them. The DX project sought to advance naval technology with many developments. The key to resolving this apparent paradox was to provide large growth margins so that new-development systems could be installed after testing was completed.

DX brought major, sometimes revolutionary, advances to the Navy in engineering, weapons, habitability, modernization, integration, and construction. Many of these advances were interrelated. For example, propulsion silencing would amplify sonar range and sensitivity. For the first time, ship system integration was required of the shipbuilder. The ships and all new systems had to work as built without a prototype phase for testing. This consideration was a further reason along with cost to avoid gold-plating.

ENGINEERING

The project office sought to install gas turbines, controllable/reversible-pitch propellers, and automated propulsion control for DX/DXG.[45] As a company-funded initiative, General Electric marinized its TF39 turbofan engine from the C-5 FDL transport aircraft (built despite the demise of the corresponding FDL ships) as the LM-2500 gas turbine. The LM-2500 promised 20 percent better fuel economy than the earlier FT4A but was untested in the naval environment. Specific fuel consumption for the LM-2500 at full power was about the same as a steam plant of similar power output (20,000–25,000 shp).[46] The design staff of the Naval Ship Engineering Center still favored steam propulsion.[47] One advantage for the U.S. Navy of going to industry to design DX under TPP was thus to avoid such areas of conservatism within its own design organization. The Navy would give the competing DD 963 design contractors the voluminous research studies on gas turbines from Project Seahawk.[48]

The Navy stocked two grades of fuel oil: black oil (NSFO) for most steam plants, and costly jet fuel (JP-5) for aviation, diesel engines, and the *Brooke-* and *Garcia*-class destroyer escorts. After his experience as Pacific Fleet maintenance officer during 1960–62, Admiral Sonenshein, Deputy Chief of Naval Material for Logistics Support after leaving the FDL project, wanted to get rid of black oil. Navy distillate fuel (NDF) was twice as expensive as black oil, but its higher quality so greatly

reduced boiler maintenance that total cost would decline. It was commercially available worldwide as diesel fuel marine (DFM). To support a new fuel policy, the project office required DX to use NDF. Competing DD 963 designers could still propose a steam plant to burn black oil, but it would be pointless because the Navy still would evaluate life-cycle costs based on NDF. A single shipboard fuel system for both viscous black oil and low-viscosity higher-grade NDF was impractical. This specification was extremely significant in a Total Package competition, because without cheap black oil, steam's life-cycle cost advantage for a destroyer's operating profile disappeared compared with gas turbines.[49] Both winning Total Package ship classes designed under the earlier fuel policy, FDL and LHA 1, used steam. Above all, it was unlikely that a steam-powered ship could be shown to meet the silencing and fast-response requirements for DX. OSD approved NDF (NATO fuel symbol F-76) as the standard shipboard fuel in 1969, and the Navy supply system quickly converted to it.[50]

HELICOPTERS

The DASH drone helicopter had been a complete failure: costly, with a very high loss rate and practically no exercise kills. It was canceled in 1967 just as the DX program was developing. A large new ASW helicopter, backed by the OpNav aviation desks and intended to carry a complex airborne ASW system such as that on the proposed VSX (S-3A) aircraft, had been disapproved. From experience in Vietnam, Admiral Weschler insisted on a versatile helicopter, not just for ASW but for reconnaissance, search and rescue, over-the-horizon targeting, cargo and passenger transfers, and if weapons were available, antiship attack. He soon took over the Ships Characteristics desk at OpNav and transferred development of the new helicopter, the Light Airborne Multipurpose System (LAMPS), there from the DX program office, which was busy enough.[51] An integrated ASW fire control system was specified to tie together the helicopter, sonobuoys, and long-range passive sonar.

In 1967 the only available ASW helicopter was the large SH-3, so a hangar and flight deck for this aircraft were required for DX, as aboard the *Iroquois* class. Canada had developed a winch, called Beartrap, to help land the big SH-3s on small ships, but Admiral Weschler and many helicopter pilots hoped not to need it. American sailors were startled to see Soviet destroyers, now appearing frequently on the high seas, keeping the sea better in bad weather than the FRAMs and other older American types. Helicopter operations and the prospect of fighting stable new Soviet ships in storms and the rough North Atlantic required DX to have good seakeeping qualities, and specifically to operate helicopters in rough seas (sea state 6: 14-foot waves, 27-knot wind). A large CH-46 cargo helicopter had to be able to land on the flight deck. This feature was to prove extremely useful in future operations.

WEAPONS

The missions for DX were an ASW-oriented subset of the original DX/DXG missions:[52]

- Operate with strike, amphibious, or ASW forces, to shield them, replenishment groups, and military and mercantile convoys against submarine threats
- Detect and destroy submarines alone or as part of a coordinated system
- Destroy shore targets at close range
- Provide naval gunfire support for ground forces
- Conduct surveillance and trailing of enemy ships and submarines
- Conduct blockades
- Provide air control for ASW, search and rescue, and patrol

By 1978 the destroyers assigned to the combat fleets had additional missions. These showed the continued importance of the original DXG cruiser missions for the DX/DD 963 class:

- Destroy surface targets out to 60 nautical miles
- Provide electronic intelligence collection and tactical deception
- Detect, decoy, and destroy attacking aircraft or missiles

Weapons and electronics requirements included ASW weapons and helicopters to attack hostile submarines, preferably beyond submarine attack range; guns and/or missiles having destructive and range capabilities to equal or exceed those of opposing units; point-defense missile systems and AAW capability for the ship's main gunfire control system; and command and control facilities to coordinate on-scene ASW units. (The squadron flagship facility specified in the Concept Formulation plan was deleted from the DX ship specification to save the cost of additional NTDS display consoles and software programming. The DXG/DXGN types were expected to have a flag plot, but ultimately it was dropped from them, too.)[53]

Contemporary surface ASW doctrine used active sonar almost exclusively. DX's advances in ASW included ship silencing, passive sonar search, LAMPS and sonobuoys, and a powerful computer system to integrate them all. The surface Navy was not noise-conscious. Captain Baylis, DX technical director, was a submarine designer and very noise-conscious. DX was required to be much quieter in terms of radiated hullborne noise than any previous surface warship. The DD 963 contract offered an incentive of $1 million per ship for better silencing than the minimum requirement. This was the only incentive fee in the contract.

The combat system was to be automated, all-digital, and integrated, using remote-control torpedo launching. The DX program office developed doctrines for designating ASW weapons to sensors so that computer programs could incorporate them as automatic procedures. OpNav approved these procedures as new combat doctrines. This was the reverse of the usual relation between OpNav and ship designers.[54]

CREW SIZE AND HABITABILITY

Standard fleet crew size for a ship of DX's complexity would be about 400. Admiral Weschler intended to cut it to 250 by emphasizing automation, such as elevators for ammunition and stores, and low-upkeep internal fittings and hull paint. Cutting the crew to 250 would save $75 million over the life cycle of each ship, a saving that was higher than the estimated contract cost of the ship. Total Package competition required the industry designers to specify the crew size and ratings required so that the Navy could estimate life-cycle costs. This favored highly automated systems such as gas turbines.

Destroyer accommodations had traditionally been spartan. Asked to describe habitability aboard his World War II destroyer, one admiral recalled, "She didn't have any." As late as 1954 the *Encyclopaedia Britannica* stated, "Although much was added to the destroyer during the course of its development, little was left off; the hull is so packed with machinery that provision for the crew appears to have been made only as an afterthought."[55] Berthing areas and mess decks were passageways. They were noisy, crowded, badly ventilated, and subject to maximum pitch and roll amplitude. Heads were inconveniently distant from berthing compartments. Ships still in service in the 1950s had troughs; sailors thought it fun to ignite a wad of toilet paper and float it downstream under their shipmates. DX was required to have living spaces with heads and showers clustered together off the passageways and located to minimize pitch and roll effects. Berthing compartments were to be larger, air-conditioned, and easy to maintain, with privacy, individual bunk lights, and lockers for each man. The Navy cited the Coast Guard *Hamilton*-class high-endurance cutter to DD 963 designers as an example of what it wanted.

MODULARITY, MODERNIZATION, AND CONVERSION

Admiral Weschler pointed out that in a way, a warship had to be paid for three times: initial construction, midcareer modernization, and the cumulative pay for the crew. The near-term need was for battle group ASW escorts, but the threat from aircraft and antiship cruise missiles might increase in the future. In a significant innovation for warship design, Admiral Zumwalt's Major Fleet Escort Study suggested reserving space, power, and weight margins if practicable in DX, to install the DXG weapons suite quickly in case the air threat did increase.[56] Similarly, development of new ASW weapons was a certainty, but the characteristics of the new systems were still indefinite. Converting six DL 2– and DD 931–class destroyers to guided missile ships took three years each. Modernization of older ships caused new equipment to intrude into living spaces, further increasing crowding, heat, and noise.

DX would have a module-based design both for anticipated initial construction savings and so that the ships could be modernized or converted to DDGs without rebuilding them. Standard design practice was to leave a 10 percent margin for equipment additions during a ship's career. DX required an unprecedented 25 percent margin, with the additional 15 percent earmarked for specific weapons systems. The ships would initially have unused capacity in their electrical generating plants, cooling systems, dry nitrogen systems for radar waveguides, and other utilities to support systems to be added later. Modernization margins would reduce the ships' contract construction cost, since the Navy would not pay to install the new weapons until a future overhaul. The margins also reduced risk by removing developmental systems from the construction schedule. The DX specification defined the interfaces needed

to mount two alternative weapons suites during modernization and conversion. This was analogous to expansion slots in a computer, but it was unprecedented in naval architecture.

Modernized DX destroyers would remain ASW ships. Growth room was reserved for future systems under development. The most significant developmental system was a passive towed-array sonar. Oceanographic research showed that a breakthrough had been made for this technology, but a deployable towed-array sonar needed to be developed. Admiral Weschler presciently recognized that a towed array would be absolutely essential for future ASW. Space and weight were reserved for a towed array and for additional computers. Those reservations were apparently sized for the SQS-35 active-only variable-depth sonar, but the DX project office never planned to install that system. Submarines began operating with a towed-array sonar (BQR-15 STASS) in 1969. It took 15 years for the DD 963s to receive a tactical towed-array sonar integrated with their combat system (the SQQ-89 ASW modernization program with the SQR-19 towed array).

During modernization a new twin-arm missile launcher with a 24-round magazine (Mark 26 Mod 0) could replace the ASROC box launcher and loader. The new launcher could support more missile types, such as Harpoon and Standard, as well as ASROC. A major challenge to the competing shipbuilders was to design an automatic ASROC loader to fit within the space reserved for the compact Mark 26. The Mark 26 was planned for DXG, DXGN, and Aegis (then the Advanced Surface Missile System fleet air defense project) so it offered commonality and simplified logistics.

Converted DX ships would become guided missile destroyers mounting the Tartar-D system. As with the modernized ships, the ASROC launcher would be replaced by a Mark 26 24-round launcher. The conversion specification further required the destroyer to support a second Mark 26 twin-arm missile launcher but with a larger 44-round magazine. Permissible sacrifices would be the point-defense missile system, one 5-inch gun, the towed sonar reservation, and the helicopter hangar, despite the loss of ASW capability. Conversion would require additional computer power and the large SPS-48 height-finding air search radar to replace the original SPS-40.

Modularity proponents took the concept a step farther to "podularity." This was an advanced notion to reserve weather deck space with clear firing arcs and standardized mechanical and electrical interfaces for prefabricated deck-mounted weapons. Admiral Weschler brought the idea from mission-specific external stores on combat aircraft. Specific weapons included an improved electronic warfare system (which entered service as SLQ-32) and close-in antiship missile defenses (Mark 15 Phalanx 20 mm guns). (The later German MEKO concept implements a type of podularity, using deck-penetrating openings for a selection of pallet-mounted weapons.) The program office suggested incorporating flag-plot electronics in vans that could be mounted on the helicopter deck and connected to the ship's combat information center to make any DX a flagship. A van-based flag plot was never developed.

Modernization and conversion margins were Navy ideas. No OSD document consulted for this study mentions the potential to add or change DX mission capabilities, although OASD(SA) appreciated the value of the modernization and conversion margins.[57] OSD officials thought that some missions must be incompatible, and thus that design features to support them could never be cost-effective, although OSD as an organization clearly was not qualified to make such judgments.[58] An example may have been a notion that an AAW ship such as DXG had more important duties than shore bombardment, therefore would never be so employed, and so did not need as powerful a gun battery as DX. As another example, Harold Brown, OSD director of research and engineering (and later Secretary of the Air Force and Secretary of Defense), once told Congress that since aircraft carriers supported amphibious assaults, he doubted that they needed to go faster than amphibious transports, about 20 knots.[59] Aside from his landlocked assumption that voyages must start at the same time and place, aircraft carriers need 30-knot speed to land aircraft. Further, they arrive well before an amphibious force for surveillance, to clear defenses, and for other tasks. The DX program office explained destroyer mission requirements to OSD many times.

SHIP INTEGRATION AND CONSTRUCTION

Several Navy weapons managers, not wanting to lose the influence that contracting authority gave them, claimed that a shipbuilder could not buy their complex new systems. The DX project office gave in on this point for 11 systems, which

was still much less than for previous ships. The project office ultimately was to find that managing these government-furnished systems for the DD 963 class was more difficult and more costly than if they had been bought by the shipbuilder as originally planned.[60]

The CD contracts would pay for three preliminary designs. The TPP production contract would be awarded to the shipbuilder whose DD 963 plans were evaluated as the most cost-effective. Costs included detail design, construction, contractor-furnished equipment (CFE) installed on board, a one-year warranty, spare parts, crew training, fuel consumption, and other operating costs.[61] The quality of the destroyer design, although it carried the largest weight in the technical evaluation, was not the only factor of interest. "The hardest thing in the whole program was to convince industry that you meant business, that you really wanted design innovation," Admiral Weschler recalled.

A systems engineering plan was required to join the destroyer design's technical details with the production plan. Shipbuilders had protested that the recent FDL competition favored shipyards owned by aerospace firms. OSD and the Navy assured them that competition for the DX contract, the largest in the history of American shipbuilding, would be open to all shipbuilders.[62] However, the same procedural base as for FDL was used, and a contemporary article reported, "While there is left open the possibility that a builder will be able to suitably modify and improve an existing yard, discussions with Defense officials leave little doubt that what is really sought is a completely new yard with the central building ways located indoors and with the variety of ships and heavy-lift equipments located along the flow lines leading into the main construction area."[63]

DX: Weapons and Performance

Congress authorized $30 million for Contract Definition of DX in FY 1968, requiring the contracts for competitive designs to be awarded by June 30, 1968. Admiral Sonenshein, while still FDL project manager at OpNav, convened a panel to draw up DX characteristics (September 16–23, 1967).[64] Admiral Mark Woods, commander of the Naval Ordnance Systems Command (NavOrd), assigned his best officers and designers to the project to contribute to the combat system.[65] The

DX ship development plan specified the DX performance requirements as statements of the warship capabilities, operating profiles, and weapons and government-furnished equipment that the design was required to support.

Sea state 4 (force 5 on the old Beaufort scale) is 4- to 8-foot waves, or moderate seas; sea state 5 is rough seas. The committee specified 30 knots in sea state 4 at full-load displacement to accompany aircraft carriers and potential high-speed merchant ship convoys.[66] Sea state 4 or better occurs about 54 percent of the time on the open oceans in the Northern Hemisphere, where the ships would expect to fight the Soviet Navy. These requirements precluded a single-shaft destroyer escort propulsion plant. The specification explicitly required two shafts and propellers for redundancy, maintenance, and reduced cavitation.[67] The main deck forward of the fantail had to be completely enclosed so that the crew could transit between compartments without going out onto the weather decks.

Older destroyers refueled every two to three days from ships in company. The new destroyer had to minimize imposition on escorted carrier battle groups, convoys, and amphibious groups for support. DX was required to operate for six days without refueling, the same schedule as a carrier, so that DX would never need to refuel from the carrier.[68] Required endurance was 6,000 nautical miles, roughly a round trip to the Norwegian Sea or one-way to Japan, to support transocean convoys and carrier raids against the Soviet littoral. OASD(SA) thought this range requirement was reasonable. Propulsion could be by steam, diesels, gas turbines, or any combination of them. Admiral Sonenshein knew that only gas turbines could provide the necessary silencing and response. The DX specification did not dictate them, because to do so would remove the shipbuilder's contractual responsibility to meet the performance requirement, as opposed to simply delivering a ship with gas turbine propulsion.[69]

Other requirements for the contractors' preliminary designs included these:

- 25-knot sonar search speed combined with silencing for long-range active and passive sonar detection
- 30-knot sonar tracking
- Good seakeeping for sonar search in rough seas and to close with and attack high-speed submarines

- Helicopter operation up to sea state 6
- Low radiated noise to minimize susceptibility to homing torpedoes and to reduce detectability by submarine sonars
- Effective radar and electronic warfare systems
- High speed and good maneuverability
- Fast-response countermeasures[70]

Despite Admiral Sonenshein's effort to tailor TPP to the realities of American shipbuilding, a significant difficulty right from the start was that no shipbuilder alone had the Navy's experience in the preliminary design of a major warship. How was industry to judge the military importance of possible design factors in making trade-off decisions? Would the additional cost of, for example, active fin stabilizers make a design less competitive? Would the Navy favor a design for a small, congested, but cheap ship, or a larger one that could offer better habitability and wider gunfire arcs? Would OSD demand mass production even of an inferior design? Gibbs & Cox warned, "Ability to construct ships efficiently and rapidly at a favorable price is also not necessarily associated with the ability to produce effective designs."[71]

Two new lightweight 5-inch/54-caliber Mark 45 gun mounts and the new all-digital Mark 86 gunfire control system were specified for naval gunfire support and antiship attack. In August 1967 NavOrd suggested designing DX to allow a future rearmament with an 8-inch gun, or a new 175 mm gun being designed in a joint Army-Navy project, for naval gunfire support.[72] OASD(SA) was surprised that ship strength and stability required, in terms of ship size, an additional 300 tons to support each 5-inch gun, about the caliber that the Army mounted on a 50-ton tank.[73]

DX was the first U.S. Navy class designed to mount an antiship cruise missile defense (ASMD) system. The original Sea Sparrow system (Basic Point-Defense Missile System) with one manual fire control director was specified for antiaircraft warfare and ASMD. It was a primitive weapon even on paper, but it was all that was available. NavOrd disliked the small antiaircraft battery but acceded to it since the ships could be converted to guided missile destroyers if the air threat worsened.[74]

Emphasis on antisubmarine warfare led the committee to begin with the armament and missions for the FY 1969 destroyer escort, the most advanced previous ASW ship design concept, as the starting point for DX.[75] A large SQS-26CX

Table 4.1 Combat systems for initial DX configuration and for modernized and converted destroyers

Systems	DX	Modernized DX	DDG Conversion
Search radar	—SPS-40B 2-D air search	—Same	—SPS-48 3-D air search to replace SPS-40
	—SPS-55 surface search	—Same	
Helicopter	—Hangar for 1 SH-3 or 2 UH-2 helicopters	—Same	—Hangar removal permitted
Guns	—2 Mk 45 5-inch/54, 1,200 rounds	—1 8-inch/55 or 175 mm replaces 1 Mk 45 5-inch/54	—1 5-inch removal permitted
	—Mk 86 gunfire control with SPG-60 and SPQ-9 radars		—CW illumination channel added to SPG-60 for Standard-MR guidance
		—2 20 mm Vulcan/Phalanx CIWS added	
	—2 remote optic sights (ROSs)		
ASW weapons	—1 8-cell ASROC with auto-loader, 24 missiles	—1 Mk 26 Mod 0, 24 ASROC/Harpoon missiles	—1 Mk 26 Mod 0 launcher, 24 ASROC/Harpoon/Standard-MR missiles
	—2 triple Mk 32 12.75-inch TT	—Space reserved for passive towed array with size = SQS-35 active IVDS	—No towed array foreseen
	—Mk 116 underwater fire control		
	—Sonar: SQS-26CX		
AAW weapons	—1 8-cell Sea Sparrow BPDMS	—Improved NATO Sea Sparrow	—1 Mk 26 Mod 1 launcher, 44 ASROC/Harpoon/Standard-MR missiles
	—Mk 115 manual director	—Mk 23 target acquisition system	
		—1 Mk 74 fire control with 1 SPG-51 illuminator	
Other weapons	—Chaff rockets	—Harpoon ASCMs	—Harpoon ASCMs
		—Fleet EW system (SLQ-32)	—Fleet EW system (SLQ-32)

sonar and the all-digital Mark 116 underwater fire control system developed for Seahawk were specified. The SQS-26-series sonar had a passive receiver. Modern convoys would probably sail faster than 20 knots, requiring a submarine to go faster than that to intercept it. At such speeds submarines of the era would make so much noise that a passive sonar would detect them easily.

The Navy-specified weapons payloads for DX and DXG are shown in table 4.1. A mission representative of the times was to recover manned spacecraft following splashdown, requiring capability to operate large helicopters such as the SH-3. A similar specification was for the combat search-and-rescue mission of picking up military aviators ditching at sea.

DXG and DXGN Merge

Advanced all-digital versions of the Naval Tactical Data System, the Tartar-D AAW system, and the new all-digital multipurpose Mark 26 missile launcher were available for DXG. OSD expected DXG to be a "single-ended" guided missile ship mounting one Mark 26 missile launcher (44-missile magazine) to fire both Tartar-D (Standard-MR) and ASROC ASW missiles. The Navy wanted an independent ASROC launcher in DXG. Transferring launchers as government-furnished equipment from old destroyers reduced cost, and just the loader would be new. Aegis used the Standard-MR missile airframe and the Mark 26 launcher, so commonality suggested savings if Aegis ships, DXGs, and converted DXs all mounted Mark 26 launchers for all missiles. DXG and converted DXs grew to two Mark 26 launchers: one to replace the ASROC launcher (24 rounds) and one for AAW missiles (44 rounds).[76]

Aegis was promising.[77] Aegis and nuclear propulsion were both government-developed technologies that would dominate any ship design. Admiral Weschler therefore recommended that DXG should be both nuclear-powered and Navy-designed for Aegis compatibility, not designed and built under TPP. A 1967 sketch depicts a notional DXG mounting Aegis (ASMS) target illumination radars, but not the primary phased-array search-and-engage radar.[78] In 1967–68 Congress had voted funds for the two nuclear-powered cruisers that Secretary McNamara refused to build. The program office thought that these ships should be built as prototype DXGs featuring mechanically scanning radar (SPS-48) and armed with a missile system compatible with Aegis.[79] Aegis might be ready for the third and subsequent DXG ships built. DXG would thus merge with DXGN. The nuclear propulsion lobby cared more about whether warships were nuclear-powered than about their armament. Both funded cruisers, the *California* (CGN 36) class, were built mounting an obsolescent weapons suite of Mark 13 missile launchers and an analog ASW system. DXGN became the *Virginia* (CGN 38) class.

That was as far as plans for DXG ever developed. The capability to convert DXs quickly to guided missile ships reduced the urgency. Resuming work toward a conventionally fueled DXG, for Total Package Procurement and with commonality with the contract DX design, remained possible. However, in 1970 Admiral Weschler concluded that the experience of the DD 963 procurement showed that TPP should not be used for a ship as complex as a guided missile destroyer, even without ASMS or nuclear propulsion.[80] (See chapters 6 and 12.) The *Kouroosh/Kidd* (DDG 993) class is not a DXG design but a converted DX plus an additional SPG-51D missile director. For Admiral Zumwalt, the successor to DXG was the *Oliver Hazard Perry*–class frigates. In terms of ship characteristics, another successor is the *Arleigh Burke* (DDG 51) class, which has the same mission requirement: armed for task force air defense, built in large numbers, and equipped to support ASW but without a helicopter facility.

The DD 963 Ship Acquisition Project Office

In 1958 the United States had bought its 962nd authorized destroyer, an old Royal Navy ship with the incongruous name of HMS *Charity,* for transfer as an article of military foreign aid to Pakistan. DD 962 never had a U.S. Navy name. She became the Pakistani Navy's *Shah Jahan,* named for the builder of the Taj Mahal, Emperor of the World until he was overthrown for waging unsuccessful wars and bankrupting the kingdom.[81] Early in 1968, while President Johnson was being similarly deposed for similar offenses, the first projected DX

was assigned the next sequential destroyer authorization number and became DD 963.

In keeping with OSD directives, the Navy set up a DD 963 ship acquisition project office within the Naval Ship Systems Command. The ship acquisition project manager prepared the request for proposals and would evaluate the competing designs and cost bids from the responding shipbuilders. This was a major effort because electronics makes any modern ship design complex, and because multiple competing designs needed to be evaluated and compared. Captain Baylis's Technical Division was the strongest part of the project office. His staff addressed all the disciplines of naval architecture (hull and fittings), marine engineering (propulsion and auxiliary machinery), electrical engineering, electronics, weapons, radar, and sonar, and could cope with problems and integration among these different systems.[82]

Rear Admiral Henning, previously FDL technical director, was the DD 963 ship acquisition project manager. He headed the Source Selection Board of working-level professionals who would evaluate the prospective shipbuilders' technical proposals in detail, and report their evaluation to the Source Selection Advisory Council (SSAC), headed by the commander of the Naval Ship Systems Command. The SSAC held the cost proposal evaluation prepared by a separate panel of cost estimators and contract specialists. The SSAC reported to Admiral Gallantin. It was clear that the huge DD 963 contract award would require political ratification by the Secretary of the Navy and the Secretary of Defense. Official selection of contractor would be made by the Secretary of Defense.[83]

The complex organization for the source selection evaluation mystified some in Congress and industry. Almost 350 civilian and military representatives from 27 naval commands participated.[84] Many officials who were not on the source selection committees still had opinions about "what the Navy really wants" in its new destroyers. Competing shipbuilders wondered throughout the design phase whether to respond to conflicting advice from officials whose preferences might be either irrelevant or significant.

The Six-Day War

During the Arab-Israeli Six-Day War (June 1967) the Soviets aligned themselves with the Arabs and reinforced their Mediterranean fleet with 10 additional ships. Soviet harassment of U.S. Navy operations in the Mediterranean was severe during the Six-Day War, with deliberate attempts to cause accidents. Leonid Brezhnev called for international pressure on the United States to withdraw the Sixth Fleet from the Mediterranean. Denying use of the eastern Mediterranean Sea to the U.S. Navy was of great interest to the Soviets because Polaris A2 submarines and Sixth Fleet aircraft carriers stationed there could launch strikes on the southern Soviet Union. Further, withdrawal of U.S. naval forces would leave the Soviets' anti-Western Arab clients relatively stronger in that region.[85]

Israel's swift military victory in the Six-Day War contrasted sharply with the hobbled and bloody campaign in Vietnam. Disenchanted and frustrated, the week the Six-Day War ended Secretary McNamara commissioned the Pentagon Papers study to find out how the United States had become so deeply involved in Vietnam. In August the Senate convened hearings on the Vietnam War, during which the Joint Chiefs of Staff privately decided to resign en masse to protest Secretary McNamara's misrepresentations of their views. Then they decided not to resign, but they ensured that the Senate hearings exposed military doubts about the mendacious defense secretary. President Johnson soon announced (if a press leak can be called an announcement) that McNamara would be dismissed. The demoralized secretary kept his post until a new position at the World Bank opened early in 1968, but his drive was gone.[86] This may have spared DX, approved thus far, from the budget axe that Alain Enthoven apparently noticed. In January 1968 OSD approved the DX development plan. At the end of the Johnson administration Enthoven returned to his Los Angeles home and became a vice president at Litton Industries.[87] His 1971 book *How Much Is Enough?* on his Pentagon career mentioned neither DX, despite his interest in its survival, nor his subsequent employment.

The Navy had long used destroyers for intelligence missions, but the new electronics systems and their crew requirements had grown beyond a small destroyer's capacity. During the 1960s low-cost surveillance ships were converted from freighters. These ships were almost defenseless and operated alone, without the air cover available to *Maddox* during the Tonkin Gulf incident. During the Six-Day War the Israeli Navy and Air Force repeatedly attacked USS *Liberty* (AGTR 5), one of the new type. They left the ship a constructive total loss, with 34 Americans dead and 75 wounded. The only American aircraft that could support her promptly were reportedly naval nuclear-armed F-4 Phantoms that were recalled to their carrier by authorities in Washington, who were more alarmed by their launch than by the incipient massacre aboard *Liberty*.[88] The feisty vice CNO, Admiral "Rivets" Rivero, ordered the other ships to be armed with at least 3-inch guns, but their operations continued before the directive was carried out. USS *Pueblo* (AGER 2) was on her final mission during the life of the DX committee. Noticing that she was unsupported, North Korean naval forces seized *Pueblo* and a wealth of intelligence material without resistance on January 31, 1968, and imprisoned her crew for a year. The DD 963 class would return surveillance missions to the destroyer force.

Discoveries about Soviet capabilities were alarming. Awareness dawned that the Soviet Navy might be capable of much worse than submarine attacks on slow transports, the basis for concentration on destroyer escort construction. The Soviets believed that they could use long-range surveillance to cue their forces into place to intercept carrier task forces. Soviet submarines could go faster than the intelligence community had estimated. A first-generation Soviet November-class submarine, until then thought to have a maximum speed under 25 knots, pursued *Enterprise* at about 30 knots, fast enough to make an intercept from most bearings.[89] Among U.S. submarines, only the *Skipjack* class (SSN 585; six built, one lost in 1968) could make 30 knots, and they were not suited for ASW. This incident probably gave decisive impetus to political approvals for the DD 963 destroyers, new towed-array sonars, and the *Los Angeles* (SSN 688)–class submarines. The new second generation of Soviet nuclear submarines included the Charlie I class, which could fire antiship missiles from underwater in a "pop-up" attack at a range of about 30 miles. That was well inside an aircraft carrier's escort screen and indicated Soviet naval confidence that Charlie I submarines would get that close. The Soviets even tested a submarine-launched ballistic missile for antiship nuclear attack.[90] Finally, any doubt that Soviet naval tactics and conventional weapons could vanquish existing Western warships vanished in the Mediterranean Sea on October 21, 1967.

During the insecure armistice between Egypt and Israel following the Six-Day War, the old British-built Israeli destroyer *Elath* was patrolling 15 miles off Port Said when a lookout sighted a large incoming low-altitude missile. With World War II–era guns *Elath* fired without effect at the missile, a Russian-built SS-N-2 Styx. The SS-N-2's guidance radar had detected *Elath*, and this missile, followed by a second, abruptly veered straight for her. Both missiles hit amidships and destroyed her propulsion spaces and electrical generators. Aboard the crippled destroyer the crew fought the fires, attended to casualties, and desperately radioed for support (the Egyptians had attacked the Israeli ship on the Jewish sabbath). Two hours later, a third SS-N-2 hit her astern, and *Elath* capsized. As she sank, a fourth SS-N-2 exploded amid her survivors. Casualties were severe: 47 killed and 90 wounded out of 199.

Elath was armed similarly to most U.S. Navy destroyers. She mounted a prototype electronic warfare system that detected her assailants, two Soviet-built Komar-class missile boats firing from inside the harbor, but she had no chaff or radar-jamming capability.[91] Shipboard guns would have been similarly useless in protecting an American destroyer caught in the same circumstances. The *Elath* incident and the new Soviet underwater-launched antiship missiles rammed home the need for warships to mount independent missile defenses. It invigorated development of ship-based weapons for fighting a conflict short of nuclear war. The immediate response was to adapt existing weapons, such as Sea Sparrow, for shipboard use. For the longer term, the NATO Sea Sparrow system, the Vulcan/Phalanx close-in weapons system, the rolling-airframe missile, LAMPS, the Harpoon medium-range antiship cruise missile, and several other weapons all had begun development by 1972.

The DD 963 Design Competition

Selecting Competitors to Design the DD 963 Class

The DD 963 design competition started in February 1968, when the Naval Ship Systems Command (NavShips) officially solicited all interested shipbuilders to outline their approaches to fulfill a multiyear contract for a class of destroyers meeting the DD 963 specification. This was the first phase of Total Package Procurement. Each prospective shipbuilder described how it would develop a preliminary design for the destroyers and its plans for a production facility. The ship design itself was not required at this stage, much less actual construction. In May 1968 six firms submitted proposals. The contenders were Litton Industries, General Dynamics, Bath Iron Works, Newport News Shipbuilding, Todd Shipyards, and Avondale Shipyards.

Naval architect J. W. Devanney described how Litton developed its proposal:

> This response took two disparate approaches. [The naval architects] were inclined to draw up a ship meeting the requirements and [to] indicate that we would design a system [shipbuilding plan, logistics support, etc.] around this ship. [Another] group wanted to look at and cost out a large number of combinations of power plant types, hull dimensions, and hull forms.
>
> Project management effected an uneasy compromise. The first group was allowed to draw up two "baseline ships," one steam and one gas turbine, and present them in the proposal with a disclaimer that these were only indications of Litton's familiarity with the state of the art with respect to destroyers and [of] its abil-

ity to perform a standard preliminary design. At the same time, the systematic search procedure was developed in considerable detail [and promised use of a] parametric program, more detailed programs related to structures, hull form parameters and seakeeping, and a series of 5′ model tests on dissimilar contenders.[1]

A Litton representative in Washington who delivered the proposal to the Navy recalled that "it read and looked like a high school project."[2] A significant item to note in the quoted passage is Devanney's exclusive focus on hull design and powering.

The Office of the Secretary of Defense usually considered two competitors to be adequate, but the Navy wanted three, to avoid a sole-source situation if one did not perform as expected.[3] In July 1968 NavShips awarded contracts to Litton Industries, Bath Iron Works, and General Dynamics for Contract Definition, the second phase of Total Package Procurement. Each winning firm had probably spent over $2 million of its own funds on the initial proposal efforts. Aerospace firms like Litton and General Dynamics were accustomed to expensive proposals, but it was new to the shipbuilders.[4] The losing firms, all shipbuilders, had probably spent about $1 million each.

The Navy would disburse the $30 million congressional authorization among the three selected Contract Definition contractors. Each firm was required to prepare the following:

- A contract design for a destroyer meeting the DD 963 performance requirements developed under the DX project, specified in terms of payload, maneuverability, speed, endurance, stability, survivability, and seakeeping

- An analysis of the trade-offs made during development of the destroyer design
- Plans for a series-production shipyard, for training the destroyer crews, for providing the initial set of spare parts and other logistics support, and for a warranty to fix any defects discovered in the ships after delivery to the Navy
- The price for a single contract to carry out the entire project, including detail design, obtaining all material, creating the shipyard, building the destroyers, and carrying out all the support plans[5]

The Navy's design guidance information to the contractors included data on gas turbines from the Project Seahawk experiments, and specifications for the government-furnished weapons and electronics that the ships must mount. The actual shipbuilding contract, optimistically foreseen as $1.364 billion, would be awarded to the firm proposing the lowest life-cycle cost for an acceptable destroyer.[6]

Litton Industries, owner of Ingalls Shipbuilding in Pascagoula, Mississippi, planned to build a highly automated new shipyard to meet the market it forecast would arise to replace the obsolete warships and merchantmen from World War II. Beset by poverty and repression of civil rights, Mississippi was the most rural state in the Union. Politically it was an alienated backwater that had not voted for a winning presidential candidate since 1944. Since its congressional delegation rarely changed, its politicians grew powerful from seniority within Congress. Senator John Stennis, Democrat of Mississippi, chaired the 1967 hearings that ended Robert McNamara's Pentagon career. Stennis was well attuned to Pentagon shipbuilding plans.

Prospects of luring the West's most modern shipyard to Mississippi attracted politicians who hoped to advance the regional economy. In November 1967 a $130 million tax-free bond backed by the state of Mississippi to build the new shipyard for Ingalls sold out in a day.[7] Litton would lease the yard from the state, at no charge for the first five years, and the state would benefit from the new jobs and concentration of technical expertise. After winning the FDL competition, Litton began building the new West Bank yard, the first new shipyard to be built in the non-Communist world since World War II. It was under construction during the DD 963 Contract Definition (CD) com-

petition. While it had only a few of the ultimate army of workers it would need on board, with its new yard Litton won 41 percent of all American shipbuilding contracts between 1966 and 1969, including the FDL class, commercial container ships, and the nine-ship *Tarawa* (LHA 1)–class amphibious assault ships, awarded under a Navy TPP contract in May 1969.[8]

Bath Iron Works had built 25 percent of the World War II destroyer fleet, and many senior Navy officers had started their careers on Bath-built ships. The legendary reputation of the World War II *Fletcher* (DD 445)–class destroyers began with the amount of combat damage that Bath-built *Fletcher*s had withstood. Bath had completed the first ship of every non-nuclear postwar DLG and destroyer class. "If you had based [the DD 963 award] on prejudice, it would have been a Bath ship hands down," Admiral Weschler recalled, "because they had never built a bad ship."[9] Bath teamed with Gibbs & Cox for the DD 963 ship design, with Lockheed for weapons systems integration, and with the National Steel & Shipbuilding Company in San Diego for West Coast support. Lockheed and National Steel later left the team, but Hughes Aircraft joined it. Hughes built the Navy's Naval Tactical Data System consoles and was an expert systems integration firm. Gibbs & Cox was the Navy's favored firm for preparing warship contract designs. The Navy DD 963 project staff eagerly awaited the Bath–Gibbs & Cox proposal, but there was concern that the firms had worked with the Navy so closely for so long that their design might not be sufficiently innovative.[10]

General Dynamics planned to capitalize on four strengths: its shipyard and design team at Quincy, Massachusetts (previously Bethlehem Steel's Fore River yard); Electric Boat's expertise at submarine construction; Convair's design of seaplanes in San Diego; and the Pomona, California, missile division's expertise in advanced naval weapon design and its design for the prototype integrated ASW combat system. Pomona hoped to develop itself as a system integrator and suggested that it should lead the DD 963 proposal effort. However, General Dynamics executives believed the Navy's assurances that it did not want the aerospace industry to usurp shipbuilding, so they chose Quincy to lead the DD 963 project. Quincy had designed the DE 1040 and DEG 1 classes of destroyer escorts for ASW, which were the only postwar destroyer-type ships whose contract design was not prepared by Gibbs & Cox, and had been active in the Seahawk project.[11] The project manager moved to DD 963 from Electric Boat in New London, after arguing with Admiral Rickover, Electric Boat's virtual sole customer. He brought Richard F. Cross III from the Pomona ASW project to design the destroyer. A naval architect educated at the Massachusetts Institute of Technology, Dick Cross had developed an advanced computer program to address the details of naval architecture. For the initial proposal he drafted a design for a new destroyer in just four days,[12] an impressive demonstration of capability.

The Litton Industries Design

Litton, a newcomer to the ship design industry, had discovered during the FDL competition that the available pool of naval architects was small. Litton recruited young naval architects abroad;

Dr. Reuven Leopold, Litton's DD 963 design manager. Later he was technical director of the Naval Ship Engineering Center for warship design during the design of the *Kouroosh/Kidd* and *Ticonderoga* classes. Using his knowledge of marine gas turbines, he has also designed commercially successful aircraft turbofan jet engines. (U.S. Naval Institute collection)

Britain, Turkey, and Israel were well represented. Reuven Leopold, Litton's director of ship engineering, was one such immigrant, age only 30. He held two naval architecture degrees from MIT and had done advanced research work there, including devising a computer-based model for the optimization of complete ship designs, before joining Litton in 1966: "my first job out of college."[13] Litton concentrated its team at Culver City, California, an embedded suburb of western Los Angeles. This team won the first Navy Total Package competition, for the FDL class, and was on its way to winning the second, for the LHA class. Leopold was busy working on the designs for the FDL and LHA projects and was not involved in the DD 963 preliminary request-for-proposals response phase. Confident in the naval architects, Litton executives told them to design the best destroyer they could conceive and not to compromise for fear that Litton could not sell a high-quality design.[14] This was exactly the approach the Navy hoped industry would take.

SYSTEM ENGINEERING

Reuven Leopold attributed the success of Litton's designs to their focus on the complete ship, as distinct from the traditional focus of naval architecture on hull and propulsion design. He characterized previous methods:

The traditional approach to ship design has often relied heavily on the indiscriminate use of past experience and on subjective opinions as to what "looks good." Critical attention to subsystems has generally been relegated downstream. Producibility aspects have rarely been considered until the design has progressed to a very detailed stage. An information network between subsystems through formal integration methods has not been fully utilized. . . .

In the past, little formal attention was given to designing availability into a ship system. All attention was directed toward assuring that all elements of the ship worked together to give the desired performance during acceptance trials. If it happened that equipment failures were few and far between during the ship's life after trials, then this was worthy of praise. However, this would be uncoordinated feedback from operations rather than design since there was no special, comprehensive analysis to predict total downtime and uptime and,

Table 5.1   Litton Industries design criteria for DD 963 destroyer class

Effectiveness Element and Subelement	Definition (variables represent classified data)
Operational capability	
Endurance	Achieve an endurance of 6,000 nautical miles at a speed of 20 knots with full normal fuel in calm seas with a clean bottom. (Maximize endurance where economically feasible.)
Ship speed	Achieve a sustained speed of 30 knots in a sea state of 4 (head seas) with bottom condition corresponding to a roughness allowance of 2 years out of dock, using full sustained power.
Readiness	Minimize the acceleration lag, the time in minutes required to develop power for 30 knots (on same basis as in above) from an endurance steaming (20 knots) condition.
Sonar capability	Maximize the optimum sonar speed, that speed in knots for which the sonar search (or sweep) rate is greatest.
	Reduce radiated hullborne acoustic noise below spectrum H_1. Desirable to reduce radiated noise below H_2. Reduce radiated noise below H_3 if economically feasible (incentive fee opportunity).
Maneuvering	Achieve a tactical diameter of Y yards at a speed of 30 knots with full rudder. (Minimize tactical diameter where economically feasible.)
Seakeeping	Maximize seakeeping, the expected fraction of time during which the DD's combat capabilities will not be significantly degraded by ship motion or green water over the deck while steaming on random course at a speed of 30 knots in state Z_2 sea.
Error budget and reaction time	Minimize ASW error as defined in RFP.
	Minimize ASUW error as defined in RFP.
	Minimize shore bombardment error as defined in RFP.
	Minimize AAW error as defined in RFP.
	Minimize ASW reaction time as defined in RFP.
Weapons coverage	Maximize weapons coverage, the sum in degrees of firing arcs for all weapons.
	Maximize sensor coverage, the sum in degrees of sensor coverage for all sensors.
Navigation	Achieve an absolute navigational error of Y_1 to Y_2 nautical miles continuously, at sea. This is defined as the mean distance between the ship's actual and navigational positions (i.e., the mean magnitude of the absolute error vector).
	Achieve a relative navigational error of A_1 to A_2 nautical miles for two DD ships at sea and separated by a distance of _ nautical miles.
Communication	Maximize voice communication capacity, the maximum number of duplexvoice VHF/UHF channels which the DD can simultaneously work.
	Achieve a communication capacity of Q messages per day.
Stealth	Minimize the radar cross-section in square meters, averaged over frequencies from AK MHz to BK MHz, all azimuths, and elevation angles between O_1 degree and O_2 degrees.
Replenishment	Minimize replenishment time, the time in hours required to take aboard and strike down the DD's full load of fuel, stores, and ammunition, using the fastest means of at-sea replenishment.
Operational flexibility	
Stability margin	Maximize reserve moment, the moment (in ton-feet) of the additional fixed load which could be added to the DD without reducing stability below standard (with no fixed ballast) at any normal steaming condition.
Manning margin	Maximize manning growth margin, the fraction by which design accommodations exceed design complement.
Modernization	Minimize modernization duration, the duration in months of the modernization overhaul.
Conversion	Minimize conversion duration, the duration in months of the DD's out-of-service period for conversion.
Availability and reliability	Achieve a mission reliability of L, that is, the availability of that portion of the total ship system essential for mission success. (Maximize mission availability where economically feasible.)
	Achieve a propulsion availability (full-power operation) of M_2. (Maximize propulsion availability where economically feasible.)

Table 5.1   (Continued)

57

The DD 963 Design Competition

Effectiveness Element and Subelement	Definition (variables represent classified data)
Survivability	Minimize vulnerability (as defined by the DX/DXG vulnerability study methodology) through: — Ship protection — Redundancy — Segregation of function — Subdivision — Shock resistance — Blast resistance
Overhaul Overhaul cycle Overhaul duration	Maximize overhaul interval, the interval in months between overhauls. Minimize overhaul duration, the duration in months of a normal overhaul.
Manning Operating and maintenance personnel ratio	Minimize the officer fraction, the fraction of total complement in commissioned and warrant grades. Minimize the rate fraction, the fraction of enlisted complement in grades E-5 and above. Minimize the skills fraction, the fraction of enlisted complement in critical skills.
Life-cycle cost	See life-cycle cost methodology supplement.

therefore, the long-term availability of the ship system and its vital elements.[15]

Many solutions were possible for almost any requirement. Time to carry out the CD contract was short and so demanded decentralized but consistent decisions. Leopold, put in charge of the DD 963 design project, established a list of explicit design principles to guide the naval architects and other designers in choosing trade-offs during the development of the design. For example, which shipboard systems should be adjacent for functionality and which should be separated for survivability? The design principles had to be understood by everyone on the team because the ship had to be designed as an integrated system, as if by a single all-knowing designer. The goal was the best-balanced design. The principles were as follows:

- Minimum cost (including construction and operating cost)
- Operational excellence (maximum operational effectiveness)
- Flexibility
- Availability of mission-essential systems
- Habitability
- Survivability and reduction of vulnerability
- Innovation to increase capability
- Nonobsolescence

Leopold's team refined these general principles into specific design criteria, which defined ship performance requirements, constraints, and design goals.

Table 5.1 lists the design criteria that the Litton team evaluated for its DD 963 design. A numerical weighting factor indicated the importance, or value, of each criterion. (In systems analysis terms, the weights and design criteria defined the objective function.) When evaluating a candidate design arrangement for the DD 963, the team graded every significant feature of the design arrangement on a scale of 1 to 4. This grade (the figure of merit) indicated the expected effectiveness of that feature in the candidate ship design. A good system would get a low grade if the design under review put it in a poor location, or a good grade if it were in a good location. The figure of merit multiplied by the weighting factor for the value of the associated design criterion measured that feature's value as it affected the overall destroyer design. The numerical grades showed where to exchange design features to achieve higher scores. Litton prepared and evaluated multiple candidate designs for the ship and the propulsion plant. In particular, the location of the helicopter deck almost amidships was chosen using the method just outlined. Litton's analysis put a high priority on reducing flight deck motion in high sea states, and designs with the helicopter deck located elsewhere scored much lower.[16]

Many calculations were performed using a computer and optimization software (linear programming), which showed where trade-off decisions would improve a design's score. This was particularly useful for quickly determining hull characteristics (length, beam, draft, displacement, block coefficient) for different propulsion plants or topside arrangements. The hull had to satisfy requirements (constraints) for stability, strength, flooding, and much else, and to provide the best combination of low cost (material, labor, protective coating, propulsion fuel) and producibility (cost, required shipyard skills, control of distortion from uneven loading during construction) within those constraints.

Leopold began with the general arrangement of the ship. Surface warships are among the most interesting of all engineering systems because they must be designed for multiple missions and for trial by endurance, storm, and combat. Unwise design compromises or a lack of operational skill can lead to disaster. "A ship," wrote merchant officer Joseph Conrad, "is a creature which we have brought into the world, as it were on purpose to keep us up to the mark."[17]

A commercial ship is designed around its payload, but the starting point for a warship design is its general arrangement. Arcs of weapons and sensor coverage, separation of equipment for survivability, motion analysis for helicopter operations, and similar considerations drive the design. Arrangement-related phenomena include radio-frequency interference, electromagnetic compatibility, hullborne radiated noise, and mechanical shock.[18] At Litton, DD 963 arrangements scoring relatively well were studied further. The naval architects studied existing warship designs to estimate the type of equipment, volume, weights, and tolerances for these subsystems, adjusting for new-technology equipment that they proposed to use in the DD 963 class.

One aspect of general arrangement was the appearance of the ship, in particular her profile. The Navy listed "naval presence" among the DD 963's missions. Leopold wondered what "naval presence" meant and, more specifically, how he might design features into the destroyer to support whatever it was. Knowing that the Navy believed that aircraft carriers impressed influential foreign officials, he concluded (logically, if not exactly correctly) that "naval presence" implied that the new destroyer must look impressive, too. He hypothesized that the Navy wished to impress foreign VIPs, who in general might not know much about ships and might simply regard any large warship such as an aircraft carrier as more imposing than a smaller ship. Standard U.S. Navy destroyers such as the *Charles F. Adams* and *Farragut* classes looked small and old-fashioned, with spindly stacks and "missing middles," visual voids between the stacks amidships, reminiscent of the days of banks of antiship torpedoes.[19]

The next problem was how to test the hypothesis. Somewhat impishly deciding that Litton's aerospace-oriented managers constituted a ready sample of very important persons who knew little about ships, Leopold conducted an experiment in psychology. Using photographs from *Jane's Fighting Ships,* he asked Litton executives which was larger, a Royal Navy County-class destroyer or a U.S. Navy *Farragut*-class destroyer. Measured by length and beam, the ships were almost identical in size. However, without exception everyone in his poll chose the modernistic County-class destroyer as the larger.[20]

Armed with this result, Leopold based the Litton DD 963 design profile on the County class: a long deckhouse without a gap amidships, low streamlined stacks, the same proportionate mast-stack-mast-stack interval, and tiered decks aft with the fantail lower than the main weather deck. A decade later, an artistic analysis of warship designs identified reasons for the visual success of the County-class design: visual "lines of force" that drew attention upward and forward, and half-height deck levels, which created a magnifying effect. Litton's DD 963 design rated better than other American warships such as the *California* class.[21]

Good appearance for the DD 963 class was one example of the kind of innovation the Navy hoped industry would provide as a matter of course. In general, ships with good seakeeping characteristics tend to look good too, although this does not mean that every handsome ship is therefore a good sea boat. Ironically, the American naval-presence mission did not require warships to be visually impressive for diplomacy, although European navies had long known the diplomatic value of good appearance, as did the U.S. Navy under President Theodore Roosevelt. "Naval presence" was Navy jargon for the peacetime forward-deployment requirements of endurance, surveillance, reliability, and readiness for action. This mission had originated in 1946, when Secretary of the Navy James Forrestal ordered the fleet to steam

Litton based the general appearance of its DD 963 design on the Royal Navy's County-class destroyers, represented here by HMS *Glamorgan* (D 19). These ships looked larger and more impressive than a tra-ditional U.S. Navy destroyer design such as USS *Farragut* (DDG 37, ex-DLG 6). (U.S. Naval Institute collection)

Litton's DD 963 design in April 1969 shows European influence in its appearance, with streamlined stacks and a profile layout adapted from the British County-class destroyers. The forward stack is offset to starboard and aft stack to port in this design. The cutouts at the corners of the broad transom stern hold large chaff rocket launchers. (U.S. Naval Institute collection)

"in any waters in any part of the globe" in peacetime so as not to cause "excitement or speculation" when crisis deployments were required.[22]

MODULARITY

For each general arrangement, computer programs calculated the design's size, subdivision requirements, and structural strength requirements. Weapon firing arcs obviously should be as wide as possible. Electronic equipment had to be positioned for antenna efficiency and so that radio-frequency signals would not interfere with each other. As noted, the flight deck had to be located to let helicopters operate in high sea states despite the ship's motion. Propulsion and auxiliary systems had to be separated for combat survivability. Internal spaces were allocated to subsystems, including propulsion, auxiliaries, weapons, and habitability, and to reserve space for new equipment to be added during modernization or conversion.

Litton's DD 963 design was for a highly modular ship, with each subsystem placed in a specific, dedicated module. The design was refined at the subsystem level by arranging equipment and accesses within each module. Dedicated work groups were made responsible for the subsystems and structural modules. A work group was not allowed to exceed the space allocation while it built up its subsystem design. The modular design approach gave Litton's subsystem designers independence (as was important, since the designers needed to work quickly to meet the Navy's deadline) and a maximum envelope into which they could fit their equipment. The maximum weight and size for each functional module were derived from the total ship system design. One of the largest modules was an empty area reserved for the Mark 26 Mod 1 44-missile launching system. This supported converting the Litton DD 963 to a double-ended DDG configuration without sacrificing the helicopter facility or a 5-inch gun mount, although the Navy specification permitted that.

The specification required the provision to swap out the ASROC box launcher and loader with a Mark 26 Mod 0 24-missile launching system. In existing ships mounting ASROC box launchers, the reload missiles were stored horizontally in a magazine behind the deck-mounted box launcher. The launchers were reloaded from deck level with the missiles traversing from the magazine at a shallow angle. The Mark 26 launcher, in contrast, carried missiles vertically on conveyors underneath the twin-arm launcher, and loaded weapons vertically onto the launcher rails. Designing one module to accommodate either of these two very different configurations was a challenge. Litton solved it with a design for a vertically loading ASROC magazine beneath the launcher.

The ship structure was analyzed for strength, buoyancy, stability, seakeeping, and damage control. The ship was designed in "damaged mode" to maintain stability with 15 percent of the hull length (three compartments) open to the sea from battle damage. A computer-based synthesis program contained mathematical parameters for the various design aspects. These numbers were put into an optimization program to develop the lowest-cost (defined as life-cycle cost) characteristics for a ship meeting all the requirements. This new set of ship characteristics described a potential new general arrangement, and the process was then reiterated. The cycle stopped when further rearrangements of the design did not add significant value as measured by the principles of effectiveness. After this, the ship design was further refined by changes within the functional modules.[23]

Thus a very wide range of considerations influenced the Litton DD 963 design. These design considerations (Leopold's design principles) made sense from the viewpoints of both economics and operational effectiveness. The resulting ship's size, propulsion, and appearance were outcomes from the design principles. It was a significant advance in the technique of designing a warship. Traditional naval architecture practice was to design the hull tightly around a propulsion plant and the payload of weapons and endurance stores. Litton's highly modular DD 963 design was unusual in that it was length-critical, fully 100 feet longer than an otherwise feasible ship design meeting the Navy's specifications for seakeeping and propulsion.[24]

The total size of the ship was driven by the sizes of the individual modules. Internal spaces were allocated to subsystems, including propulsion, auxiliaries, weapons, and habitability, and to reserve space for new equipment to be added during modernization or conversion. A consideration that did not enter into the design was an extensive

"design work study" that Litton prepared as required by the CD contract. The idea behind design work study was that the ship could be arranged to consolidate functions, reducing crew size but increasing vulnerability. The designers found the study worthless and ignored it, using instead the principles of functionality and survivability.[25]

The Navy specification required a flight deck, a hangar, and a haul-down winch for an SH-3 Sea King ASW helicopter. Sea Kings were frequently seen on TV in the 1960s and 1970s retrieving astronauts from the water following splashdown of Gemini and Apollo space flight capsules. The Canadian *Iroquois*-class destroyers had double hangars and a haul-down winch called Beartrap for landing, securing, and traversing the SH-3. Litton management learned about the Canadian destroyer-based helicopter facilities from the naval architecture firm M. Rosenblatt & Son, which suggested that DD 963 should use the same systems and even arranged for Canadian naval officials to describe them to Litton in Culver City.[26] Litton imported the design of the Canadian midships destroyer-based helicopter facility for DD 963.

PROPULSION AND AUXILIARIES

The single most vital consideration was the ship's engineering plant. In previous destroyers the weight of the engineering plant and fuel had been about a third of full-load displacement. The Navy specification for the main propulsion plant required two propellers, 30 knots' speed in sea state 4, 6,000 miles' range at 20 knots, use of the new standard distillate fuel, engine control from the bridge, survivability after attack, at-sea maintainability on some components, short stopping distance, and underwater silencing. The ships were expected to operate at 20 knots or less for 70 percent of their time under way, and at 25 knots or less for 85 percent. Rapid acceleration under way was required, although rapid start-up from cold iron was not. Steam, diesels, gas turbines, or any combination would be allowed. The provision of the Project Seahawk data revealed the DD 963 project office's appreciation of gas turbines. Silencing for ASW was vital, and the construction contract specified a bonus of up to $24 million for the 30-ship contract (not per ship) if the DD 963 underwater radiated noise was below a certain level.

Litton compared over 50 alternative plant configurations ranging from 60,000 to 120,000 hp (see

Table 5.2   Litton DD 963 propulsion plant characteristics. Over 50 plants were compared. The ship was ultimately designed to use either plant designated by an asterisk.

Plant	Configuration	Sustained shp	Weight (tons)	Crew	Maintenance (manhours/year)
Steam	2 1,200 psi 35,000 shp D-type boilers	44,000	1,248	67	10,414
CODAG	2 6-cyl. diesels, 2 FT4A2, CRP propellers	42,000	1,047	53	8,803
COSAG	2 1,200 psi D-type boilers, 2 single-casing turbines, 2 FT4A2	56,000	1,083	65	12,656
COGAG-1	2 FT3C9, 2 FT4A2, CRP propellers, no cross-connect	48,000	822		
COGAG	Same as COGAG-1, electric cross-connect	48,000	890		
COGAG	Same as COGAG-1, mechanical cross-connect	48,000	944		
COGAG-2	4 FT4A2, CRP propellers, no cross-connect	72,000	858	46	7,803
* COGAG	Same as COGAG-2, electric cross-connect	72,000	926	46	8,525
COGAG	Same as COGAG-2, mechanical cross-connect	72,000	978	46	9,307
* COGAG	4 LM-2500, CRP propellers, electric cross-connect	80,000	926		

Source: Gerald Boatwright and John Couch, "The Gas Turbine Propulsion System for the U.S. Navy *Spruance* (DD 963) Class Destroyers," paper presented to American Society of Mechanical Engineers, Mar. 1971. Sustained shp figures are estimated and did not appear in the original paper. Many sources rated the FT-4A at 25,000 hp, but the 24,500-lb. thrust of its JT4 (J-75) core engine translated to roughly 14,000 hp. The LM-2500 is rated at 22,500 hp; the thrust of its CF6 (TF-39) core engine is 41,100 lbs.

table 5.2). Each configuration was entered into a linear program to develop a conceptual design and a life-cycle cost estimate for the best ship design that could be built employing that plant. This enabled a hull form and general arrangement to be created to make the best use of each plant. The leading ship design for a candidate plant was designed in more detail. Devanney, by his own account, was working on the hull design and erred in attributing frictional drag on a towing-tank model to wave-forming drag. Wave-forming drag would apply to the actual ship, and as the alleged result, Litton overestimated the propulsive power the destroyer would require by about 5–10 percent. Such a bias also would explain the narrow beam chosen for the hull.[27]

Litton evaluated the risk that each plant might not meet performance requirements and the risk each posed to the construction schedule and cost. Two specific studies were the feasibility of silencing diesel engines, and simulation of ship maneuvering with various propulsion plants. Since only the best of the acceptable ship designs using each plant type were evaluated, the choice turned on which ship was the best design as measured by the design criteria weights. Model tests showed that the best maneuvering characteristics came from ·the use of controllable-pitch propellers with combined gas turbine and gas turbine (COGAG) or combined diesel and gas turbine (CODAG) plants. A CODAG plant was cheaper both in cost and in weight of the required fuel load. A COGAG plant provided much better hullborne acoustic silencing and was selected for that reason.[28]

Two models of marine gas turbines with the necessary power were evaluated: the FT4A and the promising but officially untested new LM-2500. The LM-2500 was 10 years more advanced than alternative marine gas turbines. It was a marinized variant of the new General Electric CF6 (TF39 in U.S. Air Force designation) turbofan jet engine for jumbo jets such as the commercial DC-10 and the Air Force C-5A. Litton designed the DD 963 to accept either engine.

To offset the high specific fuel consumption of gas turbines when only low power was required, Litton proposed to attach a large electric motor-generator to the main reduction gear on each propeller shaft. At cruising speeds, one gas turbine would simultaneously drive its propeller shaft and send electrical power to the motor-generator on the other propeller shaft through the electrical cross-connection. This would deliver about half the power of one gas turbine to the alternate shaft. An electrical cross-connect rather than mechanical gearing was necessary because the propeller shafts did not lie in a common geometric plane. The shaft from the forward engine room was longer and lay at a shallower angle to the keel than the shaft from the aft engine room. Taking advantage of the more powerful LM-2500 and the length of the evolving hull design, Litton proposed to mount only three LM-2500 gas turbines in the ship. Two were in the after main engine room and one was in the forward engine room.[29] Litton estimated that three LM-2500 gas turbines would deliver enough power for 30 knots. In an unusual and innovative touch, the stacks were offset to port and starboard. This asymmetrical arrangement placed the stacks directly above the main propulsion gas turbines. This increased the usable deck area in the superstructure and reduced the volume occupied by the uptakes.

Low-pressure steam was required for hotel services such as the freshwater evaporators, the galley, and the laundry, and for heating internal compartments and the exposed ASROC launcher. Steam-powered warships used steam for hotel services and to power the electrical ship's service turbogenerators (SSTGs). In the initial design, diesel generators would provide power if the SSTGs or the distribution switchboards were damaged. Litton designed DD 963 to use three diesel generators to provide all electrical power during both normal and emergency conditions, with two small oil-fired boilers to provide low-pressure steam. As an alternative design, Litton proposed three ship's service gas turbine–powered generators (SSGTGs). In this alternative, exhaust from each auxiliary gas turbine was ducted through a boiler to generate auxiliary steam. Litton adapted this innovation from waste-heat boilers in use on Gulf of Mexico oil rigs.[30] The Canadian *Iroquois* class used waste-heat boilers. As with the provision to use either of two main propulsion gas turbines, alternative electrical generation schemes reduced risk.

J. W. Devanney of the Litton design staff claimed in 1975 that a combined plant of one gas turbine and two low-power diesel engines would have powered each shaft at lower cost than a COGAG plant. The diesels would be identical to the models chosen for the electrical generators. He acknowledged that this plant could not achieve the silencing provided by gas turbines alone: the low-power diesel "clearly made the intermediate noise requirement" but did not meet the Navy's most stringent silencing specification.[31] A supplier proposed a complex combining gear for this plant, but there was no alternative if it did not work.

The diesel generators were an alternative design themselves. The low cost of CODAG was attractive, but ultimately Litton decided that the lower hullborne noise of COGAG was more important. Devanney asserted that the optimization

software kept steam plants from being evaluated because it had no provision to lengthen the hull enough for a low-cost, low-power steam plant to become competitive. However, a steam plant would not have met even the intermediate silencing requirement. Submarine silencing improvements would have made a steam or diesel plant a disastrous choice. The Navy program officers would have rejected such a ship, especially since they had competitive COGAG designs from Bath and General Dynamics. Devanney's article, widely cited by critics, will be analyzed later in the context of the political controversy that came to surround the *Spruance* class.

New destroyer escorts had active stabilizer fins to improve weapon and sensor performance. These stabilizers were rarely used and frequently needed repair. The Navy had not evaluated the use of active stabilizers to improve total ship performance, including tactical maneuverability, habitability, and operator and equipment effectiveness.[32] No body of research was available about the benefits from stabilizers. They would increase cost and maintenance. Litton's design did not offer them. Instead the DD 963 design provided seawater-compensating fuel tanks to keep the ships riding deep and stably. This required reliance on purifiers filtering salt water from fuel oil before burning. Without stabilizers, the destroyers would fail some seakeeping requirements, almost their only shortfall.[33]

The Bath Iron Works–Gibbs & Cox Design

Early in the DD 963 program, Gibbs & Cox designed a notional warship to mercantile specifications as an experiment to see what features might be imported for innovation and cost savings in the military-specification DD 963 design. Two lessons from this were to provide a wider hull for seakeeping and to straighten passageways for easier access to internal spaces and systems.[34] Traditional destroyer-type ships were very tight and awkward internally. Gibbs & Cox had designed the DLG 16/26–class cruisers, which displaced almost 8,000 tons full load. The DD 963 specification called for better maneuvering performance than these ships, and for greater payload weights carried higher up, all on less than 8,000 tons. Keeping displacement below 8,000 tons was a major challenge.

The Bath–Gibbs & Cox DD 963 design was for a shorter but beamier and heavier destroyer than

This was Gibbs & Cox's design for the Bath Iron Works DD 963 proposal, April 1969. Box-shaped prefabricated superstructure modules were expected to reduce construction costs. The combat information center was protected by the uptakes for the tandem forward stacks. (Courtesy Gibbs & Cox)

Litton's. Length overall in their initial proposal design was about 540 feet. Displacement was 6,900 tons at light load (no fuel, ammunition, stores, or crew). It included two pairs of active fin stabilizers, instead of seawater-compensating fuel tanks, for seakeeping stability during helicopter operations. The beam was 61 feet 6 inches, 12 percent broader than the DLG 16/26 hull. This compensated for the higher center of gravity of the gas turbine plant and supported the Navy-specified modernization margin of 300 tons high up in the ship. Use of seawater-compensating tanks, which Litton used, would have reduced the beam 3 feet. These differences made for a higher shipbuilding cost for the Gibbs & Cox design.

This design featured four LM-2500s for main propulsion. Two turbines in the forward main engine room drove the port shaft, and two in the aft engine room drove the starboard shaft. A mechanically cross-connected combining gear enabled any single gas turbine to drive both shafts at low power. The mechanical combining gear was feasible because the propeller shafts were parallel

(coplanar), not splayed as in Litton's design. It was located in an auxiliary machinery room separating the forward and after main engine rooms. Four diesel generators provided electrical power; as in Litton's design, auxiliary gas turbine generators were an alternative. Waste heat from the diesel cooling jackets and exhaust mufflers provided heating for the ship, including the seawater-distilling plant. An oil-fired heater could supplement the waste-heat system.

The ship's general layout was similar to the Litton design, with a midships helicopter deck and the 5-inch guns mounted at the furthest forward and furthest aft positions. Dimensions in the final proposal (after Navy-ordered changes covered in chapter 6) are shown in table 5.3. The design featured three stacks. The port forward stack contained the exhaust uptakes for the forward main propulsion gas turbines. The starboard forward stack contained the exhaust uptakes from the generators in the forward engine room and in the auxiliary machinery room aft of it. The combat information center was in the superstructure between the forward stacks, so the ducting structure provided some additional protection.[35] The hangar was aft of the single aft stack. The design included a Beartrap winch for the helicopter. It is not clear whether the ship could retain the helicopter hangar if she were converted to a DDG with two Mark 26 launchers. The broad beam would have made converting the design to mount Aegis easier than in the Litton design.

Overall, the Bath–Gibbs & Cox design "was excellent, highly competitive" with Litton's, in Captain Richard Henning's judgment, but "not superior."[36] It was highly modular to facilitate low-cost construction, modernization, and conversion. To compete financially, Bath used Litton's tactic with the state of Mississippi and obtained approval from Maine's legislature and voters for a $50 million bond, to build an automated shipyard for construction of the DD 963 class. Recent changes in U.S. tax laws made this bond taxable, however, so that Bath's interest payments would be higher than Litton's.[37]

The General Dynamics Design

General Dynamics had put the Quincy designers in charge of the DD 963 proposal. The loss of the FDL contract had left them frustrated and skeptical about the Navy's TPP process. They examined Pomona's capability for rapid destroyer design but were more impressed, or depressed, to discover that Pomona's engineers were drawing salaries 50 percent higher than theirs. They resented Electric Boat's influence. Pomona, Electric

Table 5.3 December 1969 characteristics for Gibbs & Cox DD 963 design

Length overall	571 feet 6 inches
Length between perpendiculars	530 feet 0 inches
Beam, molded max.	61 feet 6 inches
Displacement at full load	7,900 tons

4 LM-2500 or FT4A gas turbines, 2 shafts with 5-bladed 16.5-foot-diameter CRP propellers

Full power with LM-2500: 39,000 shp per shaft at 215 rpm

2 rudders linked to 1 steering gear; tactical diameter < 500 yards

4 1,500 kw diesel generators (alternative: gas turbine generators)

Boat, and Quincy remained separate both geographically and in their team contribution to the General Dynamics proposal.[38]

General Dynamics representatives showed their 1968 draft DD 963 design to Navy destroyer force officers, who thought it "too big."[39] General Dynamics strove for the smallest possible design. The Office of the Assistant Secretary of Defense for Systems Analysis preferred a small ship, too. The General Dynamics destroyer would have an overall length of 489 feet and a full-load displacement of 5,746 tons. It resembled the Quincy-built DE 1040–class destroyer escorts closely in layout and partly in appearance. Its initial armament and combat electronics payloads met the Navy specification. The Sea Sparrow launcher and fire control radar were both atop the bridge. General Dynamics designed an ASROC loader module located forward, which could be replaced during modernization by a single Mark 26 twin-arm missile launcher with a 24-round magazine. The compact ASROC loader was the most modular part of the General Dynamics design.[40]

For propulsion General Dynamics chose four LM-2500 gas turbines. The ship was too small for FT4As to be a fallback. Four auxiliary gas turbines drove 1,500 kW electrical generators. Two main engine rooms each contained two LM-2500s and two generators. The main engine rooms were almost adjacent. The Navy preferred that the main engine rooms should be widely separated for survivability, but General Dynamics gambled that small ship size and low cost would dominate the Navy's ultimate choice of destroyer.[41] No propulsion or auxiliary machinery was located forward of the engine rooms, thus reducing noise around the sonar dome and thus contributing to better sonar detection range. The huge intake and uptake ducts for the gas turbines occupied much of the volume of the large central deckhouse. A large stack above the after engine room contained all intakes and uptakes. The ducts for the forward engine room were trunked back through the deckhouse to the stack. The stack was the platform for the ship's single mast. The helicopter deck was on the fantail. The after 5-inch gun superfired over the helicopter deck and hangar on aft bearings. It was abutted to the deckhouse and could not fire on forward bearings.

Conversion of the ship to a guided missile destroyer was not shown in the General Dynamics executive summary of its proposed design. The

General Dynamics prepared this design in 1967 based on the Navy's draft specification for DX. It was labeled "DD 960" because the designers were unaware that Foreign Military Sales destroyers had already used hull numbers DD 960 through DD 962.

Estimating that low price and compact size might attract Navy evaluators, General Dynamics proposed a smaller design for the Total Package contract. (Courtesy Richard E. Cross III)

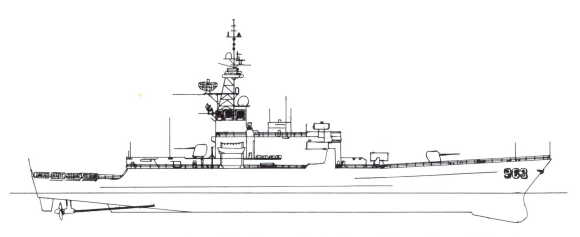

In April 1969 General Dynamics submitted this design, much shorter than Litton's and Gibbs & Cox's designs, for the DD 963 contract. Its arrangement and some details—such as the bridge wings and the sideways-venting, mast-topped stack—resembled

the GD-designed *Garcia* (DE 1040)–class destroyer escorts. The Sea Sparrow launcher was above the bridge with its director near the stack. (drawing by the author)

Tartar-D fire control radar would presumably replace the Sea Sparrow system atop the bridge. Installing a second Mark 26 launcher appeared not to be feasible. It might have replaced both the aft 5-inch/54 caliber gun and the helicopter hangar in terms of length, but reconstruction below decks would have been prohibitive. Gibbs & Cox regarded the General Dynamics design as "fictitious" for the required payload and modernization margins.[42]

The Navy, however, was very interested in General Dynamics's compact design. General Dynamics offered the lowest price among the three industry contenders. During the months following design submission, the Navy program office spent

as much effort reviewing it as it spent on the Litton and Bath–Gibbs & Cox designs put together.[43]

The Naval Ship Engineering Center Design

During the DX/DXG Concept Formulation phase the Naval Ship Engineering Center (NavSEC) had created over 1,000 computer models of feasible designs to support trade-off decisions about ship features. In these models, ship-synthesis software used alternative ship performance requirements and payload properties (weight, location, power requirements, structural support) to estimate the size of the resulting destroyer.[44] It was important to keep the destroyer's displacement below 8,000 tons to restrain Admiral Rickover and his allies in Congress from demanding nuclear propulsion, a confrontation that could kill DX/DXG as it had the FY 1967 DDG.

Under the Total Package Procurement procedures Litton, General Dynamics, and Gibbs & Cox were creating their competing DD 963 designs. Staffers at NavSEC disliked being left out of the action. They created a complete DD 963 preliminary design on their own to show that the Navy still had an in-house design capability.[45]

NavSEC's design was for a relatively small, flush-decked destroyer with a telescoping helicopter hangar. It resembled the *Leahy* and *Belknap* classes in appearance, with twin "macks" (combined mast-stacks) and the old-fashioned gap between them. Armament layout was identical to the Litton design. The ASROC installation was conventional, with a magazine and horizontal loader beneath the bridge. From illustrations in Friedman's *U.S. Destroyers* it appears that a Mark 26 Mod 0 (24-round) magazine and missile launcher could replace the ASROC launcher if below-decks changes were made. Conversion to a guided missile destroyer would have required installing missile directors above the bridge. Design variants show Aegis (ASMS) target illuminators, but installing Aegis on a 6,000-ton ship would have been impossible. There was no room for a second Mark 26 launcher and magazine. The helicopter facility was above the aft propulsion machinery room, and the hull beneath the aft 5-inch mount was too shallow.[46] The small size of the ship indicated that the modernization and conversion margins were inadequate.

NavSEC left the choice of main propulsion equipment to its propulsion office, whose civilian chief was an advocate of steam boilers. He chose steam for the NavSEC DD 963 design. This choice was unsuitable, given the known advantages of gas turbines for fast response, silencing, start-up mobility, combat survivability, and low manpower, as NavSEC's Philadelphia laboratory had found during the research for Project Seahawk.[47] Steam propulsion was suboptimal. It might permit small hull size, but the resulting destroyer would have been obsolete as delivered and expensive to operate. Small hull size provided no measurable benefit beyond lower cost for steel, a saving offset by higher costs for assembling a tighter design. The DD 963 project office had no charter to do anything with the NavSEC design and ignored it, a slight that NavSEC designers were to remember for years, as will be seen.

Design Submission

The initial designs were due in April 1969. Litton believed that its DD 963 Contract Definition package was the largest proposal ever submitted by the defense industry until then. A young officer at NavShips was told that a proposal had arrived and that he had to receive the classified material personally. He went to the loading dock expecting a box and was astounded when a fully laden truck manned by a crew of security-cleared moonlighting sailors pulled up to unload Litton's proposal. The DD 963 project office began to realize the work they faced in evaluating the competing offers.[48]

Design Selection and Completion

Contention over Gas Turbines

Early in 1969 the Defense Department requested production funds to order the first five destroyers during FY 1970 (July 1, 1969–June 30, 1970). The press revealed that all three contractor DD 963 designs featured gas turbine propulsion. Worried nuclear power lobbyists tried to block the DD 963 project with politics. Admiral Rickover asked the Chief of Naval Operations, Admiral Thomas Moorer, to stop the acquisition. The CNO backed the new destroyers.[1]

Opponents alleged that gas turbines were costly and untested and that the only Navy personnel who wanted them for destroyers were aviators. Representative Charles Mosher sent a letter to the White House charging, "These admirals have had successful experience with, and confidence in, gas turbines in their planes so they too easily assume such power plants can easily, quickly be adapted for use in ships without adequate experiment." The Project Seahawk and FY 1967 DDG episodes showed that Admiral Rickover and the nuclear power lobby intended for "adequate experiment" never to occur. Mosher's letter attacked gas turbines as an "interim tangent" from an all-nuclear fleet and demanded oil-fired steam boilers for the DD 963 class because nuclear ships also used steam. A lobbyist, probably Admiral Rickover himself, almost certainly had written the letter.[2] President Richard Nixon, a supply officer in the World War II Navy, declined to interfere.

Representative L. Mendel Rivers, chairman of the House Armed Services Committee, directed the Sea Power Subcommittee to investigate "problems" concerning the DD 963 propulsion system. The CNO himself told the subcommittee that he preferred gas turbines to steam for many reasons:

- Faster response time to maneuvering-speed orders
- Ease of maintenance
- Susceptibility of steam lines to penetration by shrapnel from bombs and shells
- Safer conditions for the crew
- Better solution for the underwater radiated acoustical noise problem
- Reduction of personnel requirements
- Ability to replace the initial gas turbine with a more efficient future engine if it were to come along ("We do this kind of thing with aircraft")

Representative Charles E. Bennett, chairman and raison d'être of the Sea Power Subcommittee, replied to Rivers (a strong Rickover ally):

The Navy is in a very difficult dilemma. It cannot formally say that it desires one or the other type of propulsion until it has examined the proposals and made its selection [of the destroyer]. On the other hand, it needs to have the money authorized at this moment so as to be able to proceed in whichever manner it decides best.

Both steam and gas turbine plants seem to be at about the same state of development—except that steam has generally been in use by the Navy for a long time and gas turbines have not. The calculated difference in cost over the lifetime spans of both steam and gas turbines is relatively slight if presently marinized gas

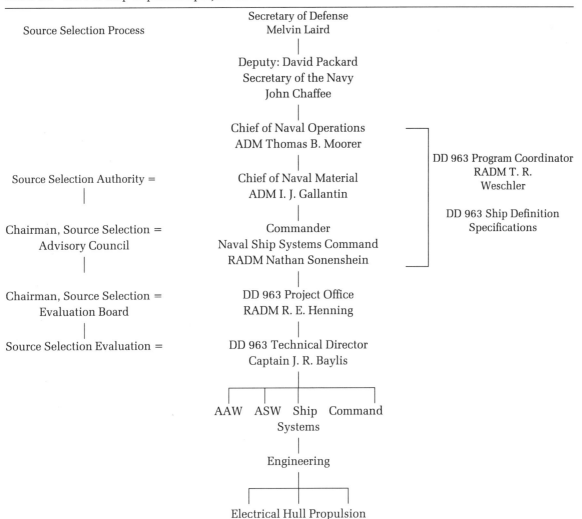

Source Selection Process

Secretary of Defense
Melvin Laird

Deputy: David Packard
Secretary of the Navy
John Chaffee

Chief of Naval Operations
ADM Thomas B. Moorer

Source Selection Authority =

Chief of Naval Material
ADM I. J. Gallantin

DD 963 Program Coordinator
RADM T. R.
Weschler

DD 963 Ship Definition
Specifications

Chairman, Source Selection =
Advisory Council

Commander
Naval Ship Systems Command
RADM Nathan Sonenshein

Chairman, Source Selection =
Evaluation Board

DD 963 Project Office
RADM R. E. Henning

Source Selection Evaluation =

DD 963 Technical Director
Captain J. R. Baylis

AAW   ASW   Ship   Command
Systems

Engineering

Electrical Hull Propulsion

turbines [FT4As] are employed. This difference could become greater if proposed second generation gas turbine engines [LM-2500s] prove successful. . . .

The decision should be made on the basis of which has the better effectiveness for the Navy, within reasonable cost differences.[3]

The Navy's earlier decision to use a single type of fuel oil (NDF) for all shipboard power plants eliminated any cost advantage for steam. Bennett's four-page letter did not mention combined diesel and gas turbine plants (CODAG/CODOG). He recommended that Congress should vote $342 million for the first five destroyers with whichever propulsion plant the Navy thought would provide the best ship. Congress agreed and in June 1969

authorized the requested funds to order the first five DD 963 destroyers during FY 1970.

The Navy Evaluates the Proposals

Admiral Sonenshein was promoted to commander of the Naval Ship Systems Command (NavShips) on July 1, 1969. That made him chairman of the DD 963 Source Selection Advisory Council, which evaluated each shipbuilder's approach to the systems management task of ship design, production, and support (see table 6.1). The Navy's evaluation of the initial proposals received in April 1969 was that none of the three was fully satisfactory as a basis for a contract award. Bath's initial bid price was $100 million

Service as Pacific Fleet maintenance officer showed Rear Admiral Nathan Sonenshein, seen here as a captain, the value of eliminating steam propulsion in warships. As Commander, Naval Ship Systems Command during 1969–72, he was the true father of the *Spruance* class. (U.S. Naval Institute collection)

per ship. Litton's price was $89.5 million. Both exceeded the Navy's FY 1970 shipbuilding authorization of $342 million for five ships.

General Dynamics (GD) offered the lowest price, and the Navy devoted half its effort to evaluating this design. It was competitive and imaginative, but its compact size had disadvantages. The aft 5-inch gun mount was abutted to the large deckhouse and could not fire far forward.[4] The living spaces were in hot parts of the ship, and the engineering spaces were neither separated nor independent enough to suit the Navy's plans for survivability.[5] The ship would be too small for FT4A gas turbines should the LM-2500 fail testing.[6] The real risk was not that the LM-2500 might fail but that proponents of the DD 963 project

might not persuade other authorities, who were already skeptical about relying on Total Package Procurement for the program, to depend on a new engine, too. The evaluators concluded that the General Dynamics design sacrificed performance that the Navy valued. The biggest concern was that the berthing spaces were too crowded and too close to the uptakes for the required habitability quality.[7]

General Dynamics also had political problems. Its Quincy yard had incurred quality complaints, cost overruns, and delays in delivery on all current contracts. The Navy despised GD–Fort Worth Division's F-111B interceptor, and Congress had killed it in 1968. GD's poor performance on the F-111 contract led to doubt that the firm was committed to the low price it quoted for DD 963.[8] Its management impressed one evaluator as being more interested in the termination fee it would collect if the Navy canceled the DD 963 contract, as had just occurred with the F-111B, than in building destroyers. The Navy rejected the General Dynamics design on September 9, 1969. GD's DX/DD 963 design effort had cost $20 million. The Navy reimbursed it for $2.5 million.[9]

Since the shipbuilding appropriation was a limiting factor, the initial acquisition price of the ships became more significant in the evaluation than life-cycle costs. The life-cycle costs were projections, while the construction price would be fixed by contract.[10] However, GD's experience proved that lowest price by itself was insufficient to win the contract. The Navy began working closely with Bath and Litton to reduce initial acquisition costs. This forced Bath and Litton to make a limited redesign of their competing destroyers.

The Navy regarded neither DD 963 design as featuring a satisfactory defensive electronic warfare system and so deleted it as contractor-furnished equipment. The helicopter haul-down system was deleted, in part in response to aviators' doubts about the safety of tethering a helicopter to a pitching ship.[11] Big ships would be stable enough to handle small helicopters without it. The Sea Sparrow antiship missile defense installation was deleted, to be replaced by the greatly improved NATO Sea Sparrow when it became available later in the decade. All these were deletions that could be, and were, restored later. As such, they differed from the irreparable reductions in capability made, for example, to the Air Force B-1 Lancer bomber, whose original Mach 2.2 dash speed was cut in

The commercially owned transport *Admiral Wm. M. Callaghan* served the Navy as the testbed for gas turbine propulsion. In an almost emergency action late in 1969, Navy gas turbine proponents had her equipped with the first LM-2500 marine gas turbine engine to test it at sea before the DD 963 production contract was decided. (L & L van Ginderen)

half in the early 1970s. Critics of the DD 963 class ignored this fact. The canard that the ships were "failures" because of "deficiencies" stemming from construction cost reduction would find its way into otherwise reputable analyses.[12]

A major specification change was to standardize the main propulsion plant on four FT4A gas turbines with the LM-2500 as an alternative. The DD 963 Source Selection Evaluation Board initially favored the FT4A, which had passed testing during the Project Seahawk research, over the new LM-2500.[13] However, the board doubted that three FT4As could provide sufficient power for Litton's DD 963 design and ordered the change to four. The Navy ordered Bath to increase its design margins to accommodate the bulky FT4A plant. Gibbs & Cox lengthened its ship 30 feet to do this. This enlarged the design significantly, but few specifics are known.[14] Nonetheless, at least some Navy evaluators still believed that the defunct General Dynamics design, which was not adaptable to the FT4A, showed that the other designs should be smaller.[15]

The standardized four-turbine plant made it easier for the Navy to compare the competing ship designs, but it narrowed the technical difference between the contractors. Bath suspected that the Navy was helping Litton to catch up in quality, since many ordered features were already present in the Gibbs & Cox design. However, alternative gas turbine plants were Litton's idea. The Navy was paying for both designs and had contractual right to use the information as it wished.[16] Bath protested Navy leaks of the initial price offers, which might affect the technical evaluators' recommendation.

In November 1969 the Navy requested that Bath and Litton resubmit their pricing proposals to reflect the changes. Bath's second bid was $86.4 million per ship, and Litton's was about $80 million. (These were target costs, to be explained shortly.) In December the evaluation panels reported that both designs were now technically adequate and that an award could be made to either contractor. Rumors put the technical rankings at 98.9 for Litton and 98.7 for Bath, on a scale of 100.[17]

Others in the Navy did not rate the designs so highly. One was Rear Admiral John Bulkeley, president of the Board of Inspection and Survey, who independently inspected all Navy ships for material fitness. He wore the Medal of Honor, won commanding the PT boats that evacuated General Douglas MacArthur from Corregidor in 1942. Admiral Bulkeley was skeptical about reliance on controllable/reversible-pitch (CRP) propellers. The Assistant Secretary of the Navy for R&D was skeptical about both the untested LM-2500 gas turbine, only one of which existed, and CRP propellers.[18] A CRP propeller able to transmit up to 50,000 shp from two gas turbines into the water had never been built. CRP propellers inherently cavitate more than fixed-pitch propellers and require an anticavitation bubbler system (Prairie) to be built in.

The Navy had identified mechanical engineering difficulties with CRP propellers during Project Seahawk and knew how to resolve them.[19] CRP design principles were well understood. Several American firms manufactured models up to 35,000 shp. Coast Guard *Hamilton*-class cutters and the Canadian *Iroquois*-class destroyers used CRP propellers. Admiral Sonenshein persuaded doubters that achieving the additional strength needed for the DD 963 CRP propellers was a manageable task.

Admiral Sonenshein had been closely involved with the development of both the LM-2500 and the FT4A marine gas turbines. Certain that the powerful and economical General Electric LM-2500 would make the DD 963 a better and less expensive ship than would the older FT4A, he authorized a crash project aboard the Military Sealift Command–chartered *Admiral Wm. M. Callaghan.* Working with NavShips and MSC, GE at its own expense replaced one of the ship's two FT4A gas turbines with the first prototype LM-2500. GE agreed to restore the FT4A afterward if the LM-2500 did not work out. Installation of the LM-2500 between voyages during December 1969 exploited the rapid swap-out feature of gas turbines. The only other change needed was a new fuel coalescer filter.[20]

The Navy gave the ship's master a histogram showing how often a destroyer typically operated at various speeds and told him to follow it for a year, starting with the next Atlantic crossing. Voyages showed that the LM-2500 was indeed 20 percent more fuel-efficient than the FT4A over a destroyer's range of power and acceleration, and as reliable. An immediate benefit was that Admiral Sonenshein persuaded doubters that an LM-2500-powered DD 963 posed little technical risk. This kept open the option to use the new engine.[21] Later, a second LM-2500 replaced *Admiral Wm. M. Callaghan*'s other FT4A. The ship operated until 1987, still doing propulsion experiments for the Navy during her voyages.

Design Selection

Litton's technical proposal gave the Navy a more attractive vision of the future than Bath's in every way. The exciting modernistic style of Reuven Leopold's European-flavored destroyer contrasted with the functional but boxy Gibbs & Cox design. Litton's proposal featured many three-dimensional illustrations. Bath's proposal relied on text and two-dimensional drawings. Bath representatives were dismayed to overhear Navy men contrasting Litton's Los Angeles offices and young, miniskirted southern California secretaries with Gibbs & Cox's bland New York headquarters and helpful but grandmotherly receptionist.[22]

What Bath did have in its favor was a more successful corporate record in shipbuilding than aviation-oriented Litton. However, Litton's existing East Bank yard at Pascagoula was one of the few that Admiral Rickover considered satisfactory for nuclear submarines. Litton's DD 963 design was competitive in quality and price. The issue before the Navy was in effect whether to penalize Litton for the undeveloped state of its new West Bank yard and over uncertainty about how it would recruit an army of skilled ship artisans. Litton had won the FDL and LHA competitions. Roy Ash, president of Litton, made a detailed presentation to the ship acquisition technical panel to show the commitment and expertise of his firm's corporate management. Evaluators were impressed that Litton had planned every detail. From the start, Litton's DD 963 design could accept alternative propulsion gas turbines and alternative auxiliary power plants; that reduced risk. Litton had arranged union agreements in advance to lock in future wage rates. Ash correctly answered even the most technical questions without once calling on his staff. Reuven Leopold's experiment regarding the design profile was not mentioned.[23]

Meanwhile, the Navy was concerned that both shipbuilders' November 1969 construction price offers still were higher than Congress would tolerate. The project office decided to replace some contractor-furnished equipment, such as life rafts, from existing Navy stocks rather than to have the shipbuilders buy new equipment. With these additional cost reductions specified, the Navy asked Litton and Bath to resubmit their "best and final" pricing proposals for production runs of 30, 40, and 50 destroyers.

The ships might have been a naval architect's dream, but figuring their costs was an accountant's nightmare. The Navy required each shipbuilder to quote its costs for ships to be built years in the future. Accelerating inflation would drive up the shipbuilder's costs for work done later in the contract. Contract terms covered wage inflation based on Department of Labor statistics, but the shipbuilders considered this inadequate. The Navy could not sign a contract adjusting for inflation above the Nixon administration's projections, lest that show doubt about the prospective success of anti-inflation policies. In theory either contractor could cover high inflation simply by quoting a higher price. In practice that risked losing the award. The competing contractor might offer a lower price and win, or might bid a high price too, in which case the Navy would kill the DD 963 project as too costly no matter who won the technical evaluation.

Litton and Bath were each required to submit a "target" price estimate based on the preliminary DD 963 designs, and a "ceiling" price for the detail design, which allowed for some cost increases. Bath quoted a ceiling price of $81.1 million per destroyer. Litton quoted $80.8 million. Target prices, for comparison with earlier figures, were $73.7 million for Bath and $71.9 million for Litton. None of these prices included weapons, sensors, or other government-furnished equipment.[24] Bath's annual financial report hinted that the firm anticipated losing the 30-ship Total Package contract but was looking forward instead to the order being split between itself and Litton. That would be contrary to the principle of Total Package Procurement but in keeping with traditional practice.[25]

The Navy had budgeted for a contract price of around $60 million per ship.[26] On February 23 Defense Secretary Melvin Laird notified Congress that the first-year (FY 1970) contract budget would cover only three rather than five destroyers. Secretary Laird was a former Naval Reserve Supply Corps officer and had been the commissioning supply officer of USS Maddox, of later Tonkin Gulf fame, or notoriety, in 1944. A contract amendment changed the financial terms, principally in how inflation adjustments would be made. Hoping that these changes would reduce the shipbuilders' prices, the Navy told Litton and Bath to submit still another best-and-final offer the following week.

The question before Bath and Litton remained whether either was willing to risk taking on a $2 billion contract with little cushion for inflation. Litton's 1969 sales were $2.177 billion, so it perhaps could take this risk; Bath's 1969 sales had totaled $189 million. Bath submitted a new ceiling price of $79.7 million per ship, about $1.3 million below its previous offer. Litton stated that on the basis of its purchases of material for the LHA-class amphibious assault ships, it could write fixed-price subcontracts for all 30 ship-sets worth of material at the start of the DD 963 project. That would reduce exposure to inflation.[27] (The Navy had awarded the Tarawa [LHA 1]–class Total Package contract to Litton in May 1969.) Litton's new ceiling-price offer was $71.3 million per ship, or $9.5 million less than it had offered a month earlier.

Skeptics later charged that this was an obvious buy-in.[28] It must also be said that multiple best-and-final requests often were an abuse of the procurement system. A frequent government tactic in a lowest-price-wins competition was to choose a favored design or contractor and then to issue repeated best-and-final requests, adjusting the specification if necessary, until the favored contractor's price was lowest. The procurement office then announced the winner and congratulated itself on how competitive the process had been and on how the government had obtained both the highest-quality design and the lowest price. Secretary of Defense Dick Cheney officially abolished multiple best-and-finals in 1989 after uncovering bribery by contractors frantic to learn competitors' prices.

This having been said, the second best-and-final call in the DD 963 procurement does seem innocent. The Defense Contract Audit Agency (DCAA) had challenged Litton's February offer as overpriced by $9.5 million per ship. Using Litton's February inflation estimate, DCAA calculated the material cost as $40.5 million per ship and did not support Litton's estimate of $50.0 million. Bath had a similar situation. The hope that hedged pricing was a source for cost reduction was a primary reason that the Navy changed its inflation adjustment policy and asked for the March 1970 repricing.[29]

The final price difference between Litton and Bath was about equal to the difference in labor costs, due to the Gulf Coast's lower wages, plus Bath's higher interest payments for its proposed new yard.[30] Litton was consistently lower in the price competition, and in the judgment of every Navy participant I interviewed, it offered an equal or superior ship design as well. The General Accounting Office later described the specific award criteria:

> [The Navy's] evaluation included five major areas of inquiry in the categories of life cycle cost, ship system, management plans, military effectiveness and corporate qualifications. In the final assessment by the Source Selection Authority, Litton was given a numerical rating superior to Bath for the following categories.

	Bath	Litton
Ship System [the actual design]	5.80	5.85
Management Plans	4.20	5.07
Military Effectiveness	6.00	6.68

> In the Navy's rating system all of these were considered in the normal range. The records also show that estimates for life cycle cost were also higher as proposed by Bath, by about 5.5%. With regard to the category of corporate qualifications, our review shows that Bath had considerable difficulty in providing a financial plan acceptable to the Navy; however, the insufficiency of the financial plan was corrected during the final stages of the evaluation. The Source Selection Advisory Council ultimately concluded that the proposal of either contractor would provide destroyers suitable for the future needs of the Navy. Based on the decision of the Source Selection Authority, as well as the lower price quoted by Litton in the best and final offer, the award to Litton Industries was made.[31]

Contract Award

Based on technical and cost factors, Admiral Sonenshein's Source Selection Advisory Council recommended an award to Litton. The contract had to be signed before the pricing proposals expired on June 26, 1970. In May Admiral Gallantin forwarded this recommendation to the Defense Systems Acquisition Review Council (DSARC), a committee of OSD officials, via Secretary of the Navy John Chaffee.

Bath learned from a Navy leak that Litton was about to win. Bath quickly enlisted the support of Maine's Senator Margaret Chase Smith, ranking Republican on the Senate Armed Services Committee, to turn a political blowtorch on the Navy to overturn or delay the award to Litton. Senator Smith signed a series of furious letters to Secretary Laird, Secretary Chaffee, and President Nixon protesting the plan to award the contract to a single shipyard. None of these letters advocated the Gibbs & Cox DD 963 design. Bath was asking to build Litton's design under a split contract.[32]

The precedent for political influence in a military source selection decision was former Secretary McNamara's 1962 award of the F-111 contract to General Dynamics over the Navy–Air Force choice of Boeing. As noted, the F-111 aircraft had severe cost and acceptance problems, leading to congressional rejection of the Navy F-111B version in 1968. Consequently, and by personal inclination, Secretary Laird preferred to trust the military's source selection. Senator Smith's letters, all written by Bath representatives, were referred to Admiral Sonenshein to draft replies for Secretary Chaffee, who upheld the Navy's decision.

The House of Representatives, however, passed an amendment to the FY 1971 defense appropriation bill requiring construction of the destroyers at two yards. Under this pressure, the Navy asked Bath and Litton for their prices for a 15–15 split and then a 22–8 split of the 30 destroyers. This amendment failed in the Senate when the responses showed that costs would increase by $200 million.[33] Under Senator Smith's verbal assault, DSARC postponed consideration of the DD 963 award until May 28. After a series of meetings DSARC then approved the award to Litton.[34] DD 963 was the first program to pass DSARC. Several sources insisted that congressional delegations from Mississippi and other Gulf Coast states must have lobbied for Litton.[35] However, except for the events described here, none of the DD 963 ship acquisition project personnel interviewed for this book heard from, or about, any politician trying to influence the Navy's decision. The Navy's source selection procedure thus withstood the political attempts to influence it.

Secretary Laird had delegated authority to sign the award document to his Deputy Secretary of Defense, David Packard, who had taken over from Paul Nitze in 1969. Packard, the industrialist cofounder of Hewlett-Packard, was skeptical both about Total Package Procurement and about the DD 963 design itself. The Navy's evaluation followed the TPP guidelines and clearly showed that the award should go to Litton, so OSD really was not being asked to choose between Litton and Bath

Iron Works as a policy decision. Secretary Packard's true decision was whether to build Litton's destroyers or nothing. After four years and three Secretaries of Defense, OSD did not have a defensible policy choice of still further delay to produce yet another DD 963 design. While the Soviet Union was building nine different classes of submarines and seven classes of surface combatants, every U.S. Navy advanced destroyer design had been canceled in favor of the DX/DD 963 class and a pair of nuclear cruisers. Four years of effort by OSD, the Navy, and industry had culminated in the favored Litton DD 963 design. Only six months had been spent on the design itself; the rest of the time was administrative. The entire American naval architecture industry had been dislocated by the effort to design the candidates for the DD 963 class, and this talent was needed elsewhere.[36]

Secretary Chaffee, Admiral Gallantin, Admiral Sonenshein, and other officials met in Secretary Packard's Pentagon office to discuss DD 963. The meeting lasted well into the night. The new destroyers would mount the most advanced ASW combat system, were designed for modernization to keep them militarily competitive for decades, and would be built under a strict fixed-price contract. The four-turbine plant guaranteed reliability, simplicity, and small crew requirements using either the specified FT4A or the alternative LM-2500. The chairman of DSARC supported the award to Litton, as did all the Navy representatives. Secretary Packard remained concerned about the risks in gas turbines and CRP propellers. In fact those were two of the least risky parts of the whole project. Admiral Sonenshein had brought with him cutaway views of internal mechanisms and used these to brief Secretary Packard in detail. Satisfied, the Secretary signed the memorandum ratifying the Navy decision for contract award to Litton.[37]

DD 963 was the biggest Total Package Procurement contract the Department of Defense ever awarded. It was also the last.[38] Secretary Packard, Admiral Sonenshein, and Admiral Weschler were satisfied that the DD 963 design was of outstanding quality. Results, however, had to be outstanding to justify the time and effort required for the TPP process. TPP did not guarantee such results. If a weapon bought under another Total Package contract proved unsuitable, the cancellation penalties paid to the contractor would be very high. Time and cost might not allow starting over

Admiral Raymond Spruance (left) meets with Admiral Chester Nimitz and Rear Admiral Forrest Sherman aboard Spruance's flagship, the battleship *New Jersey* (BB 62), during the Central Pacific campaign in 1944. Although always upstaged in the press by General MacArthur and Admiral Halsey, Spruance was a more skilled tactician, and his actions contributed more than theirs did to ultimate strategic victory in World War II. (U.S. Naval Institute collection)

to develop a replacement. In June 1970 Secretary Packard issued a new policy that required a review of operational requirements for new weapons so as to eliminate prohibitively costly functions. He further ordered that for future contracts, start-up production must minimize the government's financial commitment until all major developmental issues were resolved.[39] That ended Total Package Procurement, the most ambitious attempt to reform weapons acquisition through contractual financial incentives.

The incoming Chief of Naval Operations, Vice Admiral Elmo Zumwalt (who, before taking command of the naval forces in Vietnam, had written the 1967 Major Fleet Escort Study that justified the DX/DXG project), assented to the award.[40] On June 23, 1970, Litton and the Naval Material Command formally signed the contract to build the DD 963 class. Admiral Sonenshein, representing the Navy, announced that the first ship would be christened USS *Spruance* (DD 963). She commemorated Admiral Raymond Spruance, the taciturn victor of the Battle of Midway and the Central

Pacific Campaign, who had died in December 1969.

Design Refinement

The Naval Ship Systems Command was euphoric both about the DD 963 design and about the shipbuilding contract for series production of 30 destroyers. The *Spruance*-class destroyers would be the most modern surface warships in any navy. They could readily be modernized to keep them current. DXG had died even before the cancellation of TPP in part because the *Spruance* class could be converted to more formidable guided missile destroyers than the DXG class would have been. A converted DD 963 would mount the same Standard-MR (Tartar-D) guided missile battery as DXG while retaining the helicopter facility and a superior ASW and gun armament. The TPP contract was sound. Litton waived all rights to filing claims based on defective Navy specifications or impossibility of performance.[41] Construction was scheduled to begin in 1972 so that Litton could recruit and train workers and avoid conflicts with the LHA class and commercial construction.

The next phase was for Litton to refine the preliminary design into the detail design for production. Several design changes were made to the engineering plant during this phase. In December 1970 NavShips approved changing the DD 963 main propulsion engines from the FT4A to the LM-2500. Acquisition cost and fuel consumption were much better for the LM-2500 gas turbine than for the FT4A. Side-by-side testing for 6,000 hours aboard *Admiral Wm. M. Callaghan* showed that the engines were equal in reliability, maintainability, and operability.[42] The LM-2500 plant could use the electrical cross-connect between the reduction gear sets, but to pay for itself, the cross-connect would need to be in use 80 percent of the time between 11 and 20 knots. The cross-connect gear was heavy, it complicated propulsion control, and it reduced gas turbine acceleration. The LM-2500 plant without the cross-connection offered similar overall fuel efficiency as predicted with it. The cross-connect was deleted to simplify operation and maintenance of the propulsion system.[43]

The Royal Navy–style streamlined funnel caps were deleted, lessening the DD 963's original profile similarity to the British County class. The stacks remained rakishly offset from the centerline but now were reversed from Litton's original pro-

posal, with the forward stack offset to port and the aft stack to starboard. This improved helicopter operations when the destroyer was in formation with an aircraft carrier. During flight operations a carrier would steam on course to bring the relative wind down the angled deck. With the aft stack offset to starboard, the same relative wind would carry propulsion turbine exhaust clear of the helicopter deck.

In 1971 three ship's service gas turbine–powered generator (SSGTG) sets with waste-heat boilers replaced the original design's three diesel generators and two auxiliary steam boilers. The decision to use SSGTG/waste-heat boiler sets in the *Spruance* class drew heavily on the Canadian *Iroquois*-class destroyers, which featured two. Diesels were discarded from the DD 963 electrical plant design because they required an unproven active noise isolation system, unnecessary for the inherently quieter (as measured by hullborne noise from vibration) gas turbines. Since compressed air could be bled from the SSGTGs, two large air compressors for the Masker silencers were deleted, too.[44] This change contributed significantly to the goal of making the *Spruance* class the quietest ASW ships ever developed to that date.

The DD 963 design could carry another 350 tons at the main deck level and generated 1,000 kW more power on two SSGTGs than the weapons required for combat.[45] Its large margins of volume, power, stability, and reserve buoyancy for addition of the large Tartar-D system interested the Aegis project office. Tartar-D and Aegis were both to use the same Mark 26 launchers. The Aegis project office and Naval Ship Engineering Center staffs met after the DD 963 contract award to study whether the design could be modified to mount the entire Aegis system. Aegis was too large and complex to be incorporated as additional payload during conversion of an existing DD 963. However, it did appear feasible to modify the *Spruance* design to install Aegis during construction, with the ship and the radar system sharing common strength members. Admiral Sonenshein was interested in the Aegis–DD 963 study and in studies of the feasibility of installing Aegis on other ship designs.

Captain Wayne Meyer, the Aegis project manager, requested a meeting to go over these studies. He, Admiral Mark Woods of the Naval Ordnance Systems Command (NavOrd), and Admiral Sonenshein reviewed the study results at NavShips. All agreed that the DD 963 was the best Aegis platform. They further agreed to support the DD 963 as the best platform during Shipbuilding and Conversion, Navy (SCN) budget planning at the Naval Material Command and the Office of the Chief of Naval Operations.

The DD 963 contract design was still evolving into the detail design. Preliminary-design reservations (margins) for additional weight, space, power, and other utilities were the open slots of a ship design. Admiral Sonenshein ordered that any design changes had to preserve the stability margin, internal volume, and electrical power necessary for the Aegis system.[46] Keeping the DD 963 margins was easy, because the Total Package contract blocked other Navy program offices from asking Litton to add their new systems to the ships.

Captain Meyer's Aegis project office in NavOrd quietly began a major redesign of Aegis to fit the large Aegis components within the available DD 963 design margins. The unobjectionable official explanation was that the DD 963 was simply a convenient sample platform for sizing a physically smaller and less expensive Aegis system.[47] Apparently nobody asked why, if that was the goal, the Aegis designers were doing major work that had to be unique to the DD 963, such as shifting all the radar arrays to fit around the asymmetric uptakes. This bureaucratic white lie did not arouse Admiral Rickover and the nuclear power lobby into pressuring Congress to prohibit use of Aegis project funds to design a non-nuclear Aegis ship, the true objective of the Aegis redesign.

Naval Developments during the 1970s

In 1971 the first Soviet Kresta II large ASW ships appeared at sea. They were armed with improved antiaircraft missile batteries (SA-N-3), a helicopter, and eight bulky new canisters initially evaluated as containing a new antiship missile, designated SS-N-10. Other new Soviet ASW ships in series production, the Krivak and Kara classes, mounted the new canisters, too. Intelligence reports that this missile had barely horizon range were thought to show the Soviets' confidence that their warships would get that close to their quarry, either under cover of peace or by newfound combat endurance. In fact the SS-N-10 missile did not exist. The canisters were empty until 1975, when the slow SS-N-14 dual-purpose antiship/ ASW missile entered service.

Whatever the status of their main batteries in 1971, a growing Soviet fleet confronted Western navies. The sinking of *Elath* showed the potency of Soviet weapons. The Soviets sometimes replaced their warships after only 10 years of service. Their new ships carried new weapons and showed very good seakeeping and handling characteristics. The Soviet Navy operated progressively further out to sea, a strong trend that would culminate in the construction of fixed-wing aircraft carriers to provide Soviet naval forces with air cover beyond land-based fighter range. During OKEAN-70 in April 1970, the largest Soviet fleet exercise ever conducted up to that time, their forces practiced preemptive anticarrier strikes. There were real costs to the United States and other Western nations from Soviet naval activity, including reduced diplomatic effectiveness, protection of terrorist sanctuaries by visiting Soviet warships, support for coups d'etat by anti-Western factions, increased risk to U.S. aircrews from North Vietnamese air defenses alerted by Soviet tattletales at Yankee Station, and restrictions on ship operations in response to harassment in international waters.[48]

While U.S. ground forces disengaged in Vietnam, naval action intensified to keep pressure on North Vietnam. On shore-bombardment missions a North Vietnamese aircraft bomb hit the FRAM destroyer *Higbee* (DD 806). Shellfire from shore batteries hit the newer *Goldsborough* (DDG 20). Both ships suffered fatalities but only light structural damage.

The primary combat mission of surface warships off North Vietnam was antiaircraft warfare. Cruisers with 3-D radar guided fighters to intercept North Vietnamese aircraft. Admiral Zumwalt was concerned that the Soviet Union would arm North Vietnam with antiship cruise missiles (ASCMs), although as far as is known, no ASCM attacks occurred. Several ships scored confirmed kills of North Vietnamese aircraft with surface-to-air missiles. *Long Beach* had shot down the first two in 1967 with Talos missiles. The cruiser *Sterett* (DLG 31) shot down a MiG during the attack on *Higbee* in January 1972.[49] The total number of North Vietnamese aircraft engaged or shot down by ships is unknown. Carrier-based bombers mined Haiphong harbor in 1972 at low altitude under cover of ship-based missiles. Warships off the coast had clearance to attack any aircraft flying above 500 feet. Detecting no American fighters, North Vietnamese fighters scrambled to intercept the bombers. The cruiser *Chicago* (CG 11) promptly shot down one over land with a

Talos round. After that the others turned away whenever they detected a ship's fire control radar.

In 1972 the cruiser USS *Worden* (then DLG 18), operating off the Vietnamese coast, became the first modern warship to undergo missile attack. A U.S. Air Force Wild Weasel F-4 Phantom flying from Thailand detected her radar emissions off the coast and launched a Shrike antiradar missile. The missile homed on *Worden*'s foremast. Its 60-pound warhead exploded 80 to 100 feet above the ship. Fragments ripped through radar wave-guides and through her combat information center in the aluminum deckhouse. Spall and high-voltage electrical arcing killed 4 men, wounded 26, wrecked topside electronics, and destroyed the ship's air defense capability. This appeared to be a friendly-fire error by the Air Force crew, but a technical article in 1987 suggested that the cruiser's design as a radar platform was faulty and so contributed to the error. Repairs cost over $100 million.[50]

Two foreign naval actions were also noteworthy. Submarines, ASCMs, and old destroyers were used in the Indo-Pakistani War in December 1971. Soviet-built Indian fast attack craft twice raided Karachi. They sank a destroyer with Styx ASCMs, damaged other ships, and bombarded targets ashore. (The U.S. Navy's first ASCM, Harpoon, had not yet even flown.) A Pakistani submarine torpedoed and sank an Indian ASW frigate, the first such victim since World War II. Another Pakistani submarine was lost from an internal explosion, probably from a mine she was preparing to lay (she was not under attack when lost). The Indian Navy successfully blockaded East Pakistan and contributed to the collapse of the Pakistani military defenses there.

The second action was the Turkish amphibious assault on Cyprus in July 1974, which succeeded in occupying most of the island. Greece and Turkey both operated ex–U.S. Navy FRAM destroyers. The Turkish Air Force sank one, which turned out to be Turkish. These small operations showed the value of determined naval action in achieving strategic goals.

The largest naval activities were two confrontations between the United States and Soviet fleets in the eastern Mediterranean. The first was during the Jordanian civil war late in 1970. The Soviets were concerned that the Sixth Fleet would oppose an attack on Jordan by a Soviet client, Syria. The Soviet Mediterranean fleet outnumbered the Sixth Fleet ships, and the Soviets could reinforce their

fleet more quickly. Soviet warships followed all major Sixth Fleet warships to target them for attack. The Sixth Fleet found surveillance difficult: fighters launched to intercept approaching aircraft encountered airliners 75 percent of the time.[51]

The Soviet fleet opposed the Sixth Fleet again during the Yom Kippur War after Egypt and Syria attacked Israel on October 6, 1973. Both the United States and the Soviet Union began airlifting vast amounts of military equipment to their Israeli and Arab clients, respectively. A Sixth Fleet aircraft carrier battle group south of Crete provided surveillance, warning, and air-sea rescue services for the airlift.[52] Secretary of State Henry Kissinger ordered restrictions on battle group operations to avoid provocation.

Meanwhile, the Soviet Mediterranean fleet doubled to 96 ships. On October 24 the Soviets threatened to send in ground forces to stop an Israeli counterattack on Egypt. The United States regarded the Soviet suggestion as strategically unacceptable. As part of a large-scale military alert, two additional carrier battle groups were ordered into the eastern Mediterranean. Destroyers were sent from as far away as the Baltic. The Soviet fleet began practicing anticarrier strikes to reduce the warning time of an actual Soviet strike on the U.S. forces, if it occurred, to missile time of flight.

President Nixon called this "the most difficult crisis we have had since the Cuban confrontation of 1962."[53] Nixon was beset by the Watergate scandal, or more precisely by General Alexander Haig, who spent that weekend stirring up the "Saturday Night Massacre." Secretary Kissinger acted as de facto commander in chief. He forecast that President Anwar Sadat of Egypt would forbid Soviet forces to land, and he again ordered U.S. operations to stay nonprovocative. At sea, however, both fleets thought combat was a real possibility. The U.S. battle group prepared to counter every possible Soviet move. Shipboard confidence increased that the Americans could defeat a conventional ASCM attack from the Soviet ships and submarines. Admiral Zumwalt, however, was concerned that the Soviets could overwhelm the Sixth Fleet by deploying bombers to Egypt, Syria, the Crimea, and Yugoslavia.[54] The diplomatic confrontation eased, but U.S. ships remained on heightened alert until November 18.

During the war Israeli and Arab fast attack craft fought several battles off Arab ports. Israeli electronic warfare tactics and Gabriel antiship missiles (designed as an antidestroyer weapon to counter ex-Soviet *Skoryy*-class destroyers in the Egyptian Navy) were entirely successful. The Israelis suffered no casualties and sank several Styx-armed patrol boats.

The Yom Kippur War confrontation showed that any conflict would be at the Soviets' initiative. They could concentrate their numerically smaller forces to attack the fraction of Western naval forces that were forward-deployed. Later versions of their First-Salvo tactic had to contend with improved U.S. task force surveillance and defenses. Soviet submarines, missile ships, and bombers under command from ashore now would converge on an aircraft carrier battle group and launch nuclear and conventional ASCMs at 200–400 miles' range. Torpedo-firing submarines would attack escorts in the outer screen. Simultaneous arrival of ASCM salvos would saturate the carrier's close-in defenses. Admiral S. G. Gorshkov's tactical objectives included combat stability, in particular protection of his submarines through use of surface warships and other naval forces to obstruct Western ASW forces.

Admiral Zumwalt and the *Spruance* Class

Admiral Zumwalt became CNO a week after the Navy signed the DD 963 contract with Litton. He faced the dilemma that with the Vietnam campaign winding down, the nation generally expected the defense budget to be cut, even though the military threat from the Soviet Union was rapidly rising. Zumwalt ordered a strategic study, called Project 60, into the Navy's needs. He catego-

rized naval forces as High and Low and regarded DD 963s as High ships: "There was more than enough High, more than enough Too High, already under construction or under contract when I began Project 60 and almost no Low at all."[55] In making this often-quoted but inaccurate statement he was recalling the shipbuilding budget then being planned for FY 1971–75, not current construction. In July 1970, 35 Low *Knox*-class destroyer escorts were under construction to join 30 other modern DEs already in service. High-end ships then under contract included the first 3 DD 963s, 3 cruisers, 2 aircraft carriers, and 24 submarines. The five-year defense budget plan called for building more High ships of all these types, including 27 more DD 963s.

Admiral Zumwalt regarded the *Spruance* class as "first class . . . able to perform *nearly all* missions as well as a DLGN [nuclear-powered guided missile cruiser]," but "too expensive."[56] Facing the block-obsolescence problem and a curtailed ship construction budget, he planned an innovative fleet of low-cost but advanced ships. These included the PHM 1–class hydrofoil missile boats, a new escort aircraft carrier typed as a Sea Control Ship, and the PF 109 (later FFG 7)–class patrol frigates to meet the original inventory requirements for DX/DXG. The new CNO disliked the cost of the DD 963 class and regarded the ships as more powerful than necessary: "The trouble with them was that they were too good in the sense that the Navy had given up too much to get them."[57] It would soon become apparent that others in the fleet regarded the *Spruance* class as nowhere near good enough.

Construction and Controversy

Construction of the DD 963 Class

The DD 963 class was the largest shipbuilding program at the Ingalls West Bank yard, and indeed in the world. Commercial container freighters and the Navy's *Tarawa* (LHA 1)–class amphibious assault ships were already being built there. The new West Bank yard was built almost simultaneously with these ships. The East Bank yard was completing four ammunition ships and four nuclear attack submarines. Concerned about quality and the backlog of shipbuilding at Ingalls in 1970, the Navy scheduled DD 963 construction to begin in 1972 so that Ingalls could complete its other contracts before they might affect the DD 963 program.[1]

During this interval before construction began, the Navy provided minimal oversight to Litton. Primarily this was to avoid giving Litton any ground later to allege in court that Navy actions had damaged its contractual performance. Litton was preparing the detail design for the *Spruance* class. The Navy reviewed Litton's progress. If the Navy suggested changes, Litton usually responded that because the work being reviewed was already complete, an expensive change order would be necessary. The Navy approved changing the main propulsion plant to the LM-2500 gas turbine and the electrical generation plants to gas turbine generators. These changes reduced the destroyers' cost and radiated noise, thus increasing Litton's chance to win the $24 million incentive fee for silencing.

Training thousands of new shipyard personnel while simultaneously building advanced warships by a revolutionary shipbuilding technique on a tight schedule proved to be an insurmountable challenge. As reports of trouble came to the

Naval Ship Systems Command (NavShips), Admiral Sonenshein demanded that Litton install new managers who could control the confluence of enormous projects.[2] Between 1969 and 1973 the top four positions at Ingalls changed hands 17 times.[3]

The schedule for the LHA 1 class began to slip, in part because of slippage with the commercial container ships already under construction. In January 1971 the Navy cut the LHA program from nine ships to five.[4] The cut reflected problems at Litton and a shift in the national strategic emphasis away from power-projection forces toward sea control. Litton tried to manage the DD 963 program more effectively than LHA, since most of the destroyers had yet to be ordered. Prospects were good that the Navy would want all 30 as planned, given the new strategic emphasis on sea control.[5]

Concern grew that cost, quality, scheduling, and materials problems at Ingalls were risking disaster. The contract performance requirements were inadequate to define a complex warship, not surprising, since the Navy had no previous experience in writing such a document. The Navy's technical knowledge was superior to any shipbuilder's and was difficult to transfer in any manner that would enable the contractor to make trade-off decisions. Late in 1971 the Navy abandoned its hands-off policy and began to work more closely with Litton.[6]

Construction of the DD 963 class was a pioneering effort in shipbuilding worldwide. Fire mains, fuel lines, compressed-air pipes, ventilation ducts, electrical conduits, and many other systems are distributed throughout a ship. In traditional shipbuilding each such system was installed more or less at once in the ship under construction. Naval architects routed major piping and other systems

Litton's Ingalls Shipbuilding Division built the first
two DD 963–class destroyers (left) conventionally
from the keel up and all others by modular construction. Modules for the third and fourth destroyers are
to the right. Notice the large deck openings for the
gas turbine intakes and uptakes. When ready for
launch, the ships were moved sideways on tracks
leading to the dock where the launching platform
would be moored (*lower left*). (U.S. Naval Institute
collection)

during detail design. Designers indicated only the
general locations of thin pipes and conduits,
leaving builders to install them wherever they
might fit.[7] In modular construction, by contrast,
every subsection of every distributed system had
to be designed in detail. Every subsection in a
module obviously had to connect precisely to the
corresponding subsections in the adjacent modules.

During the final DD 963 contract negotiations
with the Navy, Litton agreed to build the first two
destroyers relatively conventionally, from the keel
up in three major sections. The ships' plans were
updated and then crosscut into designs for modules. These crosscut plans showed locations and
listed all the material needed to assemble each
module. The rest of the class was built modularly.
This dual effort was very costly for Litton. The
Spruance-class destroyers were the most complex
ships ever built by modular construction anywhere. Modular construction of tankers in Sweden and Japan was simple in comparison, since
cargo transfer piping was the only significant
system distributed across modules in a tanker
design.[8] Ingalls became, and still is, the world
leader in modular ship design.

The large area (611 acres) of empty land at the
Ingalls West Bank yard facilitated the laying out of

Modular construction of a *Spruance*-class destroyer. The prefabricated hangar is being lowered amidships. The square cutout at the end of the superstructure deck is the aft ammunition trunk. The superstructure is angled so that the aft 5-inch gun mount can fire at low trajectory on forward bearings. (U.S. Naval Institute collection)

Spruance is under way on builder's trials in February 1975, painted largely in colored primers. The ship performed extremely well and was almost 10 percent faster than the Navy required. Notice the cutout at the starboard quarter, the life rails above the bridge windows, and folded-up safety nets along the deck edges forward of Mount 51. All these features soon disappeared. (U.S. Navy photo)

material for construction of the modules. Exact lengths and quantities of all material had to be delivered to each module assembly station. Shipyard workers had easy access inside and around the modules: "The West Bank Yard, which was built on an assembly line concept, was a naval engineer's delight. . . . Geographically dispersed construction facilities brought the modules and sections together in a logical sequence from both an engineering and shipbuilding perspective. Ease of fabrication and access, coupled with optimization of the shipbuilding process, were the achievements of [the DD 963] program."[9]

After the modules were individually assembled, painted, and outfitted, 200-ton-capacity cranes hoisted them into position on the growing ship. Hull sections were assembled on rollers and joined. When the ship was ready for launching, she was rolled onto a submersible platform and lowered into the water. The ship was already about 92 percent complete at launch, compared with less than 50 percent for a conventionally built warship. The traditional shipbuilding ceremonies of keel laying and launching were casualties of the new methods, but each destroyer did get a formal christening. After launch, tugs towed the ship across the river to the older East Bank yard for outfitting.[10]

Litton staggered under the complexity of these tasks. With all its corporate knowledge, the firm could not manage the work at Ingalls. Over 25 different computer systems were in use, but they were not integrated for even such fundamental requirements as scheduling. Personnel training was inadequate. Trial-and-error was common, resulting in unneeded material purchases in some areas and omissions in others.[11] Construction flaws appeared. Alignment of the large structural modules, and of internal systems such as firemain piping that joined between modules, was intricate and reportedly imperfect in some ships.[12] Incorrect preservation of steel surfaces resulted in corrosion. Bimetallic corrosion at the connections between the steel hull and the aluminum superstructure occurred. The ship's weight was greater than expected, and the center of gravity was higher than planned.

Admiral Sonenshein was concerned about the risks of the large content of new technology in a 30-ship program. Several initiatives reduced the risks. Following the successful naval nuclear reactor projects in the U.S. Navy and Royal Navy, he started a land-based test facility at the Philadelphia propulsion laboratory. It tested propulsion and auxiliary systems before they went aboard the ships, and established training procedures for the crews.[13] Litton built another land-based facility for testing and training for the combat system.[14] Electronic systems were integrated and checked out ashore on pallets, where much more rigorous testing could be done. To install the systems aboard ship, it was necessary only to load the pallets and to connect cables. Reliability of these systems proved much better than for other warships being built at the same time elsewhere by traditional methods.

"The U.S. Navy Does Not Want the 963-Class Destroyer"

Technicians at the Naval Ship Engineering Center resented the DD 963 project office's rejection of their unsolicited destroyer design, resented the whole Total Package process, and regarded the *Spruance*-class design with enmity from the start: it was "not our ship."[15] Hostility from NavSEC surfaced in an authoritative historical study of the postwar Navy: TPP "was a disaster for the Navy because it shifted responsibility for design work from the systems commands to private firms, leaving the Navy dependent on a few major construction yards. As Norman Friedman noted in his detailed study of destroyer designs, total procurement of the DD 963 type meant that 'a company with no previous preliminary design experience had to create a new destroyer design from scratch.' "[16]

Dr. Friedman's study was based primarily on NavSEC documents. Under TPP, NavSEC regarded industry as a competitor for ship design. In the case of DD 963, industry was the winning competitor. Although TPP was dead for new projects, its perceived threat to NavSEC staff careers provided a parochial reason for denouncing it. The fact was that the industry-designed DD 963 class was superior to the NavSEC staff's obsolescent steam-powered destroyer design in every significant way, cost considerations not excluded. Improvements in Soviet submarines, due in part to the Walker-Whitworth spy ring that began operation in 1968, soon made the ship silencing obtainable from gas turbines essential, in particular for operations against multiple submarines. Without the growth margin provided in the industry DD 963 design, Aegis would have died by 1972, since no other design was cost-effective.[17] TPP was

Destroyers are outfitted at Ingalls after launching. Litton's difficulties in setting up the new shipyard erased the hoped-for savings in construction costs from series production.

USS *Spruance* (DD 963) is under way after commissioning. A windbreak has replaced the life rails on the signal bridge (the open deck above the pilot house, or navigation bridge), a change made to deflect the air blast at high speed. The cutout at the starboard quarter was plated in later for Nixie torpedo decoys. Forward of the bridge, the ASROC launcher is reversed with a 2-cell module elevated for loading from the magazine below decks. (U.S. Naval Institute collection)

imperfect, but its use for the DD 963 project was no disaster. Alternative traditional approaches might well have been. But in the early 1970s proof still lay in the future.

Resentment at NavSEC lingered. NavShips had developed a detailed list of General Specifications for Warships. These recorded the Navy's accumulated knowledge from lessons learned in ship design and construction. The General Specifications did not describe past problems encountered, just procedures and materials that would avoid them in the future, whatever they were. Under the contract, Litton did not need to use the General Specifications if it informed the Navy of its rationale for using different specifications. Litton thought that its top-down design approach addressed all the unstated problems behind the Navy's General Specifications and left it at that. Corporate experience with aviation led Litton to underestimate the hostility of the sea environment and the roughness of typical sailors in contrast with aviation personnel. Litton often selected components solely on the basis of low price.[18] Insistence on use of the General Specifications would have reduced Litton's freedom to design the destroyer. In the Navy's final judgment, that would have led to a poorer design.[19]

NavSEC technicians recognized that materials Litton specified for the waste-heat boilers and the auxiliary steam piping would not last long in service. They doubted that vinyl-covered bulkheads and other commercial fittings would withstand normal shipboard wear and tear, such as a collision with heavy equipment being manhandled down passageways or an attack by an off-watch sailor frustrated with a supervisor. NavSEC's technical staff declined to advise either Litton or the Navy's supervisor of shipbuilding at the Ingalls yard about such vulnerabilities in the design.[20] Early DD 963 crews were often restricted from taking showers (water hours) so that fresh water could replenish broken-down waste-heat boilers, and spent many hours replacing corroded valves and painting damaged bulkheads.[21] Another effect of disgruntled low-level inattention at NavSEC to the DD 963 design would be the weight problem with the first Aegis cruisers.

The Soviet Navy's clear antisurface mission dismayed U.S. Navy officers, whose ships had little ability to fight back directly. For years the fleet had expected that with the DX project American warships would finally catch up in quality with the Soviet Union's. Instead, with the announced DD 963 design the Navy appeared to have caught up with Denmark and Canada. Anger from a decade of frustration exploded within the Navy's officer corps. After OKEAN-70 and the Jordan crisis Captain Robert Smith, an ASW expert and a former destroyer commanding officer, had this to say:

> What we see today are the fruits of decades of mismanagement and inefficiency in a sluggish and ill-organized bureaucracy—all at the service of leadership that has been thin on naval professionalism, weak in imagination, and slow to perceive vital, new, emerging strategic and tactical truths. The result is the unready Fleet of today. . . .
>
> *Across the board, in a persistent pattern of grave tactical myopia, the U.S. Navy's lack of suitable weapons stands out against the darkening background of this excellence in the Soviet Navy. . . .*
>
> In a uniquely unhappy mixture of strategic and technological shortsightedness, coupled with political maneuvering to bring more hulls into being, we have wrought the 1052-class of DEs [destroyer escorts], the greatest mistake in ship procurement the Navy has known. . . .
>
> We need more such voices now [referring to Admiral Rickover], enough of them insisting, as only one example, that the U.S. Navy does not want the 963-class destroyer as presently conceived—that it makes no sense to plan to build a ship that even now, on paper (and the first of which is not to be delivered for years), is inferior to competitive Soviet ships which are already at sea. . . .
>
> It will be argued that the United States cannot design a ship to match, ton-for-ton, the capability of the Soviets, because we must have the habitability that the Soviets do not. At best this sort of deluded reasoning allows a peripheral concern—the valid need for a higher retention rate—to control profound decisions as to capability for war.[22]

The U.S. Naval Institute, whose only voting members then were military officers, published Captain Smith's turgid dissent in *Proceedings* as its prize essay for 1971. He addressed primarily the precarious effectiveness of surface ship ASW and the fact that surface ship operations relied on air cover from a declining number of aircraft carriers: "too many eggs in too few baskets." He never elaborated on trade-offs for an alternative fleet configuration to oppose the growing Soviet Navy, such as more-heavily armed (if perhaps fewer) surface warships, or more numerous (if perhaps cheaper and individually less capable) aircraft carriers.

Captain Smith's essay is of interest less for what he wrote than for the controversy he stimulated. The general press noticed Smith's article, and he was interviewed on TV.[23] For years the respected naval annual *Jane's Fighting Ships* reprinted his charge that the *Spruance* class was "inferior." *Proceedings* published over 20 pages of responding letters, the heaviest response ever from the Naval Institute's usually taciturn membership. Many officers praised Smith's outspokenness, while others challenged his analysis. Admiral Rickover was delighted with Captain Smith's attack on fossil-fueled escort programs.[24] More letters criticizing the DD 963 design appeared after a member of the General Dynamics design team recommended a return to smaller destroyers,[25] as Admiral Zumwalt was about to do with the PF 109 (FFG 7) class.

Some of the controversy was a reasonable and intelligent discussion of revolutionary changes in naval technology. The long interval in destroyer construction meant that the DD 963 design had an unusual concentration of changes, too. But the criticism was almost entirely invalid and erroneous. A retired engineering officer predicted, "The unarmored *Spruance* class could be stopped with a hunting rifle since the gas turbine does not have the inherent protection of a steam turbine." A correspondent of unknown authority attacked the new destroyers as "too large (and expensive) to be of general use" and suggested, "One possible step is to return to a more moderate hull size and work in the maximum possible capability," specifically by incorporating gas turbines in the old DDG 2 design and scrapping the helicopter facility. A serving officer wrote, "The 963 is out-gunned by the 931- and 945-class destroyers, and would be outrun and outfought in a surface action with a modern Soviet destroyer. . . . I have often wondered why, instead of the 1052, and now the 963, we did not build more DDG[2]s."[26] (Ostensibly the DD 931 and DD 945 classes were the same, but the uncoordinated shipbuilding policy described in chapter 3 made them dissimilar.)

There was in fact no measurable value to the Navy of a smaller ship other than perhaps cost, which the contract competition already covered.[27] The underlying concern, that the U.S. Navy's fleet configuration had not kept up with the Soviet menace, was accurate. Fleet commanders could not, however, rely on surface ships or even submarines to secure the seas quickly from the Soviet fleet. Only carrier air strikes could locate and reduce the Soviet fleet forces quickly enough to prevent them from interfering with allied use of the sea.[28]

The immediate issue for the Navy was not its antiship lethality but the weapons and tactics it would use to defend the carriers against a surprise massed antiship missile attack. Naval forces could attack Soviet ASCM-firing bombers, submarines, and surface warships if forewarned by surveillance and if permitted by rules of engagement. Waves of missiles would be difficult targets. Since many Soviet ASCMs carried nuclear warheads, all had to be shot down at long range lest altimeter-fuzed warheads detonate despite destruction of the missile airframes.[29] That was the mission of Aegis. Later, Tomahawk cruise missiles would disperse striking power to surface warships. The *Spruance* class would bring both these essential upgrades to the fleet.

The Side-by-Side Comparison Controversy

Criticism of the DD 963 design was due both to its appearance and to a misunderstanding of modern requirements. The extolled DDG 2 class had five prominent weapons launchers mounted along its profile. The DD 963 design as initially shown had four launchers mounted along a much longer hull, but it also had internal torpedoes, the Naval Tactical Data System, longer-range sonar, automated engine rooms, helicopters, and far better cruising endurance. It was more agile than the DDG 2 and faster in any sea state higher than calm, although proof awaited completion of the first ships.

The primary reasons for the DD 963's size were its electronics payload and its growth margin. Electronics systems are the dominant postwar development in naval architecture and are the primary cost component of a warship. Except for antennas, these systems are invisibly housed inside the ship, which in turn has a bulkier structure to enclose them. More capable power and cooling systems, and the servicing crew's need for both access and accommodation, further increase demand for shipboard internal space and volume. The weight of the enlarged ship structure requires a larger hull to float and to maintain stability despite combat damage. Large intake and uptake ducts for the gas turbines have a similar impact. The payoff of electronic systems is greater effective ordnance range and higher probabilities of killing worthwhile targets. More efficient shipboard weapons mean smaller magazines and fewer weapons launchers, since fewer targets will escape destruction.

In one of the best articles ever written on modern warships, naval architect Philip Sims explained why side-by-side comparisons were fallacious:

> The *Spruance* class carries an unusually large, volume-consuming electronic suit—so much that she could be designated a DDE to stand for electronic warfare destroyer. The large helicopter deck needs a great deal of supporting room; special silencing increases the machinery room volume. . . .
>
> U.S. Navy ships have this volume problem to a greater degree than those of other nations, because of the severe requirements for sustained at-sea performance that other navies do not face. Only American ships regularly commute to work across the major oceans and then deploy with minimal support for half a year. That requires more room for the crew, increased capability for self-repair, and more fuel and stores on board. It also means that the Navy has chosen armament systems consisting of one launcher rail with reloads below deck taking up volume. This is done instead of piling up the more ominous stacks of tubes on deck and thus keeps the deck clear for fast underway replenishments. UnReps also require large passageways to allow quick strikedown. Electronics in general form a light but high-volume payload, and the equipment with the worst impact is the Naval Tactical Data System (NTDS). . . . The NTDS electronics must be grouped in a prime, unobstructed part of the ship. A big NTDS is a by-product of the American Navy's task group philosophy—a burden other nations' ships do not carry.[30]

Still, to old salt and seaman apprentice alike, these modern warships looked swollen and lightly armed. Warship critics almost traditionally confuse ship size with cost. DD 963 is a large design not only because of electronics and gas turbines but also because of the requirements to make 30 knots in sea state 4, to carry fuel for 6,000 miles'

range, and to provide growth space for future weapons. Using hull size to attain these benefits costs little more than weight in steel, the cheapest component of a ship. The DD 963 design was a big but short-lived advance in U.S. warship design policy. After hearing all the criticism that the DD 963 was inadequately armed, that it looked "too big" and sounded too costly for its visible armament, the Navy packed its subsequent surface warship designs around their weapons. It was easier to market a congested design to inexpert decision makers than to educate them.

The general press picked up the theme that the new destroyers were inferior to Soviet warships:

> David Packard, the deputy secretary of defense, is concerned about whether the Navy's new *Spruance*-class destroyer as presently planned will be good enough to meet the Soviet threat at sea when it first joins the fleet in 1975. . . . Two developments are said to have sharpened Packard's concern over the capabilities of *Spruance*-class ships *vis-a-vis* the Soviet threat at sea.
>
> One was the appearance in the Baltic Sea of a brand-new triple-threat Soviet destroyer of the Krivak class, a 3,000-ton ship with a speed approaching 45 [sic] knots and armed with surface-to-surface [sic], surface-to-air, and anti-submarine rockets. The other was criticism within the Navy as to the *Spruance* destroyer's limitations.
>
> In this connection, a statement in the Naval Institute *Proceedings* prize essay for 1971, written by Capt. Robert H. Smith, was particularly alarming. Packard was questioned about it by interested members of Congress.[31]

Side-by-side comparisons of ships' installed weapons are popular among naval enthusiasts. (Such comparisons were the original raison d'être for Frederick Jane's *All the World's Fighting Ships* in 1897.) The comparison should be whether and how different maritime forces advance national interests throughout the strategic continuum: peacetime economic vitality, diplomacy, deterrence, combat, victory, and postconflict stabilization. Cold War strategy capitalized (differently over time, as has been seen) on the ability of aircraft carrier task forces to strike Soviet targets. Besides the immediate damage and at least local defensiveness that would result, a carrier raid

New Soviet warships such as this Kresta II–class large ASW ship showed the world that the Soviets were driving for strategic superiority at sea. Western intelligence erred in estimating that they carried dedicated antiship missiles. Nevertheless, concern grew in the early 1970s that the Soviet Union could block the U.S. Navy from carrying out wartime missions. (U.S. Naval Institute collection)

would leave targets more vulnerable to follow-up attacks. Such attacks could reduce Soviet strategic reserve forces, such as ballistic missile submarines and air defenses, and wreck support facilities such as submarine tenders, power stations, and petroleum depots. U.S. carrier aircraft and submarines could mine ports and channels and threaten ballistic missile submarines with detection.

Soviet national planners could not risk the consequences of sea-based attack. Their costly efforts to defend against it were probably futile. In contrast, there would be no penalty to a land-based enemy who simply ignored an alternative fleet of warships armed for air defense and antiship attack. They might patrol relatively small areas of ocean while Soviet combat forces concentrated at decisive points on other fronts. Any surface ship without air cover would need to stay well out at sea to avoid air attack.

Comparisons with Soviet ships were inaccurate as well as irrelevant. Soviet surface warships armed for antiship attack all dated from the early 1960s and relied on land-based aircraft and shipboard helicopters to detect contacts and on shore-based command posts to evaluate combat information and to assign targets. Soviet warships entering service during the 1970s were designed more for ASW than for antiship attack. They could fire antiaircraft and SS-N-14 ASW missiles, once those were operational, at surface ships at horizon range and torpedoes at closer range. Their primitive and very small combat data systems were susceptible to decoys and would be overwhelmed by modern air attack (long-range high-altitude detection followed by low-altitude strikes, for example).

Heavy deck-mounted weapons indicated that Soviet warships carried few if any reloads. This was consistent with the First-Salvo tactic. Impressive surface-to-air missile batteries revealed an expectation of inadequate or nonexistent air cover at sea. Duplicate launchers and radars looked formidable but probably indicated that the equipment was unreliable. Low equipment reliability, short endurance, and lack of protected replenishment ships were all consistent with a strategy of brief sorties. The central threat of Soviet warships remained their potential to strike first at forward-deployed aircraft carriers, typically two per theater. The Soviet reaction to the *Spruance* class will be seen later.

National Controversy over the *Spruance* Class

By 1972 Litton was in severe trouble. It projected that the LHA 1 class was 12–16 months behind schedule. Litton had collected 40 percent of the $970 million LHA contract price but had performed only 2 percent of the estimated production labor-hours. While some of this divergence occurred simply because Litton bought much of the necessary material up front, Litton had also nearly tripled its estimate of labor-hours to build the LHA class.[32] Litton's president, Roy Ash, visited NavShips early in 1972 and quietly asked the Navy to take over Ingalls altogether and to release Litton from its shipbuilding contracts. The Navy refused.[33] The fundamental problem was that the LHA 1 and DD 963 programs had overloaded Ingalls. The Navy refused to cancel or delay either program, so Litton unilaterally delayed the LHA class, leading to an enormous cost overrun on it.[34]

Inflation accelerated, and Litton's wage and material costs increased beyond the contractual cost escalation allowances. Further, there was no contractual allowance for cost increases resulting from schedule slippage on the LHA class. Indeed, the TPP contracts exacted penalties for late delivery. Litton asked the Navy to renegotiate the LHA contract to increase the price for that class, first by $270 million[35] and then by $375 million. The Navy rejected Litton's claims as unsubstantiated.

Coming from the Navy's get-the-job-done tradition, CNO Admiral Zumwalt blamed Litton's problems on its "optimism and lack of know-how" in modular construction. He agreed that it was a promising technique for shipbuilding.[36] Ironically, it fell to David Packard and Admiral Zumwalt, two of the least enthusiastic officials involved with the DD 963 project, to defend it to Congress. Admiral Zumwalt had not liked the DD 963 design in the first place, and he ended hopes that 50 or 75 destroyers would be built: "We can't afford to pay the unit price for more DD 963 escorts than the 30. Now what we need is a larger number of cheaper escorts, each of them less capable, but in numbers that will let us cover the convoy lanes."[37]

Admiral Zumwalt initiated the PF 109/FFG 7

Originally planners in the Office of the Secretary of Defense and the Navy's Major Fleet Escort analysts had expected that DX/DD 963 destroyers would be smaller, cheaper versions of the *Knox* design (*top*). Critics thought that both classes carried inadequate armament compared with Soviet warships. USS *Arthur W. Radford* (DD 968) and USS *Pharris* (FF 1094) operated off Chile in 1980 in the Unitas XXI exercise. (U.S. Navy photo)

design for a low-cost frigate. He hoped to build 50 at many yards. Knowing shipbuilders' costs, Admiral Sonenshein figured that even a 50-ship program would support at most three yards. The Navy selected Bath Iron Works to build the lead FFG 7 frigate. To obtain standardization without TPP, the Navy required Bath as lead shipyard to obligate its subcontractors to deliver identical components to the other two FFG 7 shipbuilders (Todd yards in Seattle and Los Angeles). Bath brought the FFG 7 class in on schedule and under budget. Bath was rewarded during the 1980s with contracts for eight Aegis cruisers, thus becoming the only yard other than Ingalls to build DD 963–type ships, and with the lead yard contract for the DDG 51–class destroyers.

Mostly because of Ingalls, Litton Industries was operating at a loss. Its stock had traded at $109 in 1967 but collapsed to $10 in 1972.[38] Commercial ship orders dried up. Without TPP, Litton had no new naval ship design work. In June 1972 Litton laid off 1,000 of its 3,400 Marine Systems personnel. It offered another 1,400 a choice: relocate from Los Angeles to Pascagoula or quit.[39] This further disrupted work at Ingalls.[40] Noting the schedule slippage at Ingalls, Congress authorized none of the seven destroyers requested by the Navy for FY 1973. It did authorize purchase of long-lead-time items. Senator George McGovern, the unsuccessful Democratic candidate for President in 1972, called for cancellation of the DD 963 shipbuilding project. He had promised a post-Vietnam welfare program but was finding few savings from reductions in the war effort. The war had been financed by a tax surcharge, by inflation, and by the forgoing of military modernization.[41]

The *Spruance* class now became an element of national political controversy. Concerned about the capability of conglomerates and multinational corporations to move jobs to cheap locations, labor leaders prodded Congress to investigate such firms. As a conglomerate, Litton Industries came under suspicion. Meanwhile, after his landslide reelection President Nixon sought to reorganize the executive branch into about eight departments that he could discipline. He intended to use his Office of Management and Budget (OMB) to carry out his plan. For director of the newly powerful OMB, he appointed none other than Roy Ash. President Nixon's opponents charged that this was an outrageous conflict of interest and that Ash would now rescue Litton from the cost overruns at Ingalls.[42] A "Mississippi Littony" of troubles at

"Pascagoula's Ash Heap" became national news.[43] Ash, of course, had resigned from the firm. Representative Les Aspin, former McNamara whiz kid and future defense secretary manqué, came to prominence on the anti-Litton bandwagon.[44]

Political uproar over Ash, Litton, and the *Spruance* class evaporated when the more dramatic Senate Watergate hearings began in 1973. Despite opposition from Aspin and others, Congress authorized all 14 destroyers remaining on the contract, 7 each during FY 1974 and 1975.[45]

The Navy Controversy Heats Up Again

The controversy within the Navy over the *Spruance* class was by no means over. Opponents of Admiral Zumwalt's courageous personnel reforms attacked his interest in shipboard habitability for all-volunteer crews. Charts of warship designs depicted the already-tainted *Spruance* class as having disproportionately more space for habitability than weaponry, compared with Soviet destroyers and previous American warships. In such accounting, air conditioning for electronics counted as habitability; so did weapons growth areas. The DD 963 design was spacious also because a larger hull was more efficient for high speed in heavy seas.[46]

Much of the debate about the *Spruance* class concerned the question of how heavily each naval mission, sea control or power projection, should weigh against the other to dominate naval war plans and ship designs. Critics looked at visible weapons and perceived the *Spruance* class as 30 more convoy escorts. In actuality, the destroyers were fast task force escorts, which was why there was so much emphasis on propulsion, seakeeping, and electronics.

Giving in to simplistic criticism was easier than educating the critics. In 1974 the new Chief of Naval Operations, Admiral James Holloway, announced that his highest priority was to increase the firepower of the fleet. (A contender for CNO, Admiral Isaac Kidd, fell victim to congressional opposition after he attempted, on Nixon White House orders, to dismiss a bureaucratic critic of Ash's appointment to OMB.) Admiral Holloway did not describe the deficiency in fleet firepower, but he implied that it existed, and after three years of loud controversy the *Spruance* class was clearly the cause of it. The story of Admiral Holloway's attempt to provide the Navy with a class of nuclear-powered Aegis-armed strike cruisers will be told in chapter 13.

A few weeks before USS *Spruance* was commissioned in 1975, the Naval Institute *Proceedings* published a flawed article by J. W. Devanney, a naval architect who had served on the Litton DD 963 design team during 1968–69.[47] His theme was that the design effort was a "dishonest exercise in salesmanship" that, he asserted, resulted from the Navy's misunderstanding of, or antipathy toward, systems analysis. Interesting details about the design effort gave the article an air of authority. Devanney wrote that a steam-propelled or CODAG design could have been cheaper than the COGAG design ultimately adopted, but that CODAG was rejected out of concern that the Navy would not "buy it."

Devanney erred in assuming that the objective of the DD 963 design effort was to optimize the traditional naval architectural interests of hull shape and propulsive power. As described previously, the objective with the DD 963 was to produce the best-balanced destroyer, as measured by a much wider range of design principles. Litton traded off propulsive efficiency to minimize underwater radiated noise. It selected the best ship design with a COGAG plant for that reason. A steam or CODAG design could not achieve the silencing of the COGAG design. Even if cheaper to buy, a CODAG plant was suboptimal: individually or in force, CODAG destroyers would be operationally inferior.[48]

Devanney said that the diesels could be the same models selected to power the electrical generators and that therefore they were obviously acceptable. He passed over the fact that those diesels were the fallback for generator prime movers, in case the gas turbines had problems. Given this choice, the Navy selected gas turbine generators primarily because of their superior underwater noise characteristics.[49] This was another decision influenced by the Canadian *Iroquois* design. "As far as I know," Devanney wrote, "the CODAG's indicated cost superiority or the reasons for rejecting it were never reported to the Navy, despite the fact that the Navy was paying for the analysis."[50] That was a comment solely on Dr. Devanney's knowledge, since in 1971 Litton and NavSEC jointly presented a paper to the American Society of Mechanical Engineers that went into this decision in detail.[51] His estimate of the CODAG design's cost superiority relied on the assumption that two gas turbines would always be on line at cruising speed in a two-shaft plant. In fleet operation just one gas turbine is adequate for 19 knots.

Devanney continued, "That was when I left Litton. Afterward, the three gas turbine concept with a combining gear was replaced with a four-turbine plant (making the ship at least 50% over-powered), an alternative which even our biased results indicated was more expensive than steam."[52] The Navy was not being deceived in this either, since it was the Navy that ordered the design change to four gas turbines. And again, Devanney's assumptions about destroyer operating profiles and future costs were invalid.

The proliferation of quieter submarines after 1970 makes it obvious that the decision to go with the quietest possible destroyer was extremely fortunate. Not only was it the best military decision; it has also proved to be the best economic investment decision. Oil prices soared after the 1973 Yom Kippur War. Steam propulsion even with black oil was more expensive than gas turbines over a destroyer's operating profile. Steam-powered frigates built simultaneously with the *Spruance* class (the last *Knox*-class frigate was only a year older than USS *Spruance*) were withdrawn from active service when even the oldest *Spruance*-class destroyers were forecast to have another 25 years of service life ahead of them.

Devanney said he hoped to spur debate about "honesty in military systems analysis."[53] Instead he merely slanted the already error-ridden public record further against the *Spruance* class. Was that his goal? Privately wealthy and obviously clever, Devanney had a record of mocking the work of those who had to work hard to get ahead. At MIT he would sit at the front of the classroom and fall asleep to annoy his classmates.[54] One contrasts him with Reuven Leopold, whose biography hints at a childhood that may have been very far from the comfort of wealth. He was born in Hungary just before the Nazi holocaust and was raised afterward in Israel by his mother. She taught him that he would have to excel if he hoped to achieve anything.[55]

Besides contributing to inaccurate criticism,[56] Devanney's misleading article has been cited in political attacks on CNO Admiral Zumwalt. Retired Vice Adm. John Hayward blamed Zumwalt for the class's alleged deficiencies. "Is the *Spruance* a 'warship,' regardless of propulsion?" he wrote, not mentioning that he had worked on GD's design. He commended Devanney's article in 1977, when Admiral Zumwalt's personnel reforms were still arousing opposition.[57] As late as 1982 another critic claimed, "The *Spruances* . . . have been called yachts masquerading as warships."[58] A rare public defense of the *Spruance* class came from Admiral W. J. Lisanby, commander of the Naval Ship Engineering Center, in 1978: "While the procurement approach used for these ships is disliked because of some inherent problems, the ships exceed by far the capabilities of previous ship classes designed to accomplish similar missions."[59]

Controversy over the *Spruance* design within the Navy was significant not because the criticism necessarily had merit, but because of the consequences of the fact that the controversy occurred at all. The Navy began publishing an internal newsletter in 1975, *Surface Warfare,* specifically to explain technological developments to the fleet. Publicly, however, Navy leaders reacted to the symptoms of controversy without addressing the causes of professional concern, which were misunderstandings about the destroyers' role in national strategy and about advances in naval engineering. Official silence about the unofficial controversy sharpened the public impression that the *Spruance* class was somehow defective. This impression crept into authoritative studies of American defense policies that damned the ships as "failures" or "a disaster" without investigating whether there was any basis for such conclusions. Perhaps the true disaster was the impact on innovation in warship design to avoid a repetition of the *Spruance* controversy. The Navy designed all subsequent surface warships with little growth margin and may have become even more cautious about implementing new technology.

Sea Trials and Shakedown

Sailors and officers in the fleet read announcements about the DD 963 design and its improvements for shipboard life with sullen suspiciousness. Gas turbines impressed them as the naval architects' new toy, while the fleet continued to suffer with the troublesome 1,200 psi boilers in which, they thought, shore-based engineers had lost interest. Underpaid sailors aboard uncomfortable and dangerous old ships suspected investment in habitability was intended to attract spoiled suburban hippies and ghetto rioters into the Navy. The apparently minimal armament that outraged senior officers was, however, thought not so bad

by junior personnel. The Sea Sparrow antiship missile defense contrasted favorably with such devices as electronic blip enhancers, aboard destroyer escorts to attract antiship missiles away from aircraft carriers and other high-value units. The new Mark 45 5-inch gun sounded reliable, and the inclusion of two of them bespoke realism about combat missions.

Two models of controllable/reversible-pitch (CRP) propellers were installed for powering, tactical, acoustic, and machinery trials on the destroyer escorts *Patterson* (during 1972–75) and *Barbey* (1973–76). Both propellers were pushed to their limits to test materials and design strength. In August 1974 *Barbey* performed a crash-back maneuver that caused all five blades to separate from the propeller hub, showing spectacularly that the alloy of that particular blade carrier was too brittle. After strength analysis and research into design loads, the propeller was rebuilt with stronger components and tested successfully. These lessons were applied to the DD 963–class CRP propellers.[60] The *Barbey* incident was incorrectly cited by critics of the *Spruance* class as indicating that the ships would be unreliable or weak. In service, the ships routinely perform crash-backs, reversing from flank speed to full-astern power in less time than it takes to read this sentence. It is a useful tactical maneuver during ASW, as will be seen.

In February 1975 USS *Spruance* went to sea on builder's trials. She performed extremely well. In particular, hull vibration was much better than required, less than 25 percent of the allowable rating. The ship was the quietest destroyer ever built. Her sonar detection range was significantly better than the design requirement. She exceeded requirements in almost every test: maneuverability, seakeeping, full-power reversals, and keeping her sonar dome wet but her decks dry in stormy seas. "The bright spot today in the *Spruance* program is the technical performance of the ship," the Navy informed Congress. Litton qualified for the full $24 million silencing incentive fee.[61]

After one of her trials *Spruance* moved back into the dock of the launch platform, a type of floating dry dock, for routine underwater hull work. As the platform rose, the Ingalls workers did not notice that one of its four caissons was still flooded. The platform fractured amidships under the uneven weight. It sank unevenly, leaving *Spruance* askew on the keel blocks and in danger of falling off them. Her stern was partially immersed, and her propellers hit the platform floor. Compressed air could not displace the floodwater inside the damaged platform because of ruptures in internal bulkheads. Pumping out individual compartments in the platform risked further structural damage from the pressure of water in adjacent compartments. This formidable engineering problem was technically Litton's to solve, since it owned both platform and ship (*Spruance* had not yet been turned over to the Navy). But the Navy had an obviously strong interest in keeping its shipbuilding program on track, so Navy officers supervised a salvage operation. A combination of compressed air and careful pumping evened the pressure across the internal bulkheads and refloated the halves of the platform in 15 days. Damage to *Spruance* was limited to dented propeller blades.[62]

Litton completed the destroyers, mounting the weapons available for installation while the ships were at Ingalls. Each destroyer went through shakedown trials and then returned to Ingalls for adjustments and repairs at Litton's expense. (The TPP contract included a one-year warranty.) Then the ship sailed to her new home port.

The Contract Controversy

Litton had gambled on its TPP contracts and lost. Inflation was far greater than the firm had predicted; ship construction was behind schedule; the cost of running the West Bank yard was high; shipyard productivity was low. But the TPP contracts were set up precisely to keep the Navy from having to cover contractor cost overruns. What saved Litton was that all shipbuilders were suffering from inflation, and by 1976 Litton was only one of several contractors seeking bailouts. Newport News Shipbuilding, beset by Admiral Rickover, made even larger claims than Litton and had to be ordered by a court to continue Navy work. By July 1976 Newport News was angrily demanding that the Navy tow an incomplete cruiser, an aircraft carrier, and a flotilla of half-finished submarines out of its yard. Deputy Secretary of Defense William Clements (David Packard's successor) offered to settle with Newport News, Litton, and two other shipbuilders for about 40 percent of their total of $1.7 billion in claims. Newport News and Litton both rejected the offer.

Emulating Newport News, Litton threatened to stop work on the LHA 1 class effective August 1,

Table 7.1   Litton claims on the LHA 1–class contract. There was no claim over the DD 963 class.

Date	Total Claim
March 1973	$270 million
March 1973	$375 million
April 1975	$500 million
June 1976	Rejected offer to settle for $240 million
July 1976	$700 million
Sept. 1977	$1.1 billion
June 1978	$1.8 billion
June 1978	Claim settled for $1.6 billion

1976. On July 29, 1976, a Mississippi court ordered Litton to continue work, saving Mississippi jobs (Ingalls employed 24,000 workers), and ordered the Navy to pay present actual costs rather than the contracted price, saving Litton. Litton promptly raised its claim against the Navy for the LHA 1 class, as shown in table 7.1. Litton never filed a claim on the DD 963 class. The closest it came was to allege cross-impact with the LHA contract, specifically that the Navy "prioritized" the DD 963 class and thus disrupted timetables for the LHA. The Armed Services Board of Contract Appeals upheld this type of claim, and the government allowed $26.735 million against Litton's claim.[63]

The Carter administration settled the LHA claim with Litton in June 1978, increasing the price of the LHA and DD 963 contracts to $4.7 billion total from the TPP total of $2.9 billion. The $1.8 billion increase included $900 million for inflation, $512 million for Navy-ordered changes and delays, and $182 million as a price increase. Much of Litton's inflation loss resulted from its own delays in delivering the ships on schedule. Litton had to pay the remaining $200 million but could recover it by delivering the ships for less than $4.45 billion.[64] The Navy explained the settlement: "The Navy believes that resolution of the Litton claims at this time is essential and in the national interest for several reasons: First, the ships being constructed by Litton are essential. . . . Second, the Ingalls shipyard is a valuable component of the industrial base for other present and planned shipbuilding. . . . Third, the Ingalls yard is a national asset [i.e., a mobilization asset that would be needed in an all-out war]."[65]

Construction cost of the 30 *Spruance*-class destroyers was $3.73 billion, an average of $124.3 million each.[66] Total procurement cost for the class was higher, since the ships had to go through extensive postshakedown overhauls to add systems such as electronic warfare, Harpoon, and NATO Sea Sparrow that were unavailable when they were built. In the shoot-the-wounded contracting environment of the 1990s the *Spruance* class would have been canceled. At least one analyst suspected that Senator Stennis deterred cancellation of the DD 963 program.[67] There is no public evidence that anyone seriously proposed cancellation after the failed 1972 congressional motion to stop at 16 ships. Deliveries of the ships picked up after the settlement. And by 1978, when the contract was settled, the first *Spruance*-class destroyers were showing their capabilities.

Fleet Introduction, Engineering, and Supply

Fleet Introduction

The Navy named most *Spruance*-class destroyers for deceased admirals. The Naval Historical Center had acquired from a New York clergyman a set of recorded interviews with World War II admirals, and it named eight destroyers from this collection. Admiral Zumwalt named *Elliot* (DD 967) and *Peterson* (DD 969) for officers killed leading small-boat actions under his command in Vietnam. *Caron* (DD 970), destined for the most violent career of the entire class, and *David R. Ray* (DD 971) commemorated young enlisted hospital corpsmen who had been posthumously awarded the Medal of Honor for rescuing wounded Marines in Vietnam.

The families of the men whom the ships memorialized often attended commissioning ceremonies. After such meetings the crew felt greater personal responsibility to take care of the ship for them. *Elliot* was adopted by her namesake's home state of Maine. Soon after commissioning, she visited Maine for a four-day clambake. There, she claimed, she became the first DD 963 to come under hostile fire when someone on the beach started shooting at the anchored destroyer one night. Standing on the fantail, the young officer of the deck disputed his Vietnam-experienced bosun's report that popping noises ashore were gunfire, until a small object whizzed past his ear in the dark. The quality of the commissioning DD 963 crews, the "plank owners," was good, in part because the Navy wanted to ensure that Litton could not use delinquency as the basis for a claim. But none of the commanding officers who commissioned the first five destroyers made admiral.

Comte de Grasse (DD 974) was named for the officer who commanded the French fleet at the Battle of Yorktown. "The Count," as the crew nicknamed their ship, had the distinction of being christened by the first lady of France, Madame Giscard d'Estaing. The christening party in 1976 was the greatest social event in years on the northern Gulf Coast, whose major cities (including Pensacola, Mobile, Biloxi, and New Orleans) all celebrate French colonial heritage. Reportedly H. Ross Perot helped to pay for it.[1] The crew invited Mme. d'Estaing back to commission the destroyer, too. She declined, but the French ambassador attended the ceremony on August 5, 1978. Invitations would have been issued during the times that Litton was filing its claims against the Navy over the LHA 1 contract. A Mississippi court decided the first suit in Litton's favor in July 1976, and the Navy settled with Litton in June 1978 because it expected the firm to sue again.[2] The ship, not the shipyard, organized these soirees, but it might be interesting to compare the guest lists with the Mississippi bench. Upon commissioning, the Francophile destroyer's first operational order was sent to her in French. Initially scheduled for the Pacific Fleet, *Comte de Grasse* was reassigned to the Atlantic instead so that she could visit France.[3] Almost the only warship in the American republic's egalitarian history to commemorate a person by a noble title, *Comte de Grasse* set a place at the wardroom table for the count's ghost every day.[4]

Each DD 963–class destroyer was first assigned to a destroyer squadron (DesRon 9 in the Pacific and DesRon 10 in the Atlantic) for training. The new destroyers' importance was their value in tactics. *Kinkaid* (DD 965) and *Hewitt* (DD 966) were the first DD 963s to operate together in a fleet combat exercise. Box 8.1 reproduces a report on this exercise. The report is clearly favorable, even though both destroyers were basically incomplete. Each destroyer went to a naval shipyard for warranty repairs funded by Litton (another innovation of Total Package Procurement) and for installation of government-furnished weapons not available when the ship was at Ingalls. The naval shipyards in effect completed the first 17 destroyers. Starting with the 18th, *Moosbrugger* (DD 980), Ingalls installed the government-furnished systems.[5] Among the more successful parts of the contract was a provision that the ships could easily initiate changes that Litton paid for as a warranty retrofit.[6] When fully armed, the destroyer joined a tactical squadron. The Navy never held a public ceremony for the turnover to the tactical squadron to highlight the true initial combat configuration of the *Spruance* class. Good publicity could have offset the rampant criticism that the ships were deficient.

The primary mission of a destroyer squadron then was antisubmarine protection for an aircraft carrier and other high-value ships. The DD 963s were the quietest and best-equipped ASW ships in the Navy. The squadron commander (a Navy captain traditionally called commodore) and his staff officers invariably noticed the new capabilities. The new destroyers did not have a flag plot or accommodations for the squadron staff, but they often became flagships because they had the Naval Tactical Data System (NTDS) and sophisticated communications handled by satellite through the WSC-3 antennas and transceivers. The commodore requisitioned the commanding officer's in-port cabin (usually to the skipper's quiet dismay) and his staff officers berthed in equipment rooms reserved for future systems.

Sailors still know that destroyers have the ancient nickname of "greyhounds," but the more common and equally ancient slang is "cans" or "tin cans." Thus a *Spruance*-class destroyer is a "Spru-can."

The Ship

Warships operate in a hostile environment of the sea and enemy threats. The ship brings mobility, range, seakeeping, maneuverability, endurance, and combat survivability to weapon and sensor performance. Unlike other weapons systems, warships house and sustain their crews and provide medical facilities for casualties. The ships fight with weapons, but planning the fight through displays of combat information is equally essential to victory. Construction and operation of a modern warship draw on almost every branch of science and engineering except mining and farming. Traditionally warships were designed around their initial set of weapons and sensors. The *Spruance*-class destroyers were innovative in that they were designed as platforms for additional weapons and sensors that future missions and tactics might demand during the ships' careers. The highly modular design assisted in construction efficiency and rapid installation of new equipment during modernization. Appendix B lists the major hull and superstructure modules.

The critical need in modern warships is for internal volume and deck area for the weapons, the crew, reserve buoyancy, a large combat information center, propulsion and auxiliary machinery, fuel, stores and strike-down routes, maintenance access to equipment, clear firing arcs, and electromagnetically sound locations for radio and radar antennas.[7] The need for internal volume drives naval architects to design large deckhouses above the hull. An aluminum deckhouse reduces top weight and thus allows the hull to be narrower while still meeting stability requirements. The narrower hull (in general) creates less resistance from wave forming and allows the ship to reach speed on less propulsive horsepower, and thus with less fuel. A warship hull must be built strong enough to resist fracturing in heavy seas despite action damage and internal flooding. The requirement for 30 knots in sea state 4 by itself dictated a large hull. The hull shape itself had to accommodate the large SQS-53 sonar.

The hull length provides topside space for the flight deck and clear arcs of fire from the gun mounts and missile launchers. The antennas for the ship's radar, radio, and electronic warfare systems needed to be spaced far enough apart that they would not interfere with each other. The crew needs clear areas to receive cargo at sea (underway replenishment, or UNREP) and clear paths to carry stores and ammunition to strike-down elevators.

Deterrence of surprise attack requires continuous operational availability of all equipment during peacetime to allow immediate action. The engineering plant of a modern warship provides propulsion, fuel storage, damage control, electrical power, interior communication circuits, and ship services. Ship services include electricity distribution, compressed air, freshwater distillation, gyroscopes, weather sensors, firefighting sys-

An impressive view of a nest of new *Spruance*-class destroyers in San Diego in 1977. USS *Elliot* (DD 967) and *Hewitt* (DD 966) are still in their contract delivery configuration, but *Paul F. Foster* (DD 964) (*left*) has completed her postdelivery overhaul. Three WLR-1 electronic warfare antenna domes are on her foremast upper yardarm, and the small SPS-55 surface-search radar antenna is on a higher platform. Other radar antennas on all the ship's foremasts are the SPG-60 height-finding radar and beneath it the SPQ-9A search radar, housed in its distinctive dome. (U.S. Naval Institute collection)

Madame Valerie Giscard d'Estaing, first lady of France, christens *Comte de Grasse* (DD 974). The ship developed such strong ties with France that the Navy reassigned her to the Atlantic Fleet so she could visit that country. (U.S. Naval Institute collection)

The commanding general of the First Marine Division briefs Colonel Leftwich's sons at the christening of the destroyer named for their father, a Vietnam casualty. In 1991 helicopter-borne commandos from USS *Leftwich* recaptured the first Kuwait territory to be taken during Operation Desert Storm. (U.S. Naval Institute collection)

Box 8.1   An early evaluation of the DD 963 class in the Pacific

97

All systems listed in paragraph 2 except for the towed-array sonar were soon added during these ships' postshakedown overhauls. Towed-array sonar was added in the 1980s.

R 101806Z Dec 76
Fm ComCarGru Three [Commander, Carrier Group 3 (*Coral Sea, Ranger*)]
To ComDesRon Nine [Commander, Destroyer Squadron 9]
Info ComCruDesGru Three [Commander, Cruiser-Destroyer Group 3]
Confidential//N03300//
DD 963 Class Assessment (U)
A. ComDesRon Nine 082219z Dec 76

1. (C) Based on orig[inator's] experience operating with *Kinkaid* during COMPTUEX 1-77 and *Kinkaid/Hewitt* during READIEX 2-77 the following is forwarded in reply to ref A.

A. The ASW function was greatly strengthened by having a NTDS-capable ship dedicated to this warfare area. The ability of *Kinkaid* to act as MPACU [maritime patrol aircraft control unit] clearly was superior to that of any other unit. Multiple air assets were well utilized and safely coordinated in air space used [by] numerous other Blue [exercise friendly] and Orange [exercise hostile] aircraft. It was obvious to orig that *Kinkaid* was able to stay on top of the AAW picture as well as her primary mission ASW picture. Noteworthy was *Kinkaid*'s and *Hewitt*'s extremely reliable Link 11 and other radio circuits. Although orig did not copy *Kinkaid*'s Link 14, numerous OPSUMS [operations summaries] indicated it was most useful to other ships in maintaining the ASW and SUW [surface] tactical

picture. A superior fire control system was demonstrated by *Kinkaid*'s high score in all gunnery evolutions. Across the board *Kinkaid* strengthened the task group's warfare posture but especially so in control of ASW air assets and reliable communications.

B. The DD 963 class added a new dimension to the task group with its capability to assume PIRAZ duties and ASW, AX [ASW control], and AS [surface control] duties simultaneously. This provided the OTC [officer in tactical command] a viable alternative when cruisers were not available to fulfill their normal duties.

2. (C) When the DD 963 class gets its full bag of systems (EW suite, acoustic processor, night helo capability, Harpoon, SAMs, and towed array, etc.) it will be a most potent weapons system. Clearly it is now a fine command and control platform.

GDS-82

[Declassified 1982]

tems, heating, ventilation, and air conditioning. An integral supporting function is the supply of food, spare parts, lubricants, consumable material, and canteen services. These ship-oriented functions of engineering and supply are essential to military effectiveness.

Main Propulsion

In the early 1960s the Navy built the *Asheville* (PG 84)–class patrol gunboats to fight Styx-armed fast attack craft off Cuba or elsewhere. The small, short-range PG 84 class had a 10,500 shp gas turbine for 40 knots and two diesel engines for cruising. This CODOG installation proved unreliable. The Navy assumed that Styx antiship cruise missiles could not attack small ships and armed

the PG 84s to fight at point-blank range with 3-inch gunfire. An Egyptian Styx attack on an Israeli fishing boat invalidated that theory. Many PG 84s went to Vietnam. Admiral Zumwalt sent several, armed with short-range Standard antiradiation missiles (Standard-ARM) to home passively on radars, to counter Soviet warships in the Mediterranean. The PG 84s were poor sea boats and did not serve long, but they gave the Navy some awareness of gas turbines. The *Spruance* class introduced successful all–gas turbine propulsion into fleet operations.

Four LM-2500 marine gas turbines drive the propellers in the *Spruance* class. The Navy's DD 963 request for proposals specified two propeller shafts for redundancy and maintenance flexibility and to reduce propeller cavitation noise (de-

All *Spruance*-family ships use controllable-reversible pitch propellers. Below 13 knots the ships control speed by changing blade pitch, and they back down by reversing blade pitch. All the ships have Prairie silencing, a system that ejects compressed air from the gas turbines through the shaft struts and the leading edges of the propeller blades. (U.S. Naval Institute collection)

scribed below). Each LM-2500 engine in the *Spruance*-class plant generates 20,000–25,000 shp (16–19 MW). Combustion gases turn two turbines: the gas generator turbine and the power turbine. The gas generator turbine turns the LM-2500 compressor. The power turbine turns the propeller shaft via the main reduction gear. The power turbine is considered part of the ship, not part of the LM-2500. Unless it has suffered a thrown blade or other foreign object damage (FOD), it usually stays with the ship when the LM-2500 is removed for replacement. *Elliot* boasted a plank-owner power turbine in 1994. Defective LM-2500s can be removed and hoisted out through the air intake and a replacement engine lowered in.

Each LM-2500 is enclosed in a gas turbine module (GTM) that provides chemical-biological-radiological containment, sound insulation, fire-fighting CO_2, engine cooling, control electronics, and vibration isolation mountings.[8] The key advantage of gas turbines is their light weight, permitting use of isolation mounts to attenuate vibration and to impede shock damage from explosions. The combination of very high turbine rotation speed and GTM isolation mountings transmits very little vibration from the gas turbines to the hull. Diesel and steam power plants are heavier, and isolation

mountings are less effective or even impractical. The port LM-2500s face forward and the starboard LM-2500s face aft. A pneumatic clutch and brake connect each engine to the reduction gear. The reduction gear steps down the high-speed rpm from two LM-2500s through two stages (double reduction) to the propeller shaft. The propeller shafts rotate inward at speeds from 55 rpm up to 168 rpm. Those are slow rotational speeds and make for quieter propellers.

Hullborne vibration radiates acoustical noise into the water. Minimizing vibration improves the ship's own sonar sensitivity, interferes less with sonar searches by other ships in the screen, and provides less information to submarines. The total weight of a gas turbine propulsion plant is about the same as that of steam, diesel, or nuclear plants of similar capability. Machinery weight is lower, but structural weight (such as for ship strength around the larger uptakes) and fuel weight are both higher.[9] But the light weight and smooth high speed of gas turbines are the keys to silencing for ASW.

Warship observers sometimes write that large rafts hold the propulsion machinery in each DD 963 engine room as a unit, as in nuclear submarines. Strictly speaking, that is not so. During the DD 963 Total Package design phase, Litton evaluated raft mounting but chose individual isolation mounts for the gas turbine modules and the reduction gears. That avoided the cost of flexible couplings to the propeller shafts.[10] The large reduction gear boxes are spring-mounted to absorb mechanical shock and vibration. Sonar detection range is much greater for the DD 963 than for steam-powered ships, which were relatively noisy because of their many pumps, even using the same basic sonar (SQS-26CX/53A). In a gas turbine plant the turbines always turn at a smooth high speed, and heavy, noisy feedwater pumps are absent.

Other advantages of gas turbines are that they are safer and require fewer personnel for operation and maintenance. Finally, gas turbines require only minutes to light off, and they accelerate to full power almost instantaneously. The LM-2500s on *Spruance*-class destroyers deliver power to the shafts in less than five minutes from cold iron. They go from idle to full power in seconds. A steam plant requires hours to light off. A steam-powered destroyer saves no fuel at low speed when she must keep steam up to respond to urgent orders for high speed.

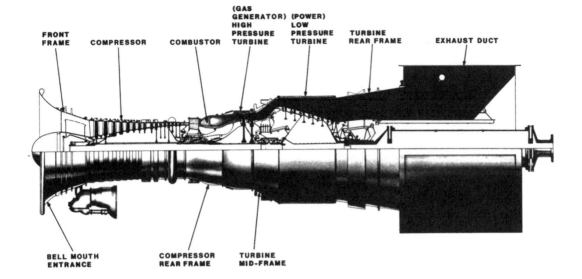

FRONT FRAME • COMPRESSOR • COMBUSTOR • (GAS GENERATOR) HIGH PRESSURE TURBINE • (POWER) LOW PRESSURE TURBINE • TURBINE REAR FRAME • EXHAUST DUCT

BELL MOUTH ENTRANCE • COMPRESSOR REAR FRAME • TURBINE MID-FRAME

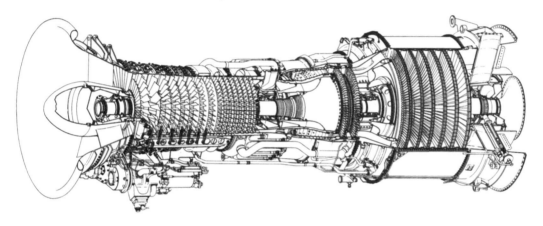

Four General Electric LM-2500 marine gas turbines power *Spruance*-class destroyers. (From David A. Blank et al., *Introduction to Naval Engineering.* 2d ed. Annapolis, Md.: Naval Institute Press, 1985).

The key technological advance in the LM-2500 was advanced blade cooling to resist corrosion and to withstand higher combustion temperature at the high-pressure turbine. This was the feature that made so many decision makers nervous about relying on the LM-2500 in the DD 963 competition. Higher temperature contains more heat energy to convert to mechanical power. Temperature at the power turbine inlet is over 2,300 degrees Fahrenheit, compared with about 1,700 degrees Fahrenheit for earlier marine gas turbines. The LM-2500's specific fuel consumption (170 grams of fuel per horsepower per hour) is 20 percent lower than earlier gas turbines with less power output, such as the Rolls-Royce Olympus and FT4A. A Naval Sea Systems Command message in 1976 coined the acronym TIT for turbine inlet temperature and extolled "high TITs."[11] Further

usage was suppressed. The LM-2500 is 40 percent smaller and lighter than the earlier gas turbines, and its airflow is less too, so the air intake and exhaust uptake are narrower.

These comparative advantages explain why Admiral Sonenshein and others made significant efforts to assure that the ultimate DD 963 design could make use of the LM-2500. A budget-constrained lack of Navy research into marine gas turbines, or other ship technologies such as sea-keeping and active fin stabilizers, as well as opposition from nuclear power lobbyists, meant that it was good fortune that the LM-2500 was available for the DD 963 class at all.

The propeller shafts rotate inward from 12 o'clock position. The shafts are hollow steel tubes. Compressed air (Prairie air) to reduce blade cavitation noise and hydraulic oil to adjust blade pitch are piped inside the shafts to the propellers. Unlike a steam plant, the LM-2500 cannot reverse its direction of spin, so to back down the DD 963 uses controllable/reversible-pitch (CRP) propellers. The propellers have five blades and are 17 feet in diameter.

Centrifugal force on the blades from the hub limits the propeller to 168 rpm. This limits the ships' highest designed speed to about 32.5 knots (see table 8.1). Gas turbines are most efficient at high power. The most economical operation is trail shaft. One engine powers one shaft, with the other engine on that shaft disengaged through the clutch. The other shaft windmills freely. This mode provides about 19 knots.[12]

The long propeller shafts from the engine rooms to the propellers provide a shallow angle of attack to the water flowing along the ship's buttock line. The ideal propeller shaft angle would be horizontal, to align the propeller with the flow of water. Given a destroyer's limited hull size, horizontal shafts would require short propeller blades and thus small, fast-rotating, and noisy propellers. The large, slow-turning propellers reduce cavitation, the generation of water vapor bubbles behind the propeller blade. Seawater absorbs the vapor, so the bubbles quickly collapse, which over time generates a continuous high-pitch acoustic noise. DD 963 propellers cavitate at all speeds. The ships eject Prairie air (compressed air) from the propeller blades' leading edges to mask cavitation. Submarines use shaft rpm to determine a target ship's speed, but this does not work against a DD 963, since the destroyer can vary ahead speed, stand still, or back down with the shafts turning at a constant 55 rpm.

High intakes and water separators are artifacts of the 1965 Project Seahawk research. The gas turbine intakes are high to avoid ingesting salt spray and litter. Gas turbine intakes are much larger than intakes for steam-powered ships, since gas turbines must pull large volumes of air through to cool themselves. Only 30 percent of air intake is for combustion; 65 percent is for cooling the gas turbines themselves; and 5 percent is for Prairie/Masker silencing and other applications. Observers[13] sometimes presume the DD 963's wide aft deckhouse must be a full-width double helicopter hangar, as in the FFG 7 or Canadian *Iroquois* class. In fact the intakes and uptakes for the aft main engine room (plus sonobuoy and aviation equipment stowage in modernized ships) entirely fill its starboard half.

Table 8.1 *Spruance*-class shaft rpm, speed, and propeller pitch

Bell	Shaft rpm	Ship Speed	Propeller Pitch (%)
All ahead full	168	32	↑
	.	.	
	100	.	
	75	.	
	55	11	+100
	55	8	+75
	55	6	+50
	55	3	+25
All stop	55	0	0
	55		−25
	55		−49
	75		
	100		
	125		
All back full	140		↓

Source: David A. Blank et al., *Introduction to Naval Engineering,* 2d ed. (Annapolis, Md.: Naval Institute Press, 1985), 258.

Uptakes duct the gas turbine exhaust to the atmosphere through six mixing tubes in each stack. Ambient air in the mixing tubes cools the exhaust gases from 900 degrees Fahrenheit to 450 degrees Fahrenheit. Uncooled exhaust gases would damage electronic antennas. Exhaust cooling reduces the ship's infrared signature to counter IR satellite cameras, shipboard optical sensors, and submarine periscopes with IR sights. Above all, it counters heat-seeking antiship cruise missiles. Soviet ASCMs were publicly reported to employ IR homing as early as 1970.[14] The presence of heat generators on Russian naval target barges indicates that some current Russian-built missiles use heat-seeking guidance for antiship attack.[15] Naval IR sensors can detect an uncooled gas turbine exhaust plume at 35 nautical miles. An enemy using IR sensors could passively detect an uncooled gas turbine ship below the visual horizon and attack her with IR-homing ASCMs.

The *Spruance* class was always intended to use cheap diesel oil (DFM, NATO fuel symbol F-76). The engineers prefer clean JP-5 aviation jet fuel, which the ships carry for their helicopters. Refineries find JP-5 expensive and difficult to produce. During the 1980s Kuwait was the only Persian Gulf refiner of JP-5. Fuel coalescer filters that would last for a month with JP-5 clogged so quickly with American DFM that crews had to replace them several times a day. DFM coated hitherto pristine fuel tanks with greasy brown sludge. The ships bought quantities of the costly coalescer filters before deploying to the Western Pacific. Once there, they found that bunker oil delivered by tankers from the Persian Gulf was as clean as JP-5. Was American oil of low quality? In fact, the sludge came not from U.S. refineries but from an additive: used motor oil drained from trucks and automobile crankcases at American gas stations. Oil recyclers sold it to marine fuel suppliers, who dumped it into their tanks.

Acoustic Silencing and Counterdetection (Stealth) Measures

The DD 963 destroyers were by design the Navy's quietest surface warships. Silencing measures prevent a ships's own equipment from interfering with sonar searches and enable the ship to approach a passive submarine contact more closely without being detected acoustically herself. Gas turbines are the quietest marine propulsion plant, without the noise of diesels or steam plant pumps. The isolation mounts further reduce the transmission of vibrations from machinery through the hull to the water and to the sonar. Even pipe hangers have rubber isolation pads.

Gas turbines give the *Spruance* class another advantage in silencing. High-pressure air bled from the gas turbine compressor stages goes to five Masker air-bubble belts around the hull and in the propeller struts. Masker belts emit streams of air bubbles that mask noise from hull drag and vibration. Masker reduces self-noise interference with the destroyer's sonar.

Prairie is a similar system that emits air bubbles through the propeller blade edges to reduce cavitation noise. Unlike water vapor bubbles from cavitation, air bubbles do not collapse. Prairie air reduces the ship noise signature that submarines detect. Bleed air reduces gas turbine engine power and can be on or off. Earlier ships such as the DE

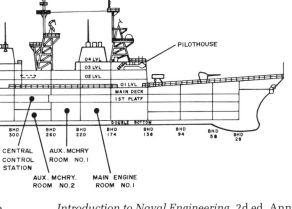

Spruance-class engineering plant layout. Wide separation of primary equipment enhances survivability against combat damage. (From Blank et al., *Introduction to Naval Engineering.* 2d ed. Annapolis, Md.: Naval Institute Press, 1985.)

1052 class had Prairie and Masker silencers, but these required additional air compressors, among the least reliable systems on a ship.

Gas turbines support sprint-and-drift passive sonar tactics, in which the ship stops to listen without flow noise (from seawater flowing past the hydrophone transducers) and then sprints at high speed to a new area or to rejoin the task group. Gas turbines provide fuel economy during these searches, since the engines can be shut down and quickly restarted. A steam-powered ship needs to keep her boilers on line because of their long light-off time, which uses fuel and increases self-noise. With towed-array sonar, silent-running discipline, and Prairie/Masker, DD 963s rarely find sprint-and-drift necessary during actual ASW operations.

Gas turbines produce a plume of hot exhaust that infrared sensors can detect up to 35 nautical miles away. Russian ASCMs from the oldest to the newest have used infrared seekers as well as radar seekers. The *Spruance*-class stacks have vents and mixing tubes to cool the LM-2500 exhaust. A different uptake configuration called the boundary-layer IR suppression system (BLISS) has been installed in the *Ticonderoga* class. Sunlight heats the ship's topside surfaces enough to make them prominent in the IR spectrum against the ocean background. The *Spruance* class has a water-wash-down system that sprays the topside surfaces with water. When in operation it would prevent chemical, biological, and radioactive contaminants from adhering to the ship, and it would cool sunlight-heated surfaces to reduce the ship's IR profile.

Technologies to reduce a warship's radar reflection, or signature, have interested the Navy for years.[16] Stealth technology complicates hostile targeting and increases the effectiveness of decoys. During overhauls the ships' superstructures are being laminated with radar-absorbent material and ceramic armor. Radar-absorbent mats cover reflective fixtures such as boarding ladders, fuel stations, and replenishment kingposts. The ships have the well-known naval stealth technology of degaussing equipment to counter magnetic-fuzed torpedoes and mines. Another measure is that during current overhauls, the ships are painted entirely in flat gray to reduce visibility. The masts above the uptakes were once painted black, a tradition from the days of boilers and black oil meant to hide smoke stains, but the masts now are uniformly gray. Even the traditional white hull numbers with black shadows and the ships' names are becoming two-tone gray.

The Engineering Control and Surveillance System

Control of maneuvering and speed from the bridge enhances the DD 963's mobility and effectiveness. A gas turbine comes equipped with a built-in control system for fuel, overspeed limitation, and other safety factors. This inner-loop control system provides interfaces for connection to the outer-loop ship control system. The relatively small number of interfaces required by gas turbines, compared with steam plants, offers the advantages of automated controls and reduced crew size. An outer-loop automated control system coordinates the engines and controllable-pitch propellers.[17] Direct throttle control from the bridge exploits the fast response of gas turbines. The engineering control system functions are to monitor and control propulsion, electrical, and damage control systems from a central location; to control steering and propulsion from the bridge; to provide single-button start-up and shut-down of main propulsion; to provide local control of propulsion and auxiliary equipment in the main machinery spaces; to ensure safe operation of equipment through hardware and software safety features; and to record propulsion orders during maneuvers (bells log).

The DD 963 engineering control and surveillance system (ECSS) consists of four consoles, which are independent digital computers linked by a digital data network:

- Two propulsion local control consoles (PLCCs), one in each main engine room. The PLCC controls the CRP propeller and the main propulsion gas turbines driving it.
- The ship control console on the bridge. It transmits engine orders (ahead, astern, speed) to the PLCCs as integrated throttle control signals.
- The propulsion and auxiliary monitoring and control equipment (PAMCE) console in the central control station, which serves as a backup for the bridge console, among other functions.

A single lever on the ship control console controls the speed and direction of each shaft, combining the old engine order telegraph and shaft rpm transmitter. The central control station (CCS) amidships provides single-location remote control and monitoring of all propulsion, electrical, and dam-

Table 8.2   DD 963 seakeeping and stability

Sea State	Ave. Wave Height (ft.)	Mission Performance Requirement
4	5	Operate embarked helicopters
5	9	Replenish and strike down under way
6	14	Continuous efficient operation other than replenishment
7	19	Limited operation and continue mission without returning to port for voyage repairs after sea subsides
8+	25+	Survivability without serious damage to mission-essential systems

Allow for following stability hazards separately:
a. 100-knot beam wind
b. High-speed turns (heel less than 10 degrees in a turn at 30 knots)
c. 27-knot beam wind with 15 percent of length open to flooding from damage

age control equipment. A damage control console in the CCS monitors fire and flooding sensors located throughout ship. An electric plant control equipment console controls and monitors the gas turbine generators and the electrical distribution system. Pressing a single button lights off a gas turbine. The gas turbine delivers propeller revolutions within five minutes of being started, about 1 percent of the time needed to light off a steam plant. The DD 963 request for proposals did not require fast light-off, which led some observers to think that the Navy still preferred steam to gas turbines.

Signal conditioners turn equipment sensor readings into digital outputs for the PLCCs. The PLCCs operate automatically under direct control from the ship control console on the bridge and can be controlled from the CCS or operated manually. The engine rooms can be locked shut and operated entirely remotely, but in practice most ships keep a watch in the engine rooms.

ECSS is similar to a present-day computer token-ring local-area network. All the consoles are specialized digital computers. The programs mostly reside in hard-wired logic circuits, not reloadable software codes. Each console in turn transmits 16-bit words over the data network to all the other consoles. Data words include orders from the ship control console, order acknowledgments, equipment status, and alarms. ECSS provides ship's speed (log) inputs to the combat systems and NTDS. The data network is redundant for survivability.

Ship Handling and Seakeeping

The U.S. Navy's 1967 discovery that Soviet destroyers were better sea boats than the overloaded American designs was a shock. Good sea-

keeping, maneuverability, and speed in heavy seas were major design requirements for the DD 963 class, and the most important single reason for their large size (see table 8.2).[18]

With high engine power and controllable-pitch propellers, a *Spruance*-class destroyer is highly maneuverable. These ships routinely reverse from full speed ahead to full speed astern (crash-back) in less than a minute. From full ahead they can stop dead almost in their own length. Steam-powered destroyers were similarly maneuverable at high speed but needed hours to get up steam, and their complex plants were delicate. The DD 963 class can put full power to the propellers on five minutes' notice if the gas turbines are cold, and without notice when they are on line. On the DD 963 a seaman at the lee helm on the bridge controls the engine throttles directly. The steam-powered assault ship I once served on attempted a crash-back twice in two years, and on the first occasion officers handled the steam valve control wheels. Another difference is that full engine power is available during low-speed maneuvers such as coming alongside a pier or replenishment ship. In general the DD 963s respond to maneuvering orders in half the time of steam-powered ships.[19]

In 1979 *Elliot* (7,800 tons, 80,000 shp) and the much lighter steam-powered destroyer *Robison* (DDG 12; 4,500 tons, 70,000 shp) held a race in the Indian Ocean. *Robison* had so much steam up before the start that her safety valves were lifting. The destroyers lined up abreast at all stop and on the signal took off for 32 knots. *Elliot*'s Prairie and Masker air-bubble sound suppressors left a clear track of the course. That event was a draw. Higher seas (the race was in a glassy calm) would have favored *Elliot*. *Elliot* next demonstrated high-speed maneuvers such as crash-backs. *Robison* watched but declined to match.

The only really notable deficiency of the class is rolling in heavy seas. During DD 963 design selection the Navy traded Gibbs & Cox's design, with its wide beam and stabilizers, for Litton's design with silencing, lower cost for fuel and maintenance, and illusory low price. During *Spruance*'s acceptance trials her rolling in sea state 6 exceeded the specified design limit.[20] The ships heel well outboard during tight high-speed turns, not a problem given their strong righting moments, but a surprise to ship riders concerned about stability.

To a naval architect the early ships had excessive stability, evidenced by snap rolls: they righted so quickly after heeling that off-watch personnel were sometimes catapulted out of their bunks. Bunks have safety belts, but at least in the early ships they were too far down. I suspected that in a bad roll, mine would not hold me down but would instead simply suspend me over the edge by the waist, so that I would drop onto the deck headfirst. One night during a storm my destroyer rolled to port more steeply than usual. I instantly awoke, aware that I might imminently be launched from my upper berth. I grabbed the bunk rail, was indeed propelled into the air as she righted, and landed on my feet like a gymnast after a vault. I quickly checked my equipment spaces and my men's berthing compartment and found that all was well. Sailors, once used to their ships, sleep through noise and tumult but will wake at a sudden silence or stillness.

When Kevlar armor was added topside, the destroyers were ballasted to help counter rolling. The ships are volume-critical, not weight-critical, so can carry extra weight low down, such as ballast, without requiring a trade-off of equipment elsewhere. Late in the 1980s an experimental autopilot system kicked rudder automatically to induce a heeling moment to counteract the rolling moment. This rudder-roll stabilization system worked well.[21] Some conservative senior officers opposed it from concern that at critical times such as underway replenishment an autopilot would reduce maneuverability.

Slamming is what happens when the keel at the bow emerges from heavy seas and pitches back into the water. The bow sonar is obviously useless when it broaches. The DD 963's antislamming performance is excellent. During *Spruance*'s acceptance trials keel slamming did not occur even in sea state 6. At very high sea states slamming can occur, but it is rare.

Good seakeeping is an important military quality. Soviet open-ocean warships were large and broad-beamed for good seakeeping. Their encapsulated missiles could work in weather that would ground Western naval helicopters. Their towed sonars could work in seas where keel-mounted sonars were unreliable because of broaching and keel slamming. Submarines at depth are unaffected by heavy seas. A submarine might escape a pursuing destroyer by steering a course to cause severe rolling and slamming aboard the ship. This was a primary reason that Soviet nuclear submarines were built to be fast, despite the noise of their early propulsion plants.[22] When designing the *Arleigh Burke* (DDG 51) class in the 1980s, the Navy chose a hull form that would allow these ships to fight in any seas in which Soviet ships could fight. This explains in part why the first DDG 51 destroyers do not have helicopter hangars.

Soviet ships and submarines relied on aircraft and other ships to classify and localize targets for their long-range antishipping cruise missiles. Heavy weather grounded their ship-based aircraft too and blinded land-based aircraft. The newest Soviet ships apparently had direct links to surveillance satellites for targeting. Heavy weather would still restrict aircraft and would require Soviet warships and submarines to identify their own targets. Soviet warships relied on radio communications with command posts ashore for this. Most *Spruance*-class destroyers are equipped to intercept such communications. They are formidably armed for close-in battle. All-weather weapons include rocket-thrown Mark 46 ASW torpedoes (ASROC), ship-launched ASW torpedoes, Tomahawk, Harpoon, Sea Sparrow, 5-inch guns, Phalanx, and decoys.

The Electrical Plant

Before the DD 963 class most American warships had two electrical turbogenerators driven by low-pressure steam drawn from the main propulsion plant. Two separate emergency diesel generators could provide electricity if the boilers were off line. In the DD 963 class three ship's service gas turbine–powered generators (SSGTGs) provide all electricity. They are independent of the main propulsion plant, so additional separate emergency generators are unnecessary. Any two SSGTGs provide power for the entire ship under battle load. The third provides survivability for electrical power. There is one SSGTG in each main engine room. The intake and uptake ducts are in the main stacks. The third SSGTG is in its own compartment aft of the torpedo rooms. Three watertight bulkheads separate each SSGTG from the next for survivability against combat damage. USS *Paul F. Foster* (DD 964) has four SSGTGs.

Each SSGTG generates 2 MW of 60 Hz alternating-current (AC) electrical power (2.5 MW in the *Ticonderoga* class). The gas turbine is the Allison 501K engine used in Navy turboprop aircraft such as the P-3, E-2, and C-130. This is a big advantage in getting spare parts. Three static frequency converters generate 400 Hz AC power for radars and some other combat systems. Previous ships carried large motor-generators that interfered with sonar performance and created distinctive sound patterns for submarines to listen to.

The gas turbine AC generators have multiple functions. Analogous to auxiliary power units in aircraft, they provide compressed air (bleed air) to start the main propulsion gas turbines. In the ships as built, the hot exhaust (waste heat) from the SSGTG passed through a waste-heat boiler to generate low-pressure steam. Steam is used for hot potable water (showers); freshwater evaporators; the galley's kettles, steam line, and scullery; the laundry; and to warm lubricating oil and fuel oil.

The early waste-heat boilers were unreliable and difficult to repair. When a waste-heat boiler was down, the attached gas turbine generator could not be operated either, since the heat might further damage dry boiler tubes. During early *Spruance*-class operations one generator was down 40 percent of the time and two were down 10 percent of the time.[23] Shortly before becoming Chief of Naval Operations, Admiral James Watkins estimated that fixing the class would cost $400 million.[24] The good news was that inexpensive modifications to the boilers let the destroyers run the generators while the tubes were dry. The project officer, Commander Ken Smith, saved the taxpayers more money than probably any other government official in years. In some cases the boilers were too damaged for repair, and the substandard valves and pipes were badly corroded. During the late 1980s the entire steam plant was removed from some ships of the class, and electric heaters replaced it.

Another electrical function is internal communication. Routine communication is through a telephone switchboard. Battle communications rely on sound-powered phones over independent point-to-point wires. Gyroscopes and wind instrumentation have communication links to navigation and fire control systems.

Auxiliaries

Auxiliary systems may seem mundane, but they are essential to shipboard operations. Evaporators produce fresh water for drinking, chilled-water cooling, boiler feedwater, and freshwater washdown to reduce topside corrosion. Potable-water tank capacity of the first DD 963s was sized for a 250-man crew, but the complement grew far above this. Another problem was that the waste-heat boilers consumed huge amounts of fresh water for refilling after a breakdown. Freshwater tankage was inadequate, and the Navy's newest ships often went on "shower hours" when the crew could not shower so that the tanks could refill. Voids in the hull were converted to freshwater tanks in early shipyard periods.

Shipboard electrical equipment converts electrical power to radio energy, mechanical motion, and mostly to heat. Chilled water is piped to larger electrical equipment for cooling. Radars, computers, static frequency converters, and other systems are water-cooled. Air conditioning cools other equipment. The large electrical load requires great air-conditioning capacity and results in a chilly ship even in the tropics. The DD 963 class has three air-conditioning plants. The DDG 993 class, intended for tropical duty and having air-conditioned main engine rooms, has four plants. Insulation mats cover the chilled-water pipes and the internal hull shell plating to prevent condensation.

Compressed air is used for launching torpedoes, for starting the gas turbines, and for the Prairie and Masker silencers. The DDG 993 class and early Aegis cruisers use compressed air to eject dud missiles from the Mark 26 launcher rails. Sources of compressed air are high- and low-pressure air compressors and bleed air from the gas turbine compressor stages for the Prairie and Masker air-bubbler silencing systems.

The Navy's first environment-friendly ships, *Spruance*-class destroyers have holding tanks for sewage so that they do not foul harbors or coastal waters. As built, the DD 963s were equipped with sewage incinerators, but these rarely worked because the ships' engineers always had more important machinery to fix. The Naval Sea Systems Command worked on this problem. More parts, better components, and better instructions reduced maintenance time by 95 percent.[25]

Supply

The Supply Department extends the operational endurance and reliability of the entire ship. Each ship carries spare parts and supplies for 90 days and food for 45 days of independent combat operations. Planning for supply was a fundamental influence in the development of the DD 963 class. A primary reason for building all the ships under a single contract was to assure that a small range of spare parts would support all 30 destroyers, thereby reducing cost and increasing the likelihood that a ship would have a needed repair part on hand.

Replaced reparable parts are sent ashore for overhaul and return to the supply system to be issued again. During fleet introduction some parts cost more to repair than to buy new, and funding was inadequate to meet the demand. Shipboard demand was high for parts needing change-out after a certain number of hours of operation, such as burners for the gas turbine generators. Spares available ashore dwindled, and many were rationed as early as 1978. The good news was that the parts were identifiable, and the shore-based supply system could issue them quickly for high-priority requirements. In 1980 the supply staff of the Pacific Fleet's Naval Surface Force commander evaluated the *Spruance* class as the best-supported ships in the entire force.[26]

The internal arrangement of the DD 963 design reflects Navy requirements for a central galley and rapid strike-down of supplies. Galley and mess decks are concentrated amidships between the main engine rooms. Their central location minimizes the effect of rolling and pitching moments on the crew. Between meals the mess decks are useful for training and recreation. The chief petty officer's mess is forward of the galley. The officers' wardroom pantry is one deck above the galley and joined to it by a dumbwaiter. One single galley thus serves all hands. Refrigerators and a dry food storeroom are beneath the galley.

As with the ship's capacity to make fresh water, food storage volume either was underestimated or was designed to support an unrealistically small crew. Early DD 963 supply officers increased food storage by taking over the bottom level of the after missile magazine growth space to stow dry boxed and canned provisions. This space was designated

Sailors take inventory in the forward dry provisions storeroom. Spring-loaded stanchions clip into the slots in the gratings on the deck and overhead, to hold the stacked boxes steady in seas. A dumbwaiter, not visible, connects it and adjacent refrigerated storerooms to the galley and the amidships replenishment station. (U.S. Navy photo)

for a Mark 26 Mod 1 44-round missile launcher but was unused in the *Spruance* class as delivered. Its upper levels became a gym, workshops, or additional berthing.

At sea the ships resupply themselves by underway replenishment. Retractable kingposts by the ASROC launcher forward and by the Sea Sparrow launcher aft receive cargo pallets by highline from a replenishment ship alongside. Another kingpost is built into the superstructure side amidships. Helicopters can land cargo pallets onto drop points on the bow and fantail. The ships prefer not to use the drop point on the bow, from concern that any helicopter malfunction might cause it to crash into the bridge. Wherever the ship lands it, cargo is brought amidships to the strike-down elevator and sent to the storerooms below the galley, or man-handled to other storerooms.

Ammunition is replenished in the same way as stores. Each 5-inch gun mount has an ammunition elevator to strike pallets quickly down to the magazines. The 5-inch shell pallets are heavy and require pallet trucks to shuttle them from replenishment stations to the strike-down elevators. Weather deck passages along the 01 level are wide enough to shuttle pallets fore and aft. The Combat Systems Department is responsible for ammunition, but Supply helps to order it.

Warships need spare parts, lubricating oil, galley supplies, medicine, engineering chemicals, office equipment, and a huge range of other materials. Supply orders them all. In commercial ports each ship charters her own port services such as telephones, office machine repairs, and tugboat hire. Other Supply activities include the payroll, the laundry, the barbershop, and canteen services. (These last three are military functions in the U.S. Navy. In the Royal Navy Hong Kong laborers do the laundry, anyone who wants the job is a barber, and civilian contractors run the canteen.)

Destroyer tenders support *Spruance*-class destroyers at advanced bases or anchorages. Seizing, building, and holding such bases is a primary mission for the Marine Corps and construction battalions (SeaBees). Tenders can replace gas turbines in a destroyer alongside by hoisting them through the air intakes. They have versatile machine shops for mechanical repairs. The destroyer crew can move aboard the tender if the galley or berthing compartments are unusable during repair.

Manpower and Habitability

For years before the DD 963 project, the Navy had seen automation as the key to solving the problem of reducing crew size. The Navy's DD 963 specification required a complicated design-work study to calculate crew size. Design-work study was a 1960s psychobabble notion that all work was done by mindless physical rote. The sum of all motions by all personnel performing work rituals defined the set of jobs and thus crew size. Such an approach might work for something like a console. Presumably the intent was to show how to design the DD 963 class to minimize crew size along the way. Underlying it was a suspicion that a lot of shipboard work was traditional rather than essential. Bosuns still sounded bells every 30 minutes, a duty from medieval times to mark the

The crew's mess is located amidships to minimize the pitching movement. The officers' wardroom is one deck above and the chief petty officers' mess forward. A common galley prepares hot meals for all messes. The mess decks are well-equipped for contemporary appetites with salad and beverage bars.

turn of a sandglass, and after eight bells the watch changed. Litton prepared a design-work study volume for its DD 963 proposal, but the naval architects found it worthless and ignored it when they designed the ships.[27] Around 1970 General Motors built a plant in Lordstown, Ohio, under a design-work study approach. Its workers went on strike over intolerable job quality. Design-work study has since disappeared even from academic job-design textbooks.

Litton provided tasteful internal furnishings such as vinyl-coated aluminum bulkheads. Navy ships inside were usually white (fading to yellow, like the Pentagon's inside ring corridors) with light green ("bilgewater green") trim and dark green tile decks. The *Spruance* class was painted with a variety of pleasant pastels: champagne yellow, light blue, and others, with deck tiles in matching hues. Estimates that the ships would need little effort to keep clean proved overoptimistic. Vessels of any size or purpose get dirty quickly. Cleaning all the berthing compartments, passageways, inside ladders, officers' staterooms, wardroom, galley, and mess decks was a full-time occupation for about 10 percent of the crew. Bulkhead coverings and deck tiles suited for aircraft needed constant repairs because of damage from scuffing, vandalism, and collisions when sailors manhandled heavy pumps and motors down passageways. The latex coverings over bulkheads tore easily. Bimetallic corrosion weakened the cosmetic bulkheads on the 01 level where the aluminum deckhouse met the steel hull. Condensation and drips from valves kept parts of the inside of the ship moist. The entire inside skin of the ship was covered in insulation to cut down on condensation. The insulation was soft and often needed to be patched or replaced. In 1979 a report on USS *Spruance* described the 4-year-old ship as looking 20 years old internally.

When the *Spruance* class entered service in 1975, the predicted crew size, whether based on design-work study or otherwise, quickly proved inadequate. As designed, or more accurately as purchased, a DD 963 destroyer rated a crew of 224 enlisted men and 19 officers.

Another manpower problem was that nuclear weapons security required a large duty section in port. The Navy authorized ships six-section duty, meaning that the crew was divided into six sections when the ship was in an American port. The six duty sections rotated, so that every sixth day a duty section stayed on the ship for 24 hours for security and damage control. With losses to leave, training, and personnel transfers, *Spruance*-class destroyers could not guard the ship with one-sixth of the crew, so they usually stood five-section or even four-section duty. This meant that if the ship tied up in her home port for a month, a sailor could look forward to a single weekend when he could leave Friday afternoon and return Monday morning. For this he drew an extra 75 cents a day. Congress approved meaningful sea-duty pay in 1980.

Meanwhile, the Navy raised the crew allowed for the DD 963 class, although much of the increase was on paper only, because of Navywide manpower shortages. More sailors were assigned to a destroyer for Harpoon, electronic surveillance, and other new systems. Berthing compartments had room to accommodate the crew after conversion to a guided missile destroyer (287 enlisted men as designed, increased to 316 when the *Kidd* class joined the fleet). Tiers of berths, or "racks" as sailors know them, are separated by 30-inch aisles in a *Spruance*-class destroyer, rather than 24-inch aisles as in previous ships. Shipyards made room for additional racks by removing tables in the lounge area of each compartment. A ship not in the yard was on her own in finding berths for more sailors. Virtually my first action on *Elliot* in 1978 was to barter 40 pounds of coffee for 12 berths from a decommissioned destroyer. They took half the recreation space in Supply berthing.

Very few previous Navy ships had mounted gas turbines. At sea, automated remote control of the gas turbines let the ships operate without men in the engine rooms. In practice, the engineers preferred to station a watch in the engine rooms anyway, to listen to equipment and to watch for obvious malfunctions such as fuel leaks that remote gauges might not report. The first gas turbine specialists, a new rating established in

Destroyer tenders are advanced bases for destroyers. They provide maintenance and rearmament facilities and can house a damaged ship's crew during repairs. If necessary, tenders can take ships alongside for extensive work on the high seas. Here, USS *Caron* (DD 970) is alongside USS *Puget Sound* (AD 38) in Malaga, Spain, in 1981. (U.S. Naval Institute collection)

1978, were drawn from diesel engine mechanics, electricians, and internal communications men. Since the ships took longer to complete at Ingalls than expected, many of the crew were not with the ship very long after commissioning before their enlistment contracts expired. The Chief of Naval Personnel described the outcome:

> When we said in the '70's that we must design to reduce manpower on ships, we didn't know what we were talking about. What we should have said is, "Let's design our systems to minimize high technical skill requirements." Now we're having to re-group. Where we previously required nine firemen, we now need three Chief IC [interior communications] electricians. The former can still be obtained; the latter cannot, because they're now working in industry.[28]

As a workplace, a DD 963 was stressful. The ships were quiet underwater but acoustically loud in the air. Alarm clocks were useless for waking officers before watch, because even at a distance of 1 foot they were drowned out by ambient noise from the gas turbines and the helicopter stabilized-glide-slope indicator. Ventilation was much better than in the stuffy older ships, but air conditioning to cool electronics cooled everything else too, so that the ships were cold even in the tropics. Crews were worked hard in every job. Stress soon showed. In 1977 a tired gunner's mate tried a shortcut to fix a 5-inch mount during a firing exercise. This worked, and the cradle instantly swung the next round toward the breech. It caught him and fatally injured him. Later, two sailors got into a savage argument in their berthing compartment. One, a Filipino cook, fatally stabbed the other with a knife from the galley. It was the first murder aboard a Navy ship in years but was kept quiet, probably because of the racial overtones. On another DD 963 a sailor murdered the disbursing officer during a robbery. Drugs, primarily marijuana, were a Navywide problem. Even if only 1 percent of the crew was involved in criminal activity, these men greatly increased the work of the rest by requiring additional disciplinary supervision. Delinquents confined to the ship in port for punishment vandalized lounges and the gym.

By 1980 the enlisted crew allowance had soared to 297, and that did not include the helicopter detachment. All warships designed during that era required larger crews in practice than as designed, but the increase for the *Spruance* class was the largest, over 30 percent.[29] Arvin Plato, the co-author of an article about this problem, was an impassioned advocate of design-work study, but the article made no mention of it as a cause of the persistently bad estimates of crew size.[30]

Poor pay and exhausting work conditions drove experienced men from the Navy. By 1980 the number of petty officers had fallen 25,000 below authorized requirements. President Jimmy Carter and his McNamara-trained Defense Secretary, Harold Brown, had already cut the Navy's shipbuilding projects by half and decided to cut even more because sailors were unavailable for crews. Some steam-powered ships were unready for operations because of a lack of skilled crewmen. Prodded by Chief of Naval Operations Admiral Thomas Hayward, in 1980 Congress raised military pay. Starting in 1981 Admiral Hayward, President Ronald Reagan, and his outstanding Secretary of the Navy, Lieutenant Commander John Lehman, USNR, quickly restored morale. The petty officer shortage disappeared.

Survivability and Damage Control

Warships must continue to perform combat missions despite damage from enemy attack. Merchant ships are designed only to let the crew and passengers abandon ship in the event of significant damage. The Navy's specification for the DD 963 design included shock resistance, blast resistance, equipment redundancy, and stability despite damage.[31]

DD 963s are expected to "fight hurt," and their design includes features intended to enable them to withstand direct torpedo explosions under at least some circumstances. DD 963s are designed to withstand flooding along 15 percent (79 feet) of their length in a 27-knot beam wind (equal to a 100-knot wind held 15 degrees off the bow).[32] In such conditions the ship will not capsize from flooding unless so much water is trapped above decks from firefighting that she loses stability. No warship is unsinkable. Certain weapons effects are designed to achieve single-shot kills of any ship of a DD 963's size, such as by warhead detonation deep beneath the hull so that an oscillating bubble hits under the keel.[33] The ship must recover from large angles of heel while in the troughs of waves generated by an underwater nuclear burst.

The primary sources of survivability are redundancy, compartmentation, hull strength, and machinery separation.[34] Machinery redundancy and separation enable functional survival despite local underwater damage; the main engine rooms, auxiliary rooms, generators, and switchboards are all redundant. The ships must withstand mechanical shock transmitted through water. Testing is realistic. USS *Spruance* (DD 963) was exposed to a series of underwater explosions, moved in progressively closer to the hull, to shock-test equipment. Concussions gave terrific jolts to the ship. These tests measured survivability to underwater nuclear blasts and influence mines. To survive mechanical shock and whipping, radars use direct-drive, direct-current servomotors instead of gears.

V-lines measure the trim, list, and immersion the ship can survive. All fittings below the V-lines must be watertight. Spaces above the V-lines need not be watertight, since the ship would capsize or sink before the water reached them. On the *Spruance* class ventilation closures and smoke curtains above the V-lines prevent the spread of smoke or chemical agents. When Kevlar armor plating was added to the upper deckhouses during modernization, internal modifications raised the V-lines. Another survivability enhancement was to add ballast for better stability.

Damage Control Central is in the engineering central control station. At general quarters, four damage control teams take posts. During flight operations a helicopter crash and rescue team is manned. The repair parties are assigned as follows:

- Repair 2: hull forward of the main engine rooms and the superstructure forward of the mainmast

- Repair 3: hull aft of the main engine rooms and the superstructure aft of the mainmast
- Repair 5: main engine rooms and adjacent engineering spaces
- Repair 8: combat system electronics

The ships carry damage control equipment for any emergency. Fire stations, emergency power cables, shoring to stiffen bulkheads, smoke curtains, and other equipment are located throughout the ship. Damage control lockers hold pipe and hull patches, drainage educators, thermal imagers to find fires through smoke, ultrahigh-frequency walkie-talkie transceivers, tools, oxygen breathing apparatus, and much else for each repair party. Medical facilities include decontamination stations, battle dressing stations, and a dispensary. Compartments are not gas-tight citadels against chemical, biological, or nuclear agents, but air conditioning pressurizes the ship. Gases will not leak in unless an explosion causes major structural damage.

Every member of the crew must pass tests of knowledge and skill at damage control. Berthing compartments might be better assigned by watch sections instead of departments so that one unexpected hit would not wipe out the entire complement of a critical rating. The U.S. Navy is sometimes criticized for putting too much emphasis on damage control training, at the expense of practicing combat tactics that might defeat an attack before weapons strike. Perhaps, for example, damage control teams should be smaller so that more sailors could serve as lookouts for submarine periscopes.

Gunnery, Electronics, and Antimissile Weapons

Radar and Radio

The first sight of a *Spruance*-class destroyer was spectacular even for experienced naval observers. The giant warships surpassed Reuven Leopold's plan that they should look impressive. The steeply raked bow led back nearly 200 feet to the sharply cornered superstructure and massive electronics-laden masts. *"Arthur W. Radford* dwarfs the DLG *Luce* lying alongside. . . . *Radford* created an immense impression as she loomed out of the mist at point-blank range in almost total silence," a British naval analyst wrote in awe.[1] *Spruance*-class destroyers look the best from the starboard quarter, a view that emphasizes the horizontal deck lines aft. From the starboard quarter and the port bow the stacks and masts appear concentrated and focused. Another type of physical appearance is the radar cross-section, which varies in intensity with the ship's target angle (i.e., relative bearing from the ship to the observer).

In naval architecture a warship has two fundamental distinctions from other ships: weapons and damage control. Guns, guided missiles, and torpedoes comprise ordnance, the destructive weapons. Weapon performance depends absolutely on fire control equipment for targeting. Fire control in turn relies on surveillance networks to detect and identify targets. Weapons, fire control, and surveillance are so closely connected that this chapter describes them together. Surveillance information comes from signals collection, search radar, sonar, communications, visual sighting, helicopters, and the tactical computer system. *Spruance*-class destroyers transmit surveillance information to other military forces and ships.

Air and surface targets operate in the atmosphere, which unlike water is penetrable over a wide spectrum by electromagnetic waves. The electromagnetic energy spectrum includes radio, radar, infrared energy, and visible light. Table 9.1 shows how shipboard sensors and communication systems exploit the electromagnetic energy spectrum, in particular the radio-frequency section. The laws of physics being neutral, the table shows electronic warfare systems to detect hostile use of the radio-frequency spectrum. Table 9.2 shows the radar (microwave) spectrum in more detail and shows where radar systems fit within it.

The Navy developed several weapons and systems for the DD 963 class on separate schedules for installation after shakedown. This additional time allowed more modern systems to be fitted than were available in the early 1970s. Delaying installation until after construction reduced the contractual scope of work, and thus the cost, of the controversial Total Package contract. The trade-off was that the ships were not fully equipped to fight at sea upon delivery. This was a frequent but specious point of criticism of the *Spruance* class. If necessary the Navy could have equipped them with contemporary deck-mounted weapons such as Standard-ARM antiship missiles and Hip Pocket antiship cruise missile defenses.[2]

The Harpoon antiship missile became available in 1978, and the SLQ-32 electronic warfare system in 1980. Early DD 963s received their full outfits of weapons and electronics during four-month postshakedown overhauls at naval shipyards. Systems added included a NATO Sea Sparrow battery; WSC-3 ultrahigh-frequency (UHF) satellite communications; WLR-1G microwave signal detectors on the first 10 ships; rapid-bloom chaff decoy rocket launchers; and facilities for one SH-2F Seasprite helicopter. The destroyers made their first forward deployments in this configuration.

Table 9.1   Military use of the electromagnetic energy spectrum

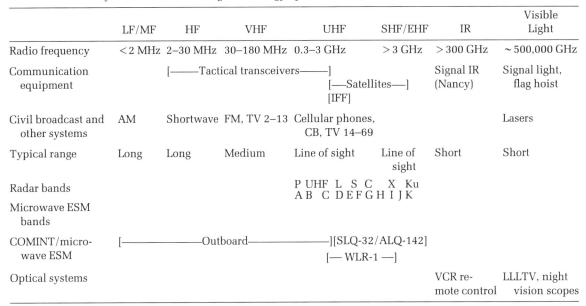

	LF/MF	HF	VHF	UHF	SHF/EHF	IR	Visible Light
Radio frequency	< 2 MHz	2–30 MHz	30–180 MHz	0.3–3 GHz	> 3 GHz	> 300 GHz	~ 500,000 GHz
Communication equipment		[——Tactical transceivers——]			[—Satellites—] [IFF]	Signal IR (Nancy)	Signal light, flag hoist
Civil broadcast and other systems	AM	Shortwave	FM, TV 2–13	Cellular phones, CB, TV 14–69			Lasers
Typical range	Long	Long	Medium	Line of sight	Line of sight	Short	Short
Radar bands				P UHF L S C X Ku A B C D E F G H I J K			
Microwave ESM bands							
COMINT/micro-wave ESM	[————Outboard————][SLQ-32/ALQ-142] [— WLR-1 —]						
Optical systems						VCR remote control	LLLTV, night vision scopes

Hewitt and *Elliot* were the first to deploy with Harpoon missiles. From *Moosbrugger* (DD 980) onward Ingalls installed these government-furnished systems during construction.

In the early 1980s most of the destroyers received two Phalanx close-in antimissile guns, a Mark 23 missile-detection radar for Sea Sparrow, and new SLQ-32 electronic warfare systems for radar detection. Seven destroyers received two quadruple armored box launchers (ABLs) for Tomahawk cruise missiles. A vulnerability assessment led to armoring the gun and missile magazines and the forward deckhouse with lightweight ceramic-laminate armor. Protection was added to radar and radio equipment, and fire-resistant insulation was added internally. These additions increased displacement by 540 tons, including 125 tons of lead ballast. Draft increased about 8 inches.[3]

During the late 1980s the *Spruance* class began a major modernization under the Warfighting Improvement Program. New systems added included the SQQ-89 ASW system, SH-60B helicopters, a 61-cell Mark 41 vertical-launch system, and SLQ-32(V)3 active radar jammers. Most of the destroyers in Operation Desert Storm fought in this configuration. Seven destroyers are armed with armored box launchers. Vertical-launch systems replaced the ASROC launcher and loader on 24 destroyers. All 31 will get SQQ-89. In 1993 the Clinton administration considered canceling the last eight modernizations and retiring those destroyers. As of late 1993, however, the decision

was to retain all 31 in active service. Fully modernized, sheathed with Kevlar armor, and ballasted, a *Spruance*-class destroyer displaces 9,250 tons full load.

Command and Control

Until surprisingly recently warships still fought like maritime stockades, with many independent weapons batteries. The final World War II battles, especially Okinawa, had saturated manually kept plots of combat data. Korean War jet aircraft operations overwhelmed manual tracking. The Navy began work on a computer-based real-time display system for aircraft carriers and guided missile cruisers to fight an air battle. Experiments conducted by the Canadian Navy showed the superior ability of digital electronics, then an infant technology, to store contact data for plotting tracks over time and to relay track data between ships over a digital data link.[4] The U.S. Navy adopted this approach and specified the requirements for the Naval Tactical Data System (NTDS) in 1955. Goals were to gather operational data, to display it in a useful format, and to coordinate a force of ships through data links. Requirements included networked operator consoles, a transistorized stored-program digital computer, and a digital data link.[5]

NTDS induced a major advance in computer science. The original NTDS computer designer was Seymour Cray, age 30, then of Univac. During

Table 9.2   Radar spectrum and *Spruance* (DD 963), *Kidd* (DDG 993), and *Ticonderoga* (CG 47) radar systems

	Radio Frequency (GHz)				
	0.4–1.5	1.5–3.9	3.9–6.2	6.2–10.9	15.0–17.2
Radar band	UHF & L	S	C	X	Ku
ESM band	B & C	F	G	I	J
Band uses	Long-range air search	Long-range air search	Air tracking	Surface search, CWI, missile seekers	Surface search, CWI, missile seekers
Search radars					
All				SPS-55, SPS-64, Furuno/Path-finder	
Spruance	SPS-40 2-D, SPS-49 2-D*				CIWS via RAIDS
Kidd	SPS-49 2-D	SPS-48 3-D			
Ticonderoga	SPS-49 2-D	SPY-1 3-D			
Fire control radars					
All				SPQ-9	CIWS, Harpoon
Spruance	Mk 23 TAS 2-D			SPG-60, Mk 95 CWI	TASM
Kidd			SPG-51 track	SPG-51 CWI, SPG-60+CWI	
Ticonderoga		SPY-1 3-D		SPG-62 CWI (= Mk 99)	TASM**
Aircraft systems	IFF, URN-25 TACAN	*LAMPS I* AKT-22: SKR-4	*LAMPS III* ARQ-44: SRQ-4	LN-66, APS-124	

Notes: * *Hayler* (DD 997) only   ** VLS ships only

the 1950s military requirements and research budgets spawned computer developments that were uneconomical for commercial industry. All subsequent NTDS computers use Cray's architecture. Cray went on to design the world's most powerful computers at Control Data and Cray Research. His commercial supercomputers still have characteristic features of NTDS, such as multiple separate processors for computation and input-output, and shared central memory.[6] NTDS became fully operational in the fleet in 1966, just as the DX/DXG project began. It was installed on aircraft carriers and guided missile cruisers (then called frigates) and was essential for PIRAZ duty off Vietnam.

The *Spruance* combat computer network is the Command and Decision System (CDS). It uses a military UYK-7 real-time computer running NTDS software with enhancements for ASW. CDS is a superset of NTDS. More than on any previous surface warship, the commanding officer (CO) of a *Spruance*-class destroyer fights her like a pilot in an airplane, controlling a completely integrated combat system. A major advance over earlier NTDS ships was automatic input of navigation data from satellite and long-range navigation (LORAN) fixes, the pit log (for speed), gyrocom-

pass, and anemometer. The CO and tactical action officer (TAO) direct all the ship's weapons from the combat information center (CIC). As built, the *Spruance* class had nine NTDS display consoles (UYA-4) in the CIC and one on the bridge. New UYQ-21 consoles might replace them.

CDS transfers target data quickly from sensors to fire control systems. It builds track files of sensor contacts (from radar, sonar, electronic surveillance, visual) and computer data (Link-11 tracks, contact identity symbols, track numbers, graphics, and other text). The UYA-4 CDS console displays simplify the workload of the CIC crew and greatly increase their ability to enter sensor contact data quickly. The computer automatically smooths out clutter. Console operators track targets in separate dimensions: air, surface, underwater. CDS records the contact data for each track, computes the target's course and speed to show its actual and expected movements, and displays it on the console as a symbol. CDS displays comprehensive plots of the tactical situation for the officers and crew to use during combat. It can show both sonar and radar contacts on the same digital graphic displays. This is highly useful since, for example, a passive sonar contact on the same bearing as a radar-tracked ship could be

The brain of the DD 963–class destroyer is Naval Tactical Data System (NTDS) software running on the Command and Decision System computers. Operations specialists use UYA-4 NTDS consoles to track targets and to set up attacks. This console is on the bridge; others are in the combat information center.

assumed to emanate from that ship, allowing ASW assets to focus on uncorrelated sonar contacts. Further, a radar contact on a sonar noise source indicates sonar range. Since all the displays are computer-generated, an operator can center the display on anything, such as a distant contact, and zoom in on the area to coordinate air strikes or other actions.

NTDS-equipped ships in an operation pass contact data to each other over the Link-11 radio data link network. Each NTDS ship transmits contact data as computer display symbols showing the target type (surface, air, submarine), target identity if known, and identity of the ship or aircraft tracking it. One ship, the network controller, polls the other NTDS ships in turn. A polled ship's NTDS computer automatically broadcasts updated contact data on all tracks for which that ship is responsible. All NTDS ships in the network record the track data broadcasts. ASW and radar sentry (AWACS) aircraft participate in the NTDS data network. The SH-2F Seasprite helicopter proved highly useful for antiship surveillance and targeting missions but could report sightings only by UHF voice radio.[7] The newer SH-60B Seahawk helicopters transmit contact data as NTDS computer symbols over a point-to-point pencil-beam data link to the destroyer. The destroyer broadcasts this information to the rest of the force in her Link-11 broadcasts. NTDS ships relay track data

over radio-teletype circuits to non-NTDS ships to help them to find targets (Link-14).[8]

The main importance of CDS is that it enables the ship to manage combat more effectively. An indicator of the importance of CDS is that destroyers and frigates, some only a year older than USS *Spruance,* without NTDS or at least Link-11 capability, have disappeared from the fleet. The ship uses CDS contact data to designate weapons systems to specific targets. An example is the provision of data to the targeting consoles for Harpoon and Tomahawk antiship cruise missiles.[9] CDS helps the ship to select the most urgent targets and to assign weapons to threats quickly. It prioritizes contacts with consideration of the console operators' tasks. At air-tracker and surface-tracker consoles, it puts the track with the oldest update at the front of the queue for updating. At the TAO and ASW consoles, it prioritizes contacts by threat, putting the fastest or closest hostile tracks at the front of the queue. Missions such as surveillance, ASW, PIRAZ, or gunfire support of landings often must be performed 100 nautical miles or farther from the main body of a battle group. CDS helps a DD 963 to continue her primary missions with less risk of disruption from enemy action. With ASW weapons, Harpoon, gunfire, decoys, and antiship defense systems, a DD 963 can defend herself in an encounter with enemy forces until air support arrives.

Gunnery and Gunfire Control

Main gun armament consists of two Mark 45 5-inch guns and the Mark 86 gunfire control system. The Mark 45 5-inch/54-caliber guns are in single turrets fore (Mount 51) and aft (Mount 52). The Mark 45 mount is modern, but the 5-inch/54 gun itself (i.e., a 5-inch-caliber gun whose barrel length is 54 times its bore) is a 1940 design. World War II ended before warships mounting it entered service. In 1947 the Navy adopted the 5-inch/54 gun for its superior antiship and shore-bombardment power, compared with a faster-firing twin 3-inch/70 mount.[10] The earlier Mark 42 5-inch/54 gun mount, which armed almost all postwar destroyers before the DD 963 class, proved unreliable off Vietnam (see chapter 2). Its complex loader was the source of the problem. The *Spruance* class's Mark 45 mount has a much simpler loader so that its reliability is better. Two Mark 45 5-inch/54 mounts weighed less than one Mark 42 and needed smaller crews.

The primary mission for the 5-inch/54 guns is fire support for ground forces. An Army maxim is that artillery is the king of battle, inflicting more enemy casualties than any other arm. During an amphibious assault, naval guns are the artillery, at least until field pieces can be securely set up ashore. Rapid fire and long range make naval gunfire highly useful for scattering and disrupting enemy defenses. For practical purposes, since 1993 the Navy's entire gunfire support capability has been concentrated in the *Spruance* class. The muzzle velocity of the 5-inch/54 gun is over 2,600 feet per second, three times that of field howitzers, and gives flat, penetrating trajectories. Time of shell flight to 10,000 yards is 15.8 seconds. Maximum range is 25,900 yards. Sustained firing rate is 18–23 rounds per gun per minute, 10 times that of field artillery. With two gun mounts, the volume of fire is further increased. The guns can engage the same or separate targets simultaneously, or one mount might fire illumination rounds during night gunfire support missions while the other fires explosive rounds.

Destroyers can watch for targets and attack them before ground forces land, both to destroy defenses and to mislead the enemy about where the actual landings might occur. They can attack targets that are dangerous for other units to approach, such as coastal fortifications and antiaircraft batteries. Outside the constraints of minefields and shoal water, they can maneuver to bring enemy units under enfilading fire and to support ground force advances during and after amphibious landings.[11] In plans for amphibious operations, a destroyer often is dedicated as direct support for a specific ground unit such as a Marine battalion landing team, in which case the ground unit designates targets. (Extensive supporting-arms coordination measures have evolved to prevent friendly-fire incidents.) Naval gunfire is useful for neutralization and interdiction missions, preventing enemy units from moving or operating in the landing areas. As will be seen in chapter 14, destroyers have saved tactical situations ashore.

The destroyers carry 600 5-inch rounds per gun. Each round consists of a projectile (the shell), the fuze, and the propellant cartridge canister. Projectile types include high explosive, white phosphorous (incendiary), and illumination (star shells). A high-explosive projectile weighs 57.3 pounds, of which 7.9 pounds is explosive, and can pierce 1.5-inch armor at 10,000 yards.[12] Fuze types include impact (point detonation), armor-

USS *John Young* (DD 973) fires her 5-inch guns. Her SPG-60 fire control radar, above the spherical SPQ-9 radar cover dome, is aimed toward the target. The SPG-60 has a TV camera mounted on it to check range safety and to assess damage. Another TV camera is on the mainmast. (U.S. Navy photo)

Table 9.3 Naval gunfire ammunition and missions

Shell and Fuze	Effect	Targets and Missions
"VT-frag": high explosive + proximity (VT, CVT)	Blast and fragmentation	Boats, aircraft and missiles in flight, personnel, vehicles, parked aircraft, missile launchers
High explosive + point detonation	Blast and fragmentation	Small ships, personnel, lightly armored vehicles, bunkers
High explosive + delay	Armor piercing, penetration, blast, cratering	Large ships, armored vehicles, bunkers, roads, runways, ranging fire
White phosphorus + proximity	Spreads incendiary phosphorus over 80-yard radius; smoke	Light structures, fuel dumps, target marking, screening
Illumination + timed	Parachute flare; backup incendiary	Surveillance, spotting, harassment

piercing (delayed or base detonation), mechanical-timed, variable time (VT, or proximity), controlled variable time (proximity-fuzed after allowing a time delay to clear friendly formations on the gun-to-target line), and others. Table 9.3 summarizes naval gunfire ammunition choices. Propellant canister types are full-charge for high-velocity low-trajectory shots and reduced-charge for arched trajectories such as for illumination rounds.

The Mark 45 gun mount is unmanned above deck. Two gunner's mates control loading the gun from consoles beneath the mount. One gunner can activate the mount to fire 20 rounds of ready-use ammunition stored in the loader drum. Four sailors in the magazine send shells vertically up a hoist tube into the loader drum. A cradle swings shells from the drum into line with the barrel and rams them into the breech. The breech block slides closed and the gun fires. The recoil energy pulls the breech open, extracts the spent cartridge, and ejects it through a port over the barrel onto the deck (i.e., it is a semiautomatic gun). Elevation range is −5 to +65 degrees. The barrel is stabilized to stay at constant elevation and train angle despite the ship's rolling. An elevator serves each mount to strike down pallets of 5-inch rounds received under way from ammunition ships and replenishment oilers (AOE and AOR types). One reason for the wide weather deck passages on the 01 level (the fo'c's'le deck) of the DD 963 is to shuttle pallets from the transfer stations to the elevators.

New Mark 45 Mod 1 mounts are replacing the original Mark 45 Mod 0 mounts on all the ships. With Mod 1 the gunner's mates can select specific rounds from the loader drum for firing. A display in the magazine tells the ammunition handlers which types of ammunition to send up the hoist into the loader drum. One Mod 1 mount can fire both illumination and explosive rounds continuously, an improvement over Mod 0. The Mark 45 Mod 1 gun mount can load and fire longer projectiles that might use booster rockets and guidance devices. The Deadeye project developed a rocket-boosted laser-guided 5-inch round for longer-range gunfire support. It was intended to home on targets that ships, helicopters, or ground forces could illuminate with laser beams. Projectile types included high explosive and a new shaped-charge antitank round. Deadeye was canceled in 1989.[13]

Guns are useful in short-of-war confrontations where a suspicious ship or aircraft may not show hostile intent until very close range. Secondary missions for the 5-inch/54 guns are close-range antiaircraft fire and antiship attack. Normally a significant target would be attacked at long range by guided missiles. Projectiles and fuzes are the same as listed above. "VT-frag" is the normal antiaircraft round: a proximity fuze on a high-explosive shell. The maximum barrel elevation of 65 degrees and the 20-rounds-per-minute rate of fire are sufficient for gunfire support but limit the guns' antiaircraft potency. Plans for a passive homing 5-inch projectile to attack incoming antiship missiles never materialized.[14] Two other gun types, 25 mm and .50-caliber machine guns, are carried for short-range counterterrorist action against boats and aircraft. Destroyers operating in the Persian Gulf and on antismuggling patrols mount one or two Mark 88 Bushmaster 25 mm chain guns, heavy machine guns for close-range shooting. These guns are mounted on the 01 level by the midships quarterdeck. The gunner aims them with an on-mount laser gunsight.

The destroyers store .50-caliber machine guns in their armories and can mount them on pintles by the bridge, aft near the Sea Sparrow launcher, or at other locations. They carry infantry small arms for landing parties.

The Mark 86 gunfire control system tracks targets and aims the 5-inch guns. It has a fire control computer, two gun control consoles in the CIC, and six sensors. These sensors are a "spook nine" (SPQ-9) high-precision surface search radar, an SPG-60 air tracking radar, two TV cameras, and port and starboard manual target designation trackers. With the Mark 86 system the ship can shoot accurately while maneuvering. It was the Navy's first all-digital gunfire control system and the first new gunfire control system developed since World War II. The Mark 86 system provides a 360-degree radar view around the ship and automatically calculates which radar contacts are approaching within range. This lets the ship engage targets by priority without having to disrupt missions such as ASW or naval gunfire support at the sudden appearance of an air or a surface threat. When a target is designated, the Mark 86 system searches for the target within a large geometric range-and-bearing "gate" or window, then quickly shrinks the gate around the target. The computer aims the guns to shoot into the small gate. Mark 86 has a counterbattery mode. This is basically an ability to interrupt a gunfire support assignment and to switch the guns to an urgent new target. Counter-battery refers to the urgency of need to silence enemy artillery. The system still tracks the original targets and can resume the original fire support mission immediately after suppressing the hostile battery. The Army and Marines have automatic artillery-locating radars that track projectile trajectories. While the Mark 86 system is not instrumented for this, a destroyer can bombard locations pointed out by ground forces.

The SPQ-9 radar is housed in a large, distinctive spherical radome on the foremast. It uses the radar X-band and a very fast spin (60 rpm) for high precision and accurate shooting. It provides surface search and low-altitude air search out to 20 miles. The radar can track aircraft or missiles flying at Mach 3 up to 2,000 feet. The SPQ-9's computer tracks up to 120 targets and can detect new targets automatically. The radar has pulse-to-pulse frequency agility to overcome enemy jamming, to reject clutter, and to allow several ships all to operate their radars without interference. Radar video and computer-tracked targets are displayed on standard CDS consoles. Target tracks are exchanged with NTDS.

For antiaircraft fire the Mark 86's computer-

Spruance-class destroyers mount 25 mm Bushmaster guns amidships for close-range counterterrorist action. This is the starboard mount, pointed aft.

driven SPG-60 radar scans the horizon sector for a target track indicated by CDS (usually from a search radar contact), finds the target aircraft's altitude and speed, and automatically locks on. SPG-60 is an X-band frequency-agile pulse-Doppler radar with a range of 65 nautical miles. It transmits four pulses simultaneously (monopulse). This has electronic counter-countermeasures value, since a hostile radar jammer must fool all four pulses to present a false image of target motion. The SPG-60 radar can keep locked on despite extreme maneuvers by a target aircraft to break out of the tracking beam.

TV cameras (remote optic sights) and target designation transmitters give some ability to shoot in good visibility while radar-silent. One TV camera is bore-sighted through an aperture in the SPG-60 antenna, and another is mounted on the mainmast above the SPS-40 air-search radar. These TV cameras have telephoto zoom lenses giving a field of view of 2.1 to 21.0 degrees. The cameras spot shell bursts for damage assessment and provide a visual safety check that unintended targets are not in the line of fire. The use of two cameras supports counterbattery fire: one continues to track the original target while the second checks the new target. Sailors manning a set of target designation transmitters near the bridge and atop the hangar can feed visual target bearing and elevation directly into the Mark 86 computer.

The Mark 86 system was designed to incorporate laser range finders and infrared or low-light-level TV gunsights for night shooting without radar. The Harpoon antiship missile, described in chapter 10, filled a similar requirement, and the Mark 86 enhancements never entered service. The Navy's new focus on littoral operations may raise the value of passive gunfire control, since it is suitable for attacking marauding patrol boats at night without revealing the destroyer to enemy radar detectors. European warships are inherently designed for littoral operations and often mount optical antisurface gunfire control systems. Another potential upgrade for the DD 963 class is to add an X-band continuous-wave emitter to the SPG-60 radar to illuminate a target for an antiaircraft missile. The *Kidd* (DDG 993)–class destroyers have this, giving them an additional missile fire control channel.

Support for the Mark 86 systems aboard early ships was, in a way, too good. Contractor personnel made warranty repairs when the destroyers returned to port from exercises. The supply system had no record of parts used so did not increase the Navy's stockpile of spares. The ships' crews did not get experience in troubleshooting. At sea, fire control technicians replaced electronic modules almost randomly with new parts in attempts to fix apparently defective equipment. Inspectors ashore found that a third of all modules received from the fleet for repair were in full operating order. Still, this rapidly depleted the scarce supply of spare modules in the fleet and left many Mark 86 systems unused pending arrival of additional spares. A more spectacular problem appeared during an early firing exercise. A DD 963 shooting practice rounds at a target suddenly shifted aim to the tug towing it. Shell splashes bracketed the alarmed tug. Back in port, experts adjusted the Mark 86 system. Accuracy improved: on the next firing exercise the incident repeated, and before firing could be stopped, shells were sent straight over the tug and barely missed her mast. (A destroyer squadron staff officer told me that a practice round actually hit the tug.) Target towlines had to be very long until this software problem was fixed around 1981.

The primary limitation of the 5-inch/54 gun is its range. As designed, the DD 963 forward Mark 45 gun (Mount 51) could be replaced by a 175 mm or 8-inch gun. The hull under Mount 52 aft is too shallow for a larger mounting. A larger-caliber gun could shoot gunfire support missions farther inland. Accuracy in long-range shooting required laser-guided projectiles, which were in development in the 1970s but were not a high priority for

the Navy. Test firing of conventional unguided rounds from the prototype Mark 71 8-inch/55 gun showed predictable shot dispersion that critics called inaccurate. In 1978 the Carter administration canceled the 8-inch gun before the detail design for the DD 963 mounting was completed. The installation design for the vertical-launch system (VLS) aboard *Spruance*-class destroyers did not require retention of the growth margin (weight and space below decks) for a larger gun, but perhaps a new version of the 8-inch gun might still be mounted. Interest continues in medium-caliber artillery, such as a vertically loaded 155 mm gun, which would fit within the space and weight margins. Cost-effectiveness probably depends on whether it would fire standard Army 155 mm shells. Another possibility is a low-cost version of the Tomahawk conventional land-attack cruise missile, using a cheaper engine and carrying fuel only for a 100–200-mile flight. Still another is to use the Army's multiple-launch rocket system. Both these missiles can dispense mines and other special submunitions. A point of caution is that some Army munitions are susceptible to a problem known as HERO, described later in this chapter. However, HERO should be a manageable problem.

Antiaircraft and Antimissile Weapons

The *Spruance*-class destroyers were the first U.S. Navy ships designed to mount point-defense weapons for antiship missile defense. Point-defense weapons defend against aircraft or missiles closing on the ship herself, as distinct from area-defense weapons, which have a longer range to protect ships in company and areas over land. Specific DD 963 point defenses include the NATO Sea Sparrow missile system and two Phalanx 20 mm Gatling guns. A *Spruance*-class destroyer might engage up to four air targets simultaneously, two with Phalanx and one each with Sea Sparrow and the 5-inch gun battery. The ship's masts, deckhouses, and other structures reduce engagements on a single bearing to two or three. Other antiship missile defenses include chaff, flares, floating decoys, radar jamming, and armor. All defenses must be continuously ready for action on a few seconds' notice. Equipment reliability and automation are vital. So are ease of maintenance and operation by young sailors.

The SWY-1 NATO Sea Sparrow missile system consists of Sea Sparrow radar-intercept missiles

119

*Gunnery,
Electronics,
and
Antimissile
Weapons*

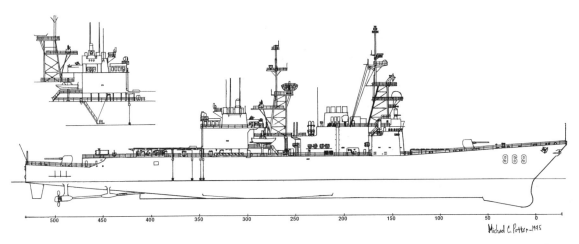

This drawing of USS *Peterson* (DD 969) depicts the fully modernized *Spruance* class. Notice the split SLQ-32(V)3 system, with the jammer section moved aft of the mainmast. A new structure supports the aft Phalanx gun.

(RIM-7M); one Mark 29 eight-cell box launcher, deck-mounted aft of the flight deck; one Mark 95 binocular-type director with a low-light-level TV camera; one Mark 91 missile fire control system; and one Mark 23 target acquisition system (TAS) missile-detection radar. The rotating box launcher aims the missiles into the missile director radar beam and permits all-weather launch. It elevates to +85 degrees and can fire a missile every two seconds. In action the missiles shoot through frangible panels on the launcher. A magazine beneath the flight deck holds 24 more rounds. Reloading is done by hand, a limitation.

The Mark 23 TAS missile-detection radar is dedicated to NATO Sea Sparrow. Its mission is to automatically detect and engage incoming missiles that are 30–60 seconds from impact. Surprise attack, submarine-launched missiles, steeply diving missiles, and saturation attacks all risk creating this situation. The rapid-scanning antenna, mounted on the mainmast, is stabilized to remain level despite the ship's rolling. It continuously surveys the area around the ship, including directly overhead. The TAS radar can track 54 targets and automatically cues the Sea Sparrow fire control system to attack incoming missiles in priority order. CDS can assign tracks to TAS to set up Sea Sparrow engagements. TAS shows which incoming missiles have been destroyed or have succumbed to decoys, and which remain threats.

A console in the CIC controls the missile system.[15] When a CDS target track is assigned to Sea Sparrow for engagement, the director scans the horizon sector for the target. The director has two 1-meter-diameter radar antennas in a binocular-type pedestal, mounted on the hangar above the main-propulsion gas turbine intake. One antenna transmits a continuous-wave radar beam, and the other is the receiver. The director is unmanned. Upon detecting the target it modulates the radar beam to determine range and bearing shift. A low-light-level TV camera on the Mark 95 director gives the console operator an optical image of the target. The launcher slews toward the target. When the target comes into range, the radar alerts the operator, who fires the Sea Sparrow missile. The missile will fire automatically if the ship has set this up.

Sea Sparrow is a Mach 1.3 missile carrying a 40 kg blast-fragmentation warhead and having greater range than the 5-inch guns. The missile receives the tracking beam from the director (sensors are on the tail fins) and the echo from the illuminated target. It distinguishes the incoming target from sea returns by the Doppler shift of the radar echo. The fire control system keeps the target illuminated and directs the missile to navigate toward the bearing where the target is moving, within the constraint that the missile seeker must keep receiving the echo from the illuminated target. The current Sea Sparrow missile (RIM-7M) has a monopulse seeker to overcome jamming, similar to the SPG-60 described above. The missile follows the echo to intercept the target and detonates.

The SWY-1 Sea Sparrow system will be added to the Rapid Antiship Missile Integrated Defense System (RAIDS) Mark 2 configuration, described later in this chapter.

Spruance-class destroyers have often practiced antimissile defense using the Mark 86 gunfire control system and Mark 91 missile fire control system to engage target drones with both 5-inch gunfire and Sea Sparrow. These systems' principal limitation is ineffectiveness against fast sea-skimming ASCMs and against simultaneous attack by multiple ASCMs.[16] NATO Sea Sparrow has an antisurface capability. In narrow waters the high-speed Sea Sparrow could disable small fast attack craft before they launched their antiship missiles.

The Mark 15 Phalanx close-in weapons system (CIWS, pronounced "sea whiz") is the last-ditch defense against antiship cruise missiles. It uses a Vulcan 20 mm six-barrel Gatling machine gun that fires depleted uranium or tungsten projectiles at 50 or 75 rounds per second to destroy or detonate incoming missiles. A high-precision Ku-band radar on the gun mount tracks an incoming target while a separate on-mount Ku-band search radar (Mark 90) automatically scans for other missiles. The gun fires automatically on computer command. The high-precision tracking radar tracks projectiles as they leave the muzzles and corrects the aim so that the stream of bullets hits the missile at the greatest range from the ship. Crews claim that the third round will hit the target. Because the engagement is so close, fragments of destroyed missiles may still spray the ship. This is one reason that the aluminum deckhouses are plated with ceramic armor, and a reason for interest in longer-range weapons such as RAM, described below. CIWS had an optically directed antisurface option, but this was deleted.

In May 1987 an Iraqi F-1 Mirage launched two Exocet missiles at the frigate USS *Stark* (FFG 31), on patrol in the Persian Gulf. *Stark*'s CIC crew detected the aircraft and noted that it was an Exocet shooter. Iraqi pilots were notorious for shooting at almost anything, but *Stark* did not illuminate the aircraft with a precision fire control radar to detect missile launch. The best radar on this ugly and poorly arranged frigate and her Phalanx gun were both masked by the mainmast. The ship did not turn to unmask these systems for a possible engagement and did not turn her Phalanx gun on. Both Exocets hit *Stark*, killing 37 men and disabling the ship. The Navy decorated the survivors for their damage control success, then punished several of them for mishandling the ship. After relieving her commanding officer, Navy authorities left *Stark* and the defenseless destroyer tender USS *Cape Cod* (AD 43), with a crew of over 600 men and women, exposed to potential Iranian terrorist attack at a Persian Gulf dockyard for weeks. *Cape Cod* could have made the damaged frigate seaworthy at any anchorage away from the danger area, or even alongside while out at sea, but area commanders did not know the tender's capabilities and imagined that dockyard service was necessary.

Search Radars

In general the trade-off in radar design is precision versus long range. As table 9.2 shows, radar efficiency depends on signal frequency (wavelengths per second). Long-range air search radar uses relatively long-wavelength signals that do not dissipate in the atmosphere or in weather clutter. Fire control radars use relatively higher-frequency shorter-wavelength signals for precision. Surface search radar range is inherently short because of the curvature of the earth.

The selection and location of shipboard radio and radar antennas must be such that signals from different radio-frequency systems do not interfere with each other. According to an engineering study, electromagnetic interference (EMI) was a factor in the Shrike attack on the cruiser *Worden* in 1972. During the Falkland Islands War in 1982, the British destroyer HMS *Sheffield* switched off her missile-warning receivers to avoid EMI from her superhigh-frequency- (SHF) band satellite communications transmitters. As shown in table 9.1, SHF radio frequencies overlap with the radar frequencies commonly used for missile guidance, including Exocets. Argentine strike aircraft were approaching and hit *Sheffield* fatally with an Exocet before she could fire chaff decoys. The electromagnetic energy field around a radio-frequency antenna is intense enough that it can disarm or trigger electroexplosive detonators. This is the HERO (hazards of electromagnetic radiation to ordnance) problem. HERO triggered a rocket launch aboard the aircraft carrier *Forrestal* in 1967, starting a flight deck fire and ordnance explosions that damaged or destroyed half her aircraft and killed 134 of the crew.[17] The Navy does not use some Army munitions because they are susceptible to HERO.

Stanchions and other topside metal equipment on warships have ground straps to prevent them from acting as antennas. Such equipment can become "slave" antennas, drawing off energy from the actual ("master") radio and radar antennas

This midships view of USS *Elliot* (DD 967) at Pearl Harbor in 1991 shows the primary electronics systems of a fully modernized *Spruance* class destroyer. *Elliot* was modernized in the late 1980s with a verti-cal launch system, Outboard, Mk 23 target acquisition radar, LAMPS Mk III helicopter facilities, and the SQQ-89 ASW system.

and reducing their performance. Absorbing radio-frequency electromagnetic energy creates a static electrical shock hazard to personnel. Aboard *Spruance*-class destroyers topside ladders, lifelines, stanchions, and other fittings are made of noncon-ductive fiberglass. This would also insulate elec-tronics against electromagnetic pulse (EMP) en-ergy transients from nuclear air bursts.

The *Spruance*-class search radar sets include the SPS-40B air search, SPS-55/64(V)9 surface search, and various fire control radars. The SPS-40B is the longest-range radar and detects air targets 200 miles from the ship. It is located on the mainmast. It is a two-dimensional radar, giving range and bearing to the contact but not altitude; the SPG-60 gunfire control radar, discussed previously, provides altitude. The SPS-40 radar can discern aircraft moving over land. The radar's video display is routed to the CDS consoles. Operators designate which contacts to track; with CDS the ship tracks many contacts simultaneously.

When the combat information center identifies contacts as hostile, the gunfire control (Mark 86) and Sea Sparrow (Mark 23 and Mark 95) systems start with track data in CDS to engage the target.

The SPS-40B has an identification-friend-or-foe unit to transmit IFF challenges to air contacts. IFF procedures were developed during World War II and the Korean War. (Pre-IFF disasters included the Army radar post near Pearl Harbor that in 1941 was unable to identify incoming aircraft and assumed incorrectly that they were friendly. At Guadalcanal in 1942 persistent IFF problems caused friendly-fire incidents and often delayed firing on enemy warships.) The challenge is an interrogation code sent on 1,030 MHz, and the response is another code to answer the specific challenge and sent on 1,090 MHz. Civil (mode 3) and military (mode 4, encrypted) modes are used. "Identification friend or foe" is really a misnomer, because all the system can positively identify are friendly contacts that so identify themselves. Damaged or clandestinely transiting friendly aircraft may be unable to do this. One aircraft might be interrogated by several ships simultaneously, but its IFF transponder will reply to only one, ignoring the others. Since IFF signals do not include error-correction codes, false data from a garbled signal may look valid. IFF unreliability was one reason for the development of PIRAZ during the Vietnam War, where the NTDS net tracked all friendly aircraft throughout their flights. PIRAZ still relies on neutral aircraft to identify themselves as such when they take off.[18]

The SPS-55 surface search radar set detects surface contacts and low-flying air contacts out to about 50 miles. It is located high on the foremast. As with the SPS-40B, the CDS consoles display this radar's video so that operators can enter contact data. CDS calculates closest point of approach, interception courses, and other navigation data.

The SPS-64(V)9 surface search radar is a high-precision short-range set for collision avoidance and for close-in navigation among obstacles and small boats in a strait, harbor, or anchorage. SPS-64 typically displays the area within about 5,000 yards of the ship. Its console tracks up to 20 contacts, automatically updates their positions on each sweep, and alerts the watch of collision danger. The Naval Sea Systems Command hopes that SPS-64 will replace the commercial Pathfinder and Furuno radars that *Spruance*-class destroyers sometimes take aboard for operational deception.[19] A destroyer navigating with a fishing-boat radar does not divulge her identity as a warship to hostile radar detection systems. This would be useful, for example, in passing through a watched strait at night. SPS-64 connects to an SLA-10 signal blanker to protect missile-warning receivers, since these operate in the same frequency range. Commercial sets are incompatible with signal blankers.[20]

Several fire-control systems have search capabilities. The SPQ-9 gunfire control radar provides surface search coverage similar to the SPS-55 within 20 nautical miles. CDS consoles display the SPQ-9 radar's video and track its contacts. The Mark 23 TAS radar for NATO Sea Sparrow is a very capable search radar. However, it does not feed tracks into CDS, although it can receive tracks from CDS. The Mark 95 illumination radar for NATO Sea Sparrow can scan the horizon continuously but again does not feed tracks into CDS. It carries a low-light-level TV with a tele-photo lens to display an optical view of contacts even at night. The Mark 90 radar on the CIWS mount scans for targets automatically and alerts the CIC to contacts.

Another surface-surveillance system is the mast-mounted sight. Originally developed as an above-rotor sensor for Army night-attack helicopters, it is a spherical turret with infrared and low-light-level TV cameras and a laser range finder. It provides a daylight-quality video display at night and in bad weather. *Elliot* and several other destroyers have a short trestle for it forward of the mainmast. A destroyer so equipped embarks a mast-mounted sight before entering the Persian Gulf. During Operation Desert Storm mast-mounted sights on ships and aircraft were used successfully for targeting, gunfire spotting, combat search and rescue, mine detection, and other missions.

Communications

Spruance-class destroyers sprout a forest of radio antennas.[21] The communications bands shown in table 9.1 have different effective ranges because of atmospheric effects. As with commercial telephone networks, shipboard radios transmit both voice and low-speed digital data communications. NTDS data links use high- and ultrahigh-frequency circuits, giving a range of about 300 nautical miles with HF and longer ranges over satellite links. DD 963 HF and UHF transmitters have antijam features. Special cryptographic devices encrypt and decrypt both voice and data transmissions.

High-frequency radio achieves long range through the reflection of broadcast waves off the ionosphere. HF transmitters tend to be very large and can broadcast for hundreds or thousands of miles. The DD 963 HF transmission antennas are a set of wire-fan antennas stretching from the forward stack to the foremast upper yardarm; two pairs of thick vertical antennas just aft of the Harpoon launchers; a pair of tall antennas on the aft stack; and a tall whip antenna on the helicopter hangar roof. HF-receive antennas include several whip antennas and a pair of tilting antennas on the fantail.

UHF is for aircraft communications, for line-of-sight transmission to other ships, and for satellite communications. Surface search radar line of sight goes about 33 percent beyond the visual horizon. The DD 963–class primary UHF transceiver is the WSC-3 ("whiskey 3") system. UHF antennas include the tall vertical cylindrical stovepipes on the yardarm wings, a small whip antenna on the foremast, and two OE-82 directional satellite communications (SatCom) antennas on the signal bridge and aft stack. UHF antennas are broadband, whereby four messages transmit simultaneously through one antenna (multiplexing). The ship aims the OE-82 directional antennas at satellites or other ships for point-to-point transmission, greatly reducing any opportunity for an enemy to detect the signal.

A recent DD 963 addition is the USC-38 extremely-high-frequency- (EHF) band SatCom transceiver for the new EHF MILSTAR satellite constellation. The USC-38 antenna is directional and is housed in a dome by the forward stack. At present most Navy SatCom voice and data transmission circuits use the UHF band. UHF satellites servicing warships at sea include Fleet Satellite Communications (FLTSATCOM) and leased commercial "gap-filler" satellite channels.

To use the analogy of commercial TV, frequency ranges in the HF and SatCom bands are the broadcast channels. Broadcast programs include CUDIXS, TADIXS, TACINTEL, OTCIXS, and Link-11 (IXS: information exchange system). CUDIXS (fleet broadcast) broadcasts general service messages from all sources to ships. A ship originating a message radios it via satellite to a shore station that broadcasts it over CUDIXS. TADIXS and TACINTEL broadcast tactical intelligence data. In particular, TADIXS supports over-the-horizon targeting for cruise missile strikes. OTCIXS and Link-11 are two-way. OTCIXS (which feeds JOTS-II [NCCS-A]) is the battle group command-and-control network. OTCIXS features a secure-voice SatCom circuit plus multiple text and intelligence data radio circuits.[22] (The British commander of the Falklands campaign chose the uncomfortable old aircraft carrier HMS *Hermes* as his flagship because he could access OTCIXS only from her.) The DD 963 automated radio system (NAVMACS) listens to all broadcast messages for codes that the ship is guarding, then automatically records those messages. It records all messages with Flash or Emergency priorities. Link-11 uses ship-to-ship HF and ship-to-ship UHF or can link through a dedicated SatCom channel.

Very-high-frequency (VHF) radios provide short-range circuits, including bridge-to-bridge voice, some aircraft control circuits, and communication with ground forces such as for shore-bombardment liaison. DD 963 VHF transceivers share the UHF stovepipe antennas and UHF whip antennas. Sonobuoy channels use VHF.

Other types of communications are visual signals and navigation signals. Visual signal methods include flag hoists, flashing lights, and flashing infrared lamps ("Nancy"). These are all slow and short-range, but an enemy cannot intercept them. DD 963s at first had an SRN-18 satellite navigation receiver on the mainmast for the primitive Transit satellites; it has been deleted. The ships now carry SRN-12 receivers for LORAN/Omega and WRN-6 receivers for the precise, jam-resistant Global Positioning System (GPS) satellites, which are replacing tactical air navigation (TACAN) beacons. The ships might receive International Maritime Satellite (INMARSAT) commercial SatCom terminals during future modernization.

Electronic Countermeasures

Electronic warfare includes systems and tactics to detect and exploit enemy use of the radio-frequency spectrum and to protect friendly use. The principal branches of electronic warfare are intelligence collection, surveillance, and countermeasures. The *Spruance* class has extensive capabilities in all three areas. This chapter focuses on tactical surveillance and countermeasures. Chapter 10 describes *Spruance*-class communications intelligence systems. Many *Spruance*-class destroyer operations have been intelligence collection missions.

Tactical electronic passive surveillance sensors and their use comprise electronic support measures (ESM, an abbreviation construed as a singular noun). ESM supports defensive threat warning, long-range situation awareness, and over-the-horizon targeting (OTH-T) for cruise missiles and tactical air strikes. Microwave ESM looks at the external characteristics of foreign radar signals, such as radio frequency and pulse-repetition frequency.[23] Electronic warfare technicians designate radar frequency bands by the letters *A–K,* as shown in table 9.2. The letter *C* occurs in both the ESM band and the radar band, which even experts find awkward. Intercepts are an extremely important source for identifying hostile activity. They are likely to be the first and often the only means to classify an enemy unit. ESM is the only method to detect a submarine beyond sonar range. Surface search radar such as SPS-55 or SPQ-9 has a surface detection range of 20–30 nautical miles. Microwave ESM can detect foreign surface radar signals at 45–60 nautical miles and some air search radar signals at 80 nautical miles.[24] Research shows that very-high-sensitivity systems can intercept radar signals at 200 nautical miles by exploiting tropospheric scattering.[25]

Surface search radar signals will propagate to greater range in conditions of atmospheric ducting. This is a weather-related phenomenon, like a mirage, that channels the radar signal energy in a surface duct for long distances. Ducting works both ways, so that the radar system will detect surface contacts at great distance, but equally sensitive microwave ESM systems will detect the signal at even greater range. This contributed to the *Vincennes* incident in 1988, as chapter 14 describes. Ducting can be worse than useless with advanced radars that track contacts automatically, because the volume of mostly useless contacts (clutter) gets so high that it can saturate the supporting computer. This was a problem for Aegis cruisers in the duct-prone Persian Gulf during Operation Desert Storm.[26]

The first 10 *Spruance*-class destroyers received WLR-1G ("whirly 1") signal intercept systems during their postshakedown overhauls. Three rotating-dish antennas were mounted on the foremast in characteristic radomes. The SPS-55 surface search radar antenna was relocated higher on the foremast for installation of the central WLR-1G antenna. The other antennas were on the adjacent yardarm wings. The WLR-1G operator could scan a wide range of frequencies for the presence of a signal and to find the bearing of the transmitter. WLR-1G scanned both communication and radar bands but did not record radio transmissions for content analysis.[27] In precision and sensitivity to weak signals, the WLR-1G was better than the early SLQ-32(V)2 systems.[28] It was more of an intelligence system than a tactical warning system.[29] *David R. Ray* (DD 971), *O'Brien* (DD 975), and possibly other *Spruances* carried both systems during the mid-1980s.

The SLQ-32(V)2 ("slick 32") microwave ESM system is installed on all DD 963s, replacing WLR-1G on the ships that had it. It is both an electronic intelligence intercept system and a radar warning receiver against antiship cruise missiles. It can detect signals from different radar transmitters simultaneously on all bearings, and it immediately warns of missile or air attack threats. The SLQ-32(V)2 antennas are roll-stabilized boxes mounted aft of the forward stack. The SPS-64 and CIWS Mark 90 radars have signal-blanking inter-

faces to protect the SLQ-32 missile warning receivers. The SH-60B Seahawk helicopter carries an airborne version of SLQ-32(V)2, the importance of which will be described shortly. SLQ-32 has a computer-based library of emitters to identify specific radars automatically.[30] SLQ-32 and its adjunct, the ULQ-16 radar signal analyzer, do not interface directly to the Command and Decision System. Instead the SLQ-32 electronic warfare technicians evaluate detected signals ("rackets") and report them to operations specialists to enter as ESM lines of bearing. The operations specialists correlate ESM data with the radar picture to identify contacts. Connecting ESM and CDS is part of the RAIDS modernization program described below.

Electronic countermeasures (ECM, another abbreviation construed as a singular noun) systems degrade enemy use of the radio-frequency spectrum. *Spruance* ECM includes radar jamming and diverse decoys. Four six-barrel Mark 36 super-rapid-blooming off-board chaff (SRBOC) mortars fire chaff and flares to create false targets. The DD 963 contract design included cutouts in the corners of the transom for large chaff rocket launchers (CHAFROC). USS *Spruance* as built had the starboard cutout, but it was plated in after the Navy canceled CHAFROC in 1974.

Decoy tactics are basically two: distraction and seduction. Decoys create highly reflective false targets to distract enemy radar operators. *Spruance*-class destroyers can launch SLQ-49 "rubber duck" inflatable floating radar decoys for distraction. Chaff clouds can create distraction decoys. In seduction chaff tactics, a chaff cloud lures a high-altitude antiship cruise missile to dive early or a low-altitude ASCM to lock on to the decoy. Chaff plumes linger for some minutes after firing. Radar-guided ASCMs home on the centroid of the cloud. Seduction chaff rounds must bloom extremely quickly. The SLQ-32 system has a chaff pattern library, presumably tailored to deceive the specific detected ASCM.[31]

Many Russian-built ASCMs use infrared homing seekers. The Mark 36 SRBOC mortar fires both chaff decoy rockets and infrared (IR) flares for seduction and distraction. Distraction rounds use a longer-range rocket.[32] Torch is a rocket-launched floating IR decoy. Ultimately a hard kill of an ASCM is preferable to ECM. During the Falkland Islands War an Argentine Exocet ASCM flew into a successful seduction bloom fired by a frigate, continued out of it, locked on to the valuable

125

*Gunnery,
Electronics,
and
Antimissile
Weapons*

USS *Briscoe* (DD 977) mounts SLQ-32(V)3 radar jammers in roll-stabilized boxes above her hangar. Part of the SLQ-32 intercept receiver remains in its original location forward of the Harpoon canisters. A combined unit would overtax chilled-water pumping capacity. The circular aperture in the transom stern is for the SQR-19 tactical towed-array sonar, and the vertical slots are for SLQ-25 Nixie antitorpedo decoys. (Jürg Kürsener)

transport *Atlantic Conveyor* 4 nautical miles farther down range, and destroyed her.[33] With hard-won experience in antiship missile defense, the Royal Navy rearmed its ships with many *Spruance* weapons, including SRBOC, Link-11, Harpoon, and Phalanx.

Recent electronic warfare additions to the DD 963 class include the SLQ-32(V)3 upgrade for antimissile jamming. These jamming transmitters use the same antenna as SLQ-32(V)2. To fit within the capacity of existing chilled-water coolant pumps servicing the superstructure, the SLQ-32(V)3 jammer units are relocated aft of the main-mast from the original SLQ-32(V)2 position by the forward stack. The aft Phalanx gun is remounted on a new platform above the port-side SLQ-32(V)3 unit. Part of the SLQ-32 ESM receiver section remains in the original location forward; it detects hostile long-range radars and gets an ESM line of bearing to them for a possible Harpoon or Tomahawk attack.

The SLQ-32 receiver detects an ASCM or aircraft attack radar. Such radars mechanically scan side to side, looking for targets. The radar antenna focuses electromagnetic energy pulses into a beam (the main lobe or axis) but produces small off-axis beams (side lobes). Any echo received is assumed to be in the main-lobe beam. SLQ-32 and ULQ-16 indicate when the hostile attack radar is scanning away. When the SLQ-32(V)3 jammer computes that a target radar's side lobe is showing, it beams out a strong spurious echo signal. The aircraft radar displays this as a contact on the main axis. If the radar is scanning dead ahead when the jamming signal from a different bearing (such as 30 degrees to the right) hits a side lobe, the radar sees a large contact dead ahead. Jamming can produce contacts at false distances. If the jammer sends out many signals, the attacker's radar screen fills with false echoes and strobes. Used along with chaff decoys, jamming greatly reduces the likelihood that the attacking ASCM or aircraft will find and hit the ship.

Electronic counter-countermeasures degrade

The highly successful LAMPS helicopter project originated as a multipurpose attack and surveillance helicopter for the DD 963 class. The LAMPS Mk I helicopter was the Kaman SH-2F Seasprite. This one is dropping decoy flares. The SH-2F was retired in 1993. (Kaman Aerospace, in U.S. Naval Institute collection)

hostile ECM. DD 963 fire control radars employ monopulse, Doppler-shift tracking, and pulse-to-pulse frequency shifting in part for ECCM.

Decoys are related to stealth. A ship can be detected in the electromagnetic, infrared, visual, and underwater acoustic environments because she emits or reflects energy that can be detected: acoustic noise, heat, visible sunlight, and radar echoes. Her sonar, radio, and radar transmissions are detectable by hostile passive ESM. The ship's effects on the environment such as her wake and magnetic field can be detected. Stealth technologies reduce the ship's detectability. To reduce the radar cross-section of the *Spruance* class, the Navy has been applying radar-absorbing material to the superstructure and grinding down reflective structures.[34] As discussed in chapter 8, the DD 963 class is extensively equipped to reduce acoustical, IR, and magnetic signatures. The oldest method of detection is visual. *Spruances* in the Persian Gulf repainted their hull numbers in subdued gray. The upper masts were originally painted black because uptake soot would stain a gray surface. But the tall masts are visible at great

distances, and around 1990 the ships started becoming all gray to reduce their visibility.

Helicopters

The *Spruances* were the first American surface combatants designed to operate manned helicopters as armament. Admiral Weschler's DX Concept Formulation team highlighted the need for a helicopter to replace the failed DASH. His experience in Vietnam showed him that an ASW-only helicopter would be inadequate. Destroyers needed helicopters to perform visual search, intelligence collection, and over-the-horizon targeting.[35] Equipment for these missions required a much larger helicopter than DASH. The *Spruance*-class hangar and 50- by 70-foot flight deck were originally designed to support the large SH-3 Sea King, with ability to land the even larger CH-46 Sea Knight transport helicopter.

Development of a destroyer-based helicopter started as part of the DX project but then became separate. The SH-3 could carry the necessary payload but was too large to operate from most American warships' flight decks. The creative solution was to put radar, sonobuoys, a magnetic anomaly detector, and torpedoes on a small helicopter and to install the sonobuoy processing electronics on the destroyer. Dividing the equipment between ship and aircraft saved the weight of the electronics and operator to allow more fuel or a second torpedo on the small helicopter. In October 1970 the Navy chose the HH-2C Seasprite search-and-rescue helicopter to convert to the Light Airborne Multipurpose System, or LAMPS Mark I helicopter, redesignated SH-2D Seasprite. Many came from duty in Vietnam, where they had been armored and armed with machine guns to fight off enemy patrols while rescuing downed airmen. Rapid progress led to the SH-2F Seasprite with stronger engines and other improvements. The Navy converted its entire inventory of H-2-series helicopters to SH-2Fs.

The Navy experimented with firing Sparrow and Sidewinder air-to-air missiles from SH-2F Seasprite helicopters for antiship missile defense. The intended targets were fast patrol boats. The heavy Sparrow was mounted on a weapons point on the Seasprite's side, and its guidance equipment displaced the copilot. In a test launch the flame from the departing Sparrow's rocket motor seared through the windshield and nearly de-

This stern view of *Merrill* in 1990 shows the original narrow helicopter hangar, sized for one SH-3 Sea King ASW helicopter or two Seasprites. The winches on the port quarter are for the SQR-15 towed-array sonar for long-range detection.

During modernization in 1992, *Merrill*'s hangar was widened by moving its entire port bulkhead outward until it was flush with the hull side, providing room for two SH-60B Seahawk LAMPS Mk III helicopters. An extension on the hangar's starboard side stores sonobuoys and other helicopter-related equipment. Her aft Phalanx antimissile gun is mounted on a new raised platform above the portside SLQ-32.

stroyed the helicopter. The pilot, Lt. Cdr. Mike Coumatos, possessed the "right stuff" of military test pilots:

> The whole cockpit was suddenly filled with smoke and wind. I thought the helicopter had exploded and this was the end. Time got slow. After a while I thought, "It still seems to be flying."[36]

He landed and discovered that all the plastic components along the starboard side had disintegrated, including the forward windshield. The missile hit the target but for obvious reasons did not enter service as Seasprite armament.

Helicopters operate as airborne extensions of the destroyer. The ship plans the mission with the Command and Decision System. On an ASW mission the helicopter flies to a designated search area, drops sonobuoy patterns, and relays the sonobuoy signals back to the destroyer. The ship analyzes the signals and issues new orders to the helicopter, including where to drop a torpedo. Chapter 11 describes helicopter ASW operations.

Inexpensive and capable, LAMPS Mark I was an immediate success when it entered fleet service in 1972. The Navy was already planning a more advanced system for the *Spruance* class, LAMPS Mark II, which required a longer-range helicopter such as the Sea King. The Seasprite manufacturer, Kaman, hopefully modified two Seasprites as candidate LAMPS Mark II helicopters (YSH-2E) with new electronic systems.[37] The Navy tested them at sea during 1972 but decided to build the next LAMPS around a new-production helicopter. This became LAMPS Mark III.

In September 1977 the Navy chose the new Army/Sikorsky UH-60A Blackhawk assault helicopter for the LAMPS Mark III airframe. LAMPS Mark III was a joint project of the aircraft designers in the Naval Air Systems Command and the ship designers in the Naval Sea Systems Command. The SH-60B was named Seahawk, obviously no relation to the Seahawk destroyer project of the early 1960s. It retains the Army's combat damage survivability features. Compared with the Army UH-60A, the SH-60B carries weapons pylons, external fuel tanks, the ARQ-44 "Hawk Link" microwave data link transceiver, a search radar,

ESM sensors, sonobuoy racks, sonobuoy receivers, and a display console similar to those in the ship's CIC. For shipboard handling the tail, stabilator, and main rotor blades fold; the tail wheel is further forward; and RAST recovery gear (described in chapter 11) is installed. SH-60B Block I upgrades, added since 1987, include Penguin ASCMs, pintles for M-60 machine guns, chaff and flare dispensers, other countermeasures, a 99-channel sonobuoy receiver, and a GPS receiver.

The SH-60B Seahawk is a far more capable aircraft than the SH-2F. It has more powerful engines, twice the range and endurance, on-board sonobuoy processing and sensor displays, and integrated systems for surveillance and communications. The backbone of the LAMPS Mark III system is the "Hawk Link," a highly directional (pencil beam) jam-resistant two-way microwave digital data link. On the destroyer the data link antenna (SRQ-4) is in a small radome on the foremast. This platform had originally held the SPS-55 radar or a WLR-1G scanner. On the SH-60B the data link antennas (ARQ-44) are in two domes on the bottom of the fuselage. The data link transmits the NTDS symbols for ESM, radar, and sonobuoy contacts from the SH-60B directly into the destroyer's CDS. The data link replaces the LAMPS Mark I one-way sonobuoy signal relay, the ship's TACAN beacon, and UHF voice radio. The data link provides navigation data for the SH-60B's return-to-ship flight. The SH-60B can relay UHF voice radio transmissions so that distant ships can communicate without broadcasting on the detectable HF band.

The SH-60B's ALQ-142 ESM receiver is an airborne version of the destroyer's SLQ-32 radar detection system.[38] Two ALQ-142 antennas are mounted at 45-degree angles beneath the nose and two beneath the engine exhausts, giving 360 degrees of coverage. Transmission of an ESM contact (racket) preempts all other data on the link. After ESM, radar contacts have priority for link use during antiship surveillance and targeting assignments, and acoustic contacts have priority during ASW assignments. Contact data are relayed between the destroyer's CDS computer and the SH-60B's mission computer as NTDS symbols. The aircrew switches between antiship and ASW modes in flight as assigned. As tactical intelligence develops, the destroyer can transmit updates to the helicopter's ALQ-142 library of threat emitters during flight. Using the SLQ-32 and ALQ-142 simultaneously, the destroyer can obtain

a real-time passive cross-fix on an enemy radar platform and launch an immediate attack with Harpoons or Tomahawks. Since the pencil-beam data link is effectively impossible to intercept by hostile ESM, before the ASCM salvo arrives the enemy ship would detect only a helicopter perhaps 100 nautical miles away, far beyond enemy antiaircraft weapons range.

Future Modernization Programs

With the shift to a littoral-oriented strategy since 1990, naval force modernization plans emphasize shallow-water warfare. Differences from open-ocean operations include reduced sea room for maneuvering, greater radar clutter and ducting, and more restrictive rules of engagement. Threats include minefields, diesel submarines, and high-speed ASCMs fired from short range by patrol boats, small aircraft, and shore batteries. Reductions in the numbers of surface warships, and the possibility that the fleet must divide to fight multiple simultaneous regional wars, mean that each warship has greater responsibility for sea control and for self-defense than in the past.[39]

Modern antiship missiles are increasing in capability and proliferation. The failure of *Stark* to respond to an Exocet threat in 1987 highlighted antiship missile defense problems. Warships will face simultaneous attack from multiple different missiles, flying lower and faster than older models, possibly using stealth features and probably with manned attack aircraft in support. Development of the Rapid Antiship Missile Integrated Defense System began in 1988 to coordinate *Spruance*-class antiship missile defense systems for faster recognition of missile attacks and for faster deployment of countermeasures. The Command and Decision System controls active-radar sensors and hard-kill weapons but does not get direct input from ESM and tactical intelligence systems. RAIDS Mark 0 adds a local-area computer network to integrate data from CDS, from the SLQ-32 and ULQ-16 ESM sensors, and from the short-range missile-detection radars (Mark 90) on the Phalanx guns. It displays a consolidated picture of the antiship missile defense system in color graphics on new consoles on the bridge and in the combat information center. The operations specialists can create and update CDS tracks with targets identified by RAIDS. RAIDS has interfaces to the Mark 86 (5-inch guns) and Mark 23 (Sea Sparrow) fire control system sensors to make target kill assessments. RAIDS Mark 2 will add the SWY-1 Sea Sparrow system radars for threat detection. Other RAIDS software displays the status of all weapons and recommends actions, such as to switch the Phalanx guns to automatic. RAIDS recommends ship maneuvers to open weapons firing arcs, to reduce EMI blind zones, and to exploit wind conditions for chaff and IR flare decoys. RAIDS is programmable aboard the ship so that the displays support command decision-making doctrines.[40]

The importance and remoteness of LAMPS missions make it vital to arm SH-60Bs for self-defense. SH-60Bs are being equipped to launch AGM-119 Penguin short-range IR-homing ASCMs and AGM-114 Hellfire antitank/antihelicopter missiles. The Navy has begun to install Penguin magazines on *Spruance*-class destroyers. An air-to-surface missile attack, or the threat of it, is likely to induce patrol boats and surface ships to switch on active radars to detect the helicopter, thus betraying themselves to detection. Naval helicopters also should carry air-to-air missiles such as Stinger or Sidewinder to defend against attack by enemy helicopters.

Candidate missile defense enhancements for the *Spruance* class include an infrared horizon-search scanner, a small Aegis-like phased-array radar, the planned RIM-7R Sea Sparrow, the rolling-airframe missile (RAM), and a cooperative-engagement data link. Rim-7R Sea Sparrow missiles would use midcourse guidance data from the new radar, and IR and antiradar homing sensors to defend against multiple, simultaneous ASCM attacks. It will have a larger, steerable rocket motor for the speed and maneuverability to attack super-

sonic ASCMs. RIM-7Rs would be loaded four per cell in the VLS.

RAM (RIM-116A) would increase the number of defensive engagements the ship can fight simultaneously. RAM is a supersonic, highly maneuverable missile with a 20-pound blast warhead. To keep it small and cheap it has only one pair of control wings. An on-board processor trims the missile's direction and trim by adjusting the wings as the missile spins. It is a fire-and-forget weapon to defend against simultaneous attacks by ASCMs beyond CIWS range. A 21-round RAM launcher will augment the SWY-1 Sea Sparrow system, providing defense in depth. RAM uses a passive radio-frequency seeker to aim itself at the active radar seeker typical of ASCMs and can switch to IR homing when close enough to detect the glint of the incoming ASCM. Its IR homing section, taken from the Army's highly successful Stinger infantry antiaircraft missile, is precise enough to intercept even a small ASCM. Because RAM homes on an ASCM's own energy emissions, it does not use a shipboard radar guidance channel. The ship aims the launcher toward a target, fires two RAMs (assuming a shoot-shoot-look doctrine), and goes on to the next target.[41]

Finally, an intrinsic element in coastal warfare is the commando raid for reconnaissance, sabotage, rescue, and other missions. *Spruance*-class destroyers are useful platforms for raids by SEAL (sea-air-land) teams and other special-forces units. The destroyers' advantages include the large flight deck to operate long-range helicopters, room to stage commando equipment, the large array of radios, and mast-mounted sights when installed.

Strike Weapons

Destroyers as Offensive Weapons

Spruance-class destroyers carry weapons for long-range offensive action against high-value ships and targets ashore. Strike weapons include Tomahawk and Harpoon cruise missiles. The Outboard communications intelligence system supports strikes, blockades, amphibious assaults, and other offensive operations. The large growth margin and modular design of the *Spruance* class enabled these destroyers to take Tomahawk and Outboard to sea.

Many operations by *Spruance*-class destroyers have been conducted in order to bring these systems into theaters of interest. The most publicized operations have been Tomahawk strikes against Iraq in 1991 and 1993. Known Outboard missions have targeted the Soviet Union, the Nicaraguan Sandinistas, Liberian rebels, South American drug lords, Libyan forces, Iraqi blockade runners, and no doubt Iran. Strike warfare missions are national strategic tasks, not traditional destroyer missions. An important side effect of the destroyers' new strategic capability has been the institutional rise of the surface warfare officer community within the Navy.

Harpoon

The Harpoon antiship cruise missile is described here because of its relation to Tomahawk and over-the-horizon targeting. In open-ocean operations Harpoon is more of a defensive than offensive weapon, in that the destroyers would most likely fire it at targets discovered relatively close to the ship. One early reason for installing Harpoon on DD 963s was to attack Soviet Echo II–

and Juliett-class submarines that surfaced to launch and control their large, long-range antiship cruise missiles. Another reason was to immobilize large surface warships and to sink small ones.

Harpoon has a 488-pound warhead carrying the explosive yield of four 16-inch shells. The DD 963 and DDG 993 classes carry eight Harpoon missiles in two quadruple sets of canisters mounted amidships. The missiles launch across the ship so that the booster exhaust plume vents over the side. The ship evaluates contacts, Naval Tactical Data System (NTDS) tracks, and electronic support measures (ESM) data to decide whether to fire. NTDS is vital for keeping accurate tracks of contacts and friendly ships to avoid "blue-on-blue" disasters. Early Harpoon missiles had a range of about 60 miles. The destroyer programmed the missile from the Harpoon control console in the combat information center to fly out on a given bearing toward a target, and optionally gave the missile a minimum or expected range to the target. Early Harpoons made an almost straight-in approach to the target, with an optional pop-up-and-dive maneuver to dodge target defenses.

The new Harpoon targeting console (SWG-1A) and updated, longer-range missiles provide better offensive capability. New Harpoons (Harpoon Block I-C) have longer range, over 67 nautical miles, and can fly dogleg courses to conceal the firing ship's location and to arrive on target from different directions. The new console calculates missile courses, way points, and arrival times so that missiles arrive simultaneously from different bearings to saturate the target's defenses. The console uses NTDS data about other ships near the target to plot courses that improve the missile's chance of finding the intended target and of avoiding defenses. Multiple ships can coordinate launches so that their Harpoons arrive on target simultaneously.

Harpoon can be fired at an ESM contact when the firing destroyer has her radar off for stealth, or when the target is over the radar horizon. The Harpoon missile seeker will attack the first target it finds on the assigned bearing beyond the seeker turn-on range.

A recent variant of Harpoon is the standoff land-attack missile (SLAM, AGM-84E). A carrier-based bomber launches SLAM, but a destroyer-based helicopter can control it. An infrared seeker and a Global Positioning System receiver replace the Harpoon radar seeker. SLAM races to the target area and relays a video infrared image of the target to a controller (which in the first demonstration test was an SH-60B LAMPS III Seahawk helicopter). The controller guides SLAM to impact. The data link is active for a very short time and is therefore difficult to jam or counter.[1] The Aegis cruiser *Lake Champlain* launched the first surface-launched test SLAM, but there are no plans to install it as shipboard armament.

Harpoon is an offensive weapon in coastal operations. Harpoon-armed *Spruance*-class ships patrolled for Libyan warships in the Gulf of Sidra (Operation Prairie Fire, 1986), and a supporting Navy A-6 Intruder bomber sank a Libyan missile boat with Harpoon. Destroyers actively hunted Iranian warships in the Persian Gulf (Operation Praying Mantis, 1988) and sank two with Harpoons and other weapons.

Tomahawk Land-Attack Missiles

The *Spruance* and *Ticonderoga* classes are the Navy's predominant surface ship Tomahawk platforms. All *Spruances* and vertical-launch-system-equipped *Ticonderogas* will carry this family of weapons. The Tomahawk system has three components: the missiles themselves, the weapons control system, and the launcher. There are several models of each component. Table 10.1 lists the versions of Tomahawk missiles. There is no special shipboard sensor for Tomahawk. Instead, the shipboard Tomahawk weapons control system correlates real-time intelligence data and narrows the target ship's location into an area that the Tomahawk antiship missile can search. Present land-attack Tomahawks are entirely self-guided.

At first the shipboard installation was to be

A DD 963 fires a Harpoon antiship cruise missile. (U.S. Air Force photo)

open racks such as used for Harpoon. After fire nearly destroyed the cruiser *Belknap* in 1975, there came to be increasing interest in warship survivability. The armored box launcher (ABL), a deck-mounted casing for four elevating tubes for Tomahawk and Harpoon, resulted. The *Spruance*-class destroyers and *Virginia*-class cruisers had sufficient weight and moment margins to carry ABLs. The *Spruance* class was an attractive platform since 31 ships were available.[2] The prototype deck-mounted armored box launcher went aboard USS *Merrill* (DD 976), the test ship for Tomahawk, in 1979. Diversion of ABLs to recommissioned battleships left most of the destroyers available to mount higher-capacity vertical launch systems.

In 1970 Chief of Naval Operations Admiral Elmo Zumwalt started research for a cruise missile with longer range than Harpoon for shipboard launch to disperse naval strike capability beyond aircraft carriers. He revived the Regulus nuclear land-attack cruise missile concept. He called the cancellation of Regulus in 1958 "the single worst decision about weapons [that the Navy] made during my years of service."[3] In 1971 the Navy suggested that a new land-attack cruise missile could exploit advances in small jet engines and microprocessors for precision guidance. In June 1972 the Office of the Secretary of Defense (OSD) authorized designing a sea- (submarine or surface) launched cruise missile for land attack. It could carry either conventional or nuclear warheads.

Table 10.1   U.S. Navy ship-launched antiship and land-attack cruise missiles. Despite imaginative reports, no missile carries a "microwave warhead" or dispenses wires to short-circuit power transmission lines.

Type	Mission	Range (nm)	Warhead	Cruise Guidance	Terminal Guidance
Harpoon RGM-84A	Antiship	67+	488.5 lb HE SAP	Inertial	Active seeker
SLAM AGM-84E	Land attack	67+	488.5 lb HE SAP	GPS + inertial	IR image relay to controller
Tomahawk sea-launched cruise missile series:					
TLAM-N BGM-109A	Land attack	1,350	Nuclear	TERCOM + inertial	TERCOM
TASM BGM-109B	Antiship	250 + search	1,000 lb HE SAP	Inertial	Direction finding + radar
TLAM-C TERCOM BGM-109C Block II-A	Land attack	700	1,000 lb HE SAP	TERCOM + inertial	DSMAC; point detonation or programmed fuze
TLAM-C GPS BGM-1090C Block III	Synchronized land attack	1,000	Improved 700 lb HE; delayed fuze option	GPS + inertial; TERCOM backup	Same as Block II-A with improved DSMAC
TLAM-D BGM-109C Blocks II-B, III	Land attack, multiple bomblets	472 (II-B)	24 bomblet packages ([22 × 7] + [6 × 2] 3.4 lb incendiaries)	Same as corresponding TLAM-C block	Same as corresponding TLAM-C block; attacks multiple targets
TMMM BGM-109C Block IV	Land attack and antiship	1,000?	HE penetrator	GPS; retargetable via data link	IR image relay to controller

The initial installation plan for Tomahawk cruise missiles used elevating deck-mounted armored box launchers (ABLs). Two 4-cell ABLs are mounted aft of the forward 5-inch gun on *Merrill* and six other *Spruance*-class destroyers. This was *Merrill*'s appearance during the Praying Mantis raid in 1988. The ASROC box launcher between the ABLs was removed in 1993.

The nuclear version generated little enthusiasm in the Navy. Nuclear weapons required awkward security and restricted ships' operations. Officers doubted that the nuclear cruise missiles would ever be used and expected them to be sacrificed in arms talks anyway.

The nuclear-warhead sea-launched cruise missile is the Tomahawk land-attack missile–nuclear (TLAM-N). Unlike other naval nuclear weapons, Tomahawk incorporates a permissive action link (PAL) to prevent launch without political authorization. It was not assigned to SIOP (single integrated operational plan) targets; instead, it was reserved to retaliate against Soviet bases in the event that Soviet forces attacked U.S. naval forces with nuclear weapons.[4]

The ship programs the overwater flight path for the missile to a designated landfall, where the missile will orient itself by matching actual terrain contours with a stored digital map. From the landfall the missile flies a course at very low altitude around enemy defenses and terrain obstacles. It updates its position en route by terrain contour matching (TERCOM) with digital maps.

Initial contour-matching fields are larger, so that the missile is more sure to orient itself. When the on-board guidance system computes that the missile has arrived at the target, the warhead detonates.[5] Tomahawk Mission Planning Centers ashore program nuclear-warhead Tomahawk missions. Before launch, the ship can reassign the aim point under a flexible targeting option (FLEX) to attack an updated target position, such as a target that is discovered to have relocated or to have survived an earlier attack.[6]

The best-known Tomahawk version is the conventional-warhead land-attack missile (TLAM-C). Early TERCOM-guided block II-A TLAM-C missiles were used against Iraq in 1991 and 1993. Each carries a 1,000-pound high-explosive semi-armor-piercing conventional warhead. It became operational in March 1986. Its warhead was taken from the old Bullpup air-to-ground missile. The weight of this warhead reduces the fuel load for the conventional-warhead Tomahawk, so that its range is about half that of the nuclear-warhead version (700 vs. 1,350 nautical miles). Strike route planning is the same as for the nuclear version, except that the conventional version uses digital scene-matching area correlation (DSMAC) for terminal guidance. A war correspondent in a Baghdad hotel in 1991 watched a Tomahawk below the level of his room fly straight down a street, and was almost surprised that it did not stop at intersections.

On approach to the target, the missile photographs the area and matches the scene with a stored digital image for final course adjustment toward the specific target. Strike planners choose landmarks that are not themselves targets, since these would be less likely to be demolished during early strikes.[7] Tomahawk flies straight into the target (point detonation fuze), or explodes at a programmed point over it, or pops up and dives on it to evade terminal defenses. In 1993 Iraqi air defense forces, which were gaining experience on the subject, observed that Tomahawks roll over before diving on targets.[8]

Tomahawk TLAM-D carries 166 bomblets to scatter over airfields, buildings, or other soft targets during its flight. It can attack multiple targets during one mission and detonates at the last target.

Since the success of the Tomahawk strikes on Iraq, national military commanders have kept Tomahawk-equipped ships forward-deployed for contingencies. *Spruance*-class destroyers launched retaliatory Tomahawk strikes on Iraq twice in

Tomahawk conventional land-attack cruise missiles (BGM-109C TLAM-C) locate and identify their targets autonomously and need no guidance from the ship after launch. This one dives on a concrete target building, which it destroyed on impact (point detonation fuze). Notice it is inverted, the result of terminal maneuvering to evade target defenses. (U.S. Navy photo)

1993. The presence of Tomahawk-equipped *Spruance*-class destroyers lets aircraft carriers depart an area without leaving potential aggressors uncovered.

Current-production (blocks III and IV) Tomahawks have a Global Positioning System (GPS) receiver and a time-of-arrival control to synchronize strikes with each other and with air strikes. Other improvements are described below. All these missiles have land-attack capability. Their missions can be planned quickly in theater aboard aircraft carriers and command ships. Since they can be aimed quickly, they can strike time-urgent targets.[9]

Tomahawk Antiship Missiles

In November 1972 design began for the Tomahawk antiship missile. It carries the same 1,000-pound high-explosive semi-armor-piercing warhead as the conventional land-attack Tomahawk. It can penetrate a submarine's pressure hull or a large warship's hull plating.[10] Originally it was planned for a range of 140 nautical miles, which was adequate for underwater launch, since submarines find targets by sonar and are invulnerable to hostile cruise missiles during approach to a target. Surface warships needed a longer-range weapon to attack Soviet warships, which could fire their antiship missiles from 250 nautical miles. Surface search radars cannot be used for long-range antiship targeting (over-the-horizon targeting) because of their inherent horizon-limited range, about 20–50 miles.

The challenge was not to build a 250-nautical-mile-range missile but to plan a targeting system for it. Critics doubted that a task group would fire weapons over the horizon at targets that task group sensors could not locate, since the probability of missing the targets would seem to be high, making an over-the-horizon attack an act of unilateral disarmament. The same critics accepted that the large Soviet antiship cruise missiles were threats from 250 nautical miles' range.

The cruise missile project officer, Rear Admiral Walter Locke, recalled that the task force commanders at the battles of Coral Sea and Midway in 1942 had launched heavy air strikes without knowing precisely where the Japanese forces were.[11] In particular, at Midway Admiral Spruance held a single sighting report from a Midway-based aircraft, which he recognized might be inaccurate (it was off by 40 miles) and would be two hours old before his own aircraft could arrive in the reported vicinity of the enemy. When Japanese aircraft bombed Midway, Spruance figured that in two hours those aircraft would be refueling on deck after the raid. He immediately launched his carriers' strike aircraft, which searched the area of ocean where the Japanese carrier group might have moved during the time of flight. This expanding area of ocean was the "area of uncertainty." Spruance's bombers flew an expanding-box search pattern through the area of uncertainty, found three Japanese carriers with aircraft on deck, and sank them.[12]

Admiral Locke asked the Johns Hopkins University Applied Physics Laboratory and the Harpoon seeker contractor, McDonnell Douglas, to investigate whether a Tomahawk could search for a target inside an area of uncertainty. Replacing the Tomahawk antiship missile turbojet with the costlier land-attack Tomahawk turbofan engine could increase TASM range to over 300 nautical miles to allow the missile to fly search patterns. In 1975 the Applied Physics Laboratory developed search patterns so that the Tomahawk antiship missile was capable of autonomous scouting and strike mis-

sions. The next problem was how to get surveillance information to the launching ship for over-the-horizon targeting. Opponents within OSD refused to authorize funds for research into over-the-horizon targeting, but the Naval Electronics Systems Command funded an experiment called Outlaw Shark.[13]

For this experiment, a computer database was set up at the Submarine Operational Command Center in Naples, Italy, and another computer was installed aboard a submarine. The Naples Outlaw Shark system copied operational intelligence data being collected for later transmission to a Sixth Fleet aircraft carrier, condensed the data, and relayed it without delay to the submarine over a computer-to-computer encrypted radio data link. Sometimes the submarine received intelligence data only six minutes after the occurrence of the event being described.[14] The submarine's computer correlated the intelligence data with her own contact data and prepared search patterns adequate for an immediate Tomahawk antiship attack. In December 1976 the submarine used this system to generate search patterns for actual ships not held by her own sensors. Analysis showed that the search patterns would result in Tomahawk hits on those ships. The first test of a Tomahawk antiship missile was a launch at a target hulk 224 nautical miles away. The Tomahawk flew 175 nautical miles toward the target and began searching. It then flew 173 nautical miles in search patterns, found the target, and overflew it (as programmed). This was the first long-range antiship cruise missile flight not to use a data link between the missile and a controller. Ex-Soviet long-range cruise missiles can be disrupted by jamming their data links, or by attacking the ships and aircraft in the link. Armed with these results, the Navy obtained OSD authorization early in 1977 to build the Tomahawk antiship missile.[15]

Tomahawk antiship missiles became operational in 1982. In contrast to the procedure for early Tomahawk land-attack missiles, for the antiship mission the ship controls all targeting and plans the entire strike mission. Tomahawk weapons control systems are SWG-2 for armored box launchers and SWG-3 for vertical-launch systems. Each provides track-control and launch-control functions. The ship uses the launch-control group to track the status of the Tomahawk missiles on board, to program missions into the antiship missiles, and to fire them. The Tomahawk antiship missile uses the Harpoon active radar guidance

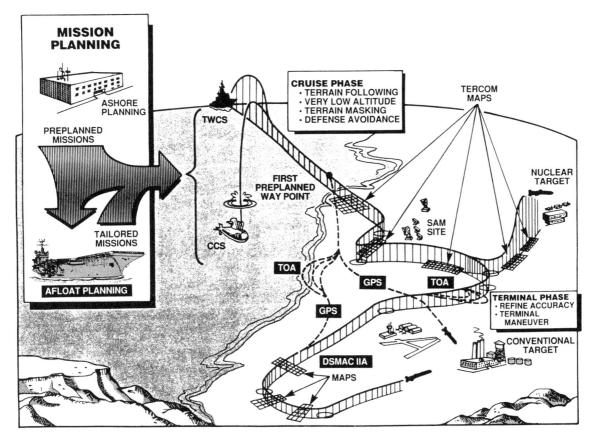

Tomahawk land-attack and antiship cruise missile flight profiles. New land-attack Tomahawks use the Global Positioning System to eliminate the need to generate digital terrain contour maps.

section with a passive seeker for identification and direction finding. The missile classifies and prioritizes targets to attack the most valuable target. As with Harpoon, Tomahawk antiship missiles can fly dogleg courses so that they attack from unexpected directions, and so that missiles from one or more ships arrive on target simultaneously.

The ship keeps a database of surface ships in her assigned patrol area on the track-control group computer. In the initial Tomahawk weapons control system, each ship kept her own track database. With newer versions, one ship is the force over-the-horizon track coordinator (FOTC) for a task group. Track data come from the ship's own sensors, NTDS data links, and intelligence sources. General surveillance information comes from surveillance satellites and aircraft, from SOSUS, and from shore-based (Bullseye) and ship-based (Outboard) radio direction finding.[16] Each ship's Tomahawk weapons control system is linked to tactical information exchange satellite radio broadcasts (TADIXS and OTCIXS) over the ship's radios.

A new system, JOTS (officially the Joint Operational Tactical System; unofficially the Vice Admiral Jerry O. Tuttle System, after its sponsor), records data from the OTCIXS and TADIXS broadcasts and maintains a local tactical picture as seen by off-ship sensors. JOTS, which uses commercial computers, is entirely passive. The ship can get into Tomahawk range without revealing her presence to the target. A Tomahawk antiship missile can be fired as a "wake-up call" or "screaming meemie" toward a radio-silent enemy battle group. If the enemy ships switch on their radars to target the actively searching Tomahawk, a follow-on strike can exploit the newly available locating information.[17]

Spruance-class destroyers and *Ticonderoga*-class cruisers can use their long-range passive towed-array sonar to locate targets over the radar horizon. Submarines have long used the same tactic. USS *Hewitt* (DD 966) used her SQR-15 towed-array sonar for over-the-horizon targeting exercises in 1986.[18]

The Vertical-Launch System

Radar-homing missiles without inertial guidance units must be fired more or less toward the target along the axis of the fire control radar beam. Mechanically complex launchers point the missiles into the beam (box launchers for small missiles such as Sea Sparrow, or large trainable rail launchers for larger missiles such as the Standard family). The ship cannot fire such missiles on bearings blocked by the ship's own superstructure (launcher cutout). Launchers for large missiles require complex loading equipment in the magazines. Launchers are vulnerable to enemy weapons, severe weather, and mechanical breakdown.

Vertical-launch systems (VLSs) had long been in use for large ballistic missiles such as Polaris and Minuteman. By the early 1970s inertial navigation instruments were small enough to fit inside small tactical missiles, making it feasible to launch such missiles from vertical cells.[19] In 1972 the Navy and General Dynamics–Pomona modified the autopilots on Standard missiles to pitch over automatically and successfully test-fired them from vertical-launch canisters. By 1977 the Navy was considering vertical-launch systems for medium-size missiles (Standard medium-range, ASROC, and Harpoon) for *Spruance*-class destroyers and *Ticonderoga*-class cruisers. In 1981 Secretary of the Navy John Lehman canceled development of a longer-range Harpoon to force the Navy toward Tomahawk and successfully pushed to deepen the vertical-launch system to fire these large missiles.[20]

This became the Mark 41 vertical-launch system. Each missile is delivered, loaded, stored, and fired in a steel canister. All canisters are identical externally. The vertical-launch system holds the canisters in cells in groups of eight, with one group of five canisters plus a three-cell-wide crane. Two Mark 41 61-cell systems were installed on new-construction Aegis cruisers starting with USS *Bunker Hill* (CG 52). One Mark 41 61-cell system was installed on *Spruance*-class destroyers during modernization; USS *Spruance* was the first ship to mount it. It replaced the ASROC box launcher and its 16-round magazine. Advantages of the vertical-launch system include the following:

- Larger magazine capacity
- Much faster firing rate (a 1-second interval between shots, as opposed to about 30 seconds to reload a rail launcher such as Mark 26)
- Ability to launch several missiles of the same or different types almost simultaneously, including Tomahawk, Standard SM-2, VL-ASROC, and the RIM-7R Sea Sparrow
- Better protection (deck-level armor for the VLS is equal to three-quarter-inch steel)
- Absence of blind arcs and launcher cutouts (the autopilot rolls the missile into the illumination beam, so that a missile from the forward VLS can pursue a target dead astern)
- The fact that dud rounds are bypassed and do not incapacitate the launcher while being cleared
- Mechanical simplicity and high reliability, since the only moving parts are the cell hatch and the rocket motor exhaust uptake hatch (which will open under exhaust pressure alone if necessary)
- Lower cost for procurement and maintenance
- Smaller crew requirements

The VLS is unmanned during firing. The ship sets the missile autopilot; the cell hatch and uptake vent hatch open; the VLS weapon control computer ignites the missile; and the missile breaks through the top cover of its canister on its way up. The autopilot pitches the missile to the ordered trajectory. Exhaust from the missile rocket motor blows out the bottom panel of the canister and vents through the uptake duct. A water-deluge system and the uptake vent handle missiles that may go into a restrained burn. Explosive shock testing of VLS-equipped Aegis cruisers showed that missile mountings had to be improved inside the canisters to avoid damage. VLS hatches have de-icing heaters.

The vertical-launch system was designed to be reloaded at sea. A collapsible crane for striking down missile canisters was sized to handle Standard medium-range missiles. Tomahawk, vertical-launch ASROC, and the booster-equipped Aegis Standard extended-range (SM-2 Block 4) missiles are heavier, and the crane handled them at only 40 percent of expected speed. Use of the VLS cranes was discontinued. The destroyer tender *Acadia* (AD 42) with her large cranes reloaded Tomahawks into the VLS for the *Spruance*-class destroyer *Paul F. Foster* (DD 964) during the Iraqi campaign, and unloaded weapons from the Aegis cruiser *Princeton* (CG 59) before she went into dry dock for repair of mine damage.[21]

The Outboard Tactical Intelligence System

Electronic support measures (ESM) use passive surveillance systems to intercept radio and radar emissions. ESM is very similar to, and often the same as, intelligence collection activity; the distinction is that if an activity focuses on providing immediate combat information to operational commanders, it is considered ESM. The electromagnetic characteristics of a radio transmission indicate the direction to the transmitter, and assist in identifying the platform operating it. A command post sending orders to a ship thousands of miles away, or a ship reporting back, obviously must transmit signals with enough power to carry over that distance. At that range atmospheric effects require use of either the high-frequency (HF) radio band or satellite communications.

HF radio circuits are much easier to set up than satellite circuits, especially by mobile ground forces, and the world's military forces have a large inventory of HF equipment. Since an HF radio transmission is omnidirectional, other receivers can detect it, too. World War II destroyers carried high-frequency direction-finding (HF/DF) equipment to detect radio transmissions between U-boats. It was an early form of radio communications intelligence (COMINT, a type of signals intelligence, or SIGINT; the other primary type of SIGINT is microwave ESM, discussed in chapter 9). For warships, the advantages of communications intelligence activities are that the intercepting ship does not need to broadcast signals herself; the detection range against surface targets can be much greater than with radar or sonar; and analysis of intercepted messages reveals much about the source, such as its identity and what it is doing. On early deployments some DD 963s carried a transportable deck-mounted communications intelligence system operated by a detachment of Navy communications technicians.

Many *Spruance*-class destroyers carry an extremely powerful communications intelligence installation, designated Outboard (SSQ-108). Most details about this system are classified. It is an advanced version of a prototype shipboard communications intelligence collection system, Classic Outboard (SSQ-72), developed starting in 1967. Classic Outboard never went aboard any *Spruance*-class destroyers.

World War II shipboard HF/DF receivers intercepted radio ground waves with a maximum range of about 180 nautical miles.[22] HF radio sky waves

travel thousands of miles. A published description of a shore-based intercept station describes a large land-based antenna to detect faint radio signals and to draw a line of bearing toward the distant transmitter.[23] Multiple stations intercepting the same signal draw simultaneous lines of bearing toward it, locating the transmitter at the intersection. The Navy's appropriate name for the shore-based direction-finding network is Bullseye. Outboard's functions are probably similar to those reported for Bullseye:

- Communication intercepts (voice, telegraphic, and probably data)
- Decryption and signal analysis to identify transmitters and their hosts (ships, aircraft, shore units, etc.)
- Radio direction finding and correlating intercepts
- Monitoring security compliance by U.S. military communications[24]

Advances in electronics made it feasible to fit Outboard, a shipboard extension of the Bullseye network,[25] aboard large warships. Outboard passed technical and operational evaluation tests and was approved for service in 1974.[26] In 1975 the Navy planned to install Outboard on 18 cruisers and *Spruance*-class destroyers. The Navy selected the *Spruance* class for Outboard for several reasons: the ships would operate with aircraft carriers; they had NTDS; they were adaptable to Outboard and had the necessary space, electrical power, and other utilities.[27] Space had been reserved in the DD 963 design for an electronic countermeasures system called Shortstop that was canceled for cost reasons in 1974 (later replaced by SLQ-32). Outboard went into this space. It was economical: one installation design would support potentially 30 installations.

The first ship design alteration (SHIPALT 1) approved for the DD 963 class was Outrigger, the project to install Outboard. Ultimately 36 Outboard systems were built. They were installed at a test site and aboard 35 warships: 5 *Belknap* (CG 26)–class cruisers; 8 nuclear cruisers; 6 Royal Navy *Broadsword*-class (Type 22 Batch 2) large frigates; and 16 *Spruance*-class destroyers. The large physical size of the Outboard system prevented it from being installed on the *Kidd* and *Ticonderoga* variants of the DD 963 design.

The Navy disclosed in 1975 that one Outboard mission was new to surface warships, and that its primary mission was tactical.[28] The new (in 1975) surface ship mission may have been intelligence collection to support nuclear targeting of Soviet command-and-control posts. President Richard Nixon revised targeting policy during 1972–74 to hold such targets hostage during war for coercion of Soviet political authorities, a strategy that advanced throughout the 1970s.[29] To the extent that the new strategy required communications intelligence capabilities, Outboard may have been a strategic intelligence collection asset.

The tactical mission was described in 1991 as "real-time over-the-horizon passive detection, localization, and targeting of hostile emitters."[30] All Outboard ships were to operate with aircraft carriers, suggesting that the Navy expected that in combat, aircraft would attack the targets. At least initially, Outboard was not connected with the Tomahawk cruise missile project, which was several years behind it in development and had an uncertain future.

The primary Outboard sensors are 24 radio-frequency sensors linked by miles of dedicated cabling within the ship. These sensors are mounted on the deck edges near the capstans, around the fantail, and on the deckhouse sides. They form an interferometer to detect radio signals throughout the low- through ultrahigh-frequency (LF–UHF) radio spectrum. Other antennas include standard radio communications antennas and a distinctive direction-finding array on the foremasthead. The antennas form the SRD-19 direction-finding subsystem.[31] Outboard receives signals from any direction.

Outboard's primary advances are local signal analysis and real-time data links to other Outboard ships. It automatically selects signals of interest and determines direction to the emitters through the array on the foremast.[32] Outboard's precision in range and bearing is accurate enough to track distant aircraft in flight and to target cruise missiles against ships that are too far away (over the horizon) to be detected by radar from any ship in a battle group.[33] An upgraded version (Outboard-II) further improves local classification capability. Outboard is a real-time tactical intelligence system, showing where and what a target is. It is probably tuned to detect and intercept even extremely brief transmissions, which is important because, for example, former Soviet command posts ashore broadcast operational orders as single-word codes to their ships to execute particular plans.[34] Outboard supports amphibious assaults, naval gunfire missions, and over-the-horizon tar-

geting.[35] It passively tracks aircraft in flight and tracks ships far beyond surface radar range.[36] An Outboard-equipped ship can substitute for airborne early-warning aircraft on many missions. This frees aircraft carriers for other assignments.

Multiple simultaneous lines of bearing are needed to fix the location and identity of a transmitter. To achieve this, Outboard transmits data via the ship's directional WSC-3 satellite communications transceivers into a dedicated tactical intelligence (TACINTEL) radio net, which other Outboard ships and the shore-based Bullseye HF/DF net monitor.[37] The WSC-3 UHF satellite communications antenna transmits data to an orbiting Fleet Satellite Communications satellite over a directional beam, which hostile forces can neither detect nor copy. Analyzed intercept data are rebroadcast over the TADIXS and OTCIXS radio networks in a form usable by tactical commanders and by the new JOTS terminals.[38]

For security reasons Outboard is not interfaced directly to the destroyer's NTDS computers. Outboard signal cables are segregated from other shipboard communication cables. However, an NTDS display console in the Outboard space shows the communications technicians the tactical situation as seen by the combat information center (CIC). Outboard operators provide data on contacts being tracked by CIC and cue CIC about new contacts. At general quarters the Outboard division officer stands watch in CIC.[39] Outboard checks friendly forces' emission control (EMCON) discipline, signal security, and electronic counter-countermeasures (ECCM) status. Operating passively, an Outboard ship does not reveal her presence to enemy surveillance measures.

No *Kidd*- or *Ticonderoga*-class ships carry Outboard, since their large missile guidance systems consume the necessary shipboard space. Of the U.S. Navy installations, only the 16 *Spruance*-class destroyers will be active after the mid-1990s. Military forces are increasing their use of satellite communications with directional superhigh/extremely-high-frequency (SHF/EHF) radio and satellite communications system to defeat intercepts by Outboard and other surface COMINT systems. However, foreign navies often do not have access to satellite communications, and ground forces still find HF radio transmitters cheap and easy to set up.

Some tactical commanders were slow to appreciate Outboard's power. Early in 1982 USS *Merrill* (DD 976), the test ship for Tomahawk, was assigned to the Third Fleet in the Pacific for a series of three war-at-sea exercises spread over two months. She brought other advanced new combat systems to the surface force, including Outboard for passive over-the-horizon targeting. *Merrill* correctly claimed to be the most powerful destroyer in the world. Here was the first opportunity in decades for the surface fleet to experiment with offensive tactics. During the second exercise a real war at sea erupted in the South Atlantic between Great Britain and Argentina. Instead of developing new tactics, however, the best use the exercise commanders could think of for *Merrill* amid this stimulus and opportunity was for her to simulate a Soviet warship in all three exercises. She earned a message from the exercise commander awarding her a bogus Order of Lenin. Afterward she went into the yards to get more Tomahawk-related systems installed and was lost to the fleet for the rest of the year.[40]

On other occasions *Spruance*-class destroyers equipped with Outboard and SQR-19 towed-array sonar, ships that could profitably have been stationed 100 miles away, have been assigned as plane guards, following aircraft carriers to pick up aircrews who might eject during launch or recovery accidents. However, in heavy seas the *Spruances* might be the only ships able to keep up with the carriers.

Soviet Responses to the *Spruance* Class

Contrary to the predictions of American critics, the Soviets regarded the *Spruance* class with alarm. They had armed themselves to make saturation attacks on perhaps two to four concentrated aircraft carrier battle groups. Through the Walker-Whitworth spy ring they were probably aware that *Spruance*-class destroyers and other Outboard ships would confront the Soviet Union with a dispersed capability to spy on their command posts and to call in strikes. Tomahawk land-attack missiles, under development for probable, eventually certain, deployment aboard the *Spruance* class, could evade most Soviet air defenses.

In 1980 the Soviet destroyer *Sovremennyy*, the first of a new class, appeared at sea. *Sovremennyy* went on sea trials visibly incomplete, hinting that the Soviets felt some urgency to put the new ship into service. The *Sovremennyy*-class destroyers are armed for antisurface attack, but their SS-N-22 (P-80) antiship cruise missiles have shorter range than the Russian Navy's large anticarrier cruise

missiles, and the ships lack the Punch Bowl satellite data link associated with the anticarrier weapons.[41] Thus the new destroyers were not part of the first-strike anticarrier force. According to a Soviet Navy admiral commanding a Northern Fleet destroyer group in 1990, the mission of the *Sovremennyy* class was specifically to counter the *Spruance* class, perhaps meaning to ambush them during their remote operations.[42] Another theory, not inconsistent with the anti-*Spruance* mission, is that the *Sovremennyy* class was built to attack surviving ships of a carrier task group decimated by a first strike from Soviet long-range cruise missiles.[43] Noncarrier warships whose destruction would be a matter of strategic urgency to the Soviets would be those with Tomahawk land-attack cruise missiles or Outboard. Those warships were predominantly *Spruance*-class destroyers and, later, their sisters in the *Ticonderoga* class.

Ironically, the *Sovremennyy* class was a rearmed version of the Kresta I/II design that had so impressed *Spruance* critics in 1970. The Soviets evidently regarded a much stronger gun and missile battery than on the Krestas to be a necessity to fight a DD 963. Use of an existing design again suggested urgency in deploying the ships; or perhaps the basic design was good enough to continue in production, despite its obsolescent steam propulsion plant.

The Soviets not only built *Spruance* counters, they also built *Spruance* clones. The latter are the very handsome *Udaloy*-class large ASW ships. The *Udaloy*s are about the same size as the *Spruances*, and they are similarly armed for ASW, although much less efficiently. The *Sovremennyy* and *Udaloy* classes both were designed and first built during the 1976–80 five-year plan approved in 1975, when the first *Spruance*-class destroyers were being completed. They are both good-quality warships designed for blue-water operations. Obviously costly, they contributed to the Soviet Union's economic collapse. In 1994 they comprise the two largest classes of large warships in the new Russian Navy, and both remain in production.

The Revival of the Navy's Role in National Strategy

In event of war in the mid-1970s, Admiral Thomas Hayward, commander in chief of the Pacific Fleet, planned a carrier-based offensive against Russian bases in Siberia. This "horizontal escalation" would divert Soviet forces away from

The Soviet Union built both *Spruance* counters and *Spruance* clones: the *Sovremennyy* (*top*) and *Udaloy* destroyer classes, respectively. They are the two largest classes of large surface warships in the independent Russian Navy, although many are laid up and are probably deteriorating. (U.S. Navy photos)

Since the late 1980s, 24 *Spruance*-class destroyers have been or are being armed with vertical launch systems for Tomahawk, vertical-launch ASROC, and potentially other missiles. USS *Paul F. Foster*

fired 40 Tomahawks during Desert Storm. A fully modernized unit, she has the SRQ-4 Hawk Link, Outboard, and the Mk 23 target acquisition system (TAS) radar, but not SLQ-32(V)3 jammers.

Europe. In 1978 Admiral Hayward became Chief of Naval Operations and brought his strategy with him. After Ronald Reagan defeated President Jimmy Carter in 1980, the new Secretary of the Navy, John Lehman, adopted this strategy as the basis for all Navy plans. President Reagan and his staff looked at the Islamic militant takeover of Iran in 1979 as a strategic disaster, the result of a "decade of neglect" of the military that had left the United States too weak to intervene. Secretary Lehman exploited this concern so as to get a major expansion of the fleet, although he ridiculed the situation as fear of the ghost of the shah of Iran. (Other still-influential ghosts include Admiral Rickover, whom Secretary Lehman fired in 1981, and General George Marshall, the 1940s advocate of service integration, a failed attempt to eliminate the Marine Corps.) Looking ahead rather than back, Secretary Lehman used the Maritime Strategy to determine the configuration of the new fleet.

Secretary Lehman ordered that all large new surface combat ships were to be armed with vertical-launch systems. The VLS design was deepened to permit all VLS ships to fire Tomahawks. In 1983 the Navy suggested building six more *Spruance*-class destroyers as a fast way to replace the

discarded DD 931–class ASW-modernized destroyers.[44] However, all plans for new destroyer construction shifted to the *Arleigh Burke* (DDG 51) class.

Soviet strategy to surge submarines into the Atlantic to split NATO made it imperative for U.S. naval forces to advance toward Soviet bases in a crisis to track Soviet warships and submarines for prompt destruction, to signal determination, and to prevent them from setting up a first-salvo ambush. The Maritime Strategy presumed that war would start conventionally in the Mideast or along a NATO flank and that both sides would hesitate to use nuclear weapons. The aggressor might plan to seize a limited strategic objective area and thus to fight only in that theater. To coerce the Soviets, the Navy planned "horizontal escalation" to force them to fight a multiple-front world war. Attacks from naval forces would annihilate the Soviet Northern Fleet and its bases, keeping the Atlantic safe for NATO. The Navy made it clear that if fighting began, its highest-priority targets would be Soviet ballistic missile submarines. That threat denied the Soviets the option to counter the U.S. naval offensive simply by ignoring it. Concern arose that if a fleet arrived close offshore in a crisis, the Soviets might attack it with nuclear weapons.[45] Deterring such an attack became the major reason for nuclear Tomahawk land-attack missiles.

Soviet diplomats at the United Nations forswore first use of nuclear weapons. That might have accurately reflected Soviet intentions, but it came across as a threat to commit further aggression using conventional forces and chemical weapons, as in Afghanistan. President Reagan dismissed the Soviet pledge as worthless and deceitful. "Deeds, not words," he told the United Nations, were needed from the Soviet Union to reduce tension. Without nuclear weapons the Soviets did not see how they could defend against the new naval threat. "It is hardly possible to imagine anything worse," groaned a Soviet commentator.[46]

The poor man was not very imaginative. A classified supplement to the published Maritime Strategy reportedly planned an amphibious assault by 13,000 Marines on Russia's Kola Peninsula, probably on the coast west of Murmansk, to set up an airfield for 150 aircraft. With their own bases unusable from attacks, Soviet submarines at sea, including ballistic missile submarines, would need to reach other ports to replenish before their crews starved. They faced running long gauntlets of mines, attack submarines, ASW aircraft, and destroyers.

The authoritative Senate Armed Services Committee and President Reagan's National Security Council never accepted the complete Maritime Strategy, but they funded the 600-ship fleet that could implement it.[47] Exercises supporting the Maritime Strategy were elements of psychological warfare intended to demoralize Soviet leaders. The Navy practiced the initial contingency operation, the immediate advance of ASW forces to threaten Soviet submarine bases and transit lanes. Western fleets practiced strikes from the Norwegian Sea and the northern Pacific. Clearly concerned about this, Soviet leaders focused on negotiating arms treaties at a time when the Soviet Union's economic survival needed all their efforts.

The *Spruance* Class and Arms Limitation

Tomahawk and the *Spruance* class were key elements of the aggressive Maritime Strategy. Secretary Lehman trumpeted that conventional Tomahawks would destroy Soviet nuclear missile–carrying bombers at their bases.[48] Tomahawk strikes could come from any coast or direction.

Most Soviet air defenses could not detect Tomahawks, and indeed air defense facilities would themselves be targets, as they were in Iraq, to clear routes for bombers to penetrate inland. Other strategic targets included ballistic missile submarine bases and ammunition depots. With Tomahawk the United States could launch strategically destabilizing raids without nuclear weapons.

Spruance-class destroyers and *Ticonderoga*-class cruisers might proceed as a surface action group far ahead of other forces to attack Soviet barrier ships, such as the large *Kiev* and *Kirov* classes of cruisers, with Tomahawk antiship missiles. Surface action groups could attack ground targets from 700 miles' range, forcing the Soviets to dilute their surveillance and attack forces away from aircraft carriers.[49] Probably more important, by maintaining a constant Outboard patrol for days, weeks, or months, before or after an outbreak, they could detect command-and-control targets based on even brief radio transmissions.

Nuclear Tomahawk land-attack missiles became operational in June 1984. As a national strategic asset, Tomahawk was a reserve weapon for transattack and postattack coercion of surviving Soviet authorities. As a theater reserve nuclear weapon in NATO and the Far East, it was a deterrent against any Soviet hope of preemptively destroying all U.S. nuclear forces. Most important, since carrier battle group escorts could attack Soviet bases and command posts with nuclear Tomahawks, this weapon removed any perception on the part of the Soviets that they might attack aircraft carriers with nuclear weapons and get away without nuclear retaliation. It was not included in the SIOP strike packages for initial nuclear attacks, and no ships carrying nuclear Tomahawks were deployed solely because of that capability.[50] The primary mission of nuclear Tomahawk was to keep a war conventional. The adoption of this strategy explains the cancellation of the nuclear-warhead air defense Standard missile in 1986, a weapon that had been intended to defeat attacks by nuclear cruise missiles.[51]

As the predominant Tomahawk platforms, the *Spruance* and *Ticonderoga* classes were an element in negotiation of the strategic arms limitation and strategic arms reduction treaties (SALT and START). In 1973 Secretary of State Henry Kissinger had supported the development of nuclear land-attack cruise missiles to get the Soviets to bargain seriously at the SALT II negotiation. The SALT II treaty proposed limiting cruise missile range to 500 kilometers until 1981, to induce further negotiation to develop new constraints. Kissinger's idea was that the United States would trade cruise missiles, an almost paper weapon without a target assignment, for actual Soviet weapons.

Tomahawk did not fit with two fundamental tenets (or flaws, depending on one's politics) of SALT II: (1) SALT II proposed to restrict not weapons, but launch systems, and (2) each side had to be able to count the other's launchers. Before Tomahawk, SALT-accountable launchers identified unique weapons, but now VLS-equipped ships could launch any version of Tomahawk or other missiles. Treaty proponents criticized Tomahawk as a "weapon in search of a mission," a weapon the Navy had not wanted but would not give up. SALT II was never ratified.

During the START negotiations in the mid-1980s, the Soviets demanded restrictions on nuclear Tomahawk land-attack missiles as their price for cutting their arsenal of intercontinental ballistic missiles. They began work on a new submarine-launched nuclear cruise missile, the SS-N-21 (Soviet RKP-55), a throwback to the old SS-N-3 (P-5) land-attack cruise missiles of the 1950s. Only a few Soviet submarines could launch the new weapon. The Soviets had no conventional potential to retaliate against the United States south of Alaska. In 1986 their demands included limiting the number of each side's submarine-launched nuclear land-attack cruise missiles to 400 and banning surface-launched versions altogether. In 1988 they proposed allowing each side to deploy nuclear cruise missiles on one class of surface warship, still within an overall limit of 400, and to limit conventional sea-launched cruise missiles to 600. The Reagan administration flatly rejected all the Soviet proposals:

> We will not eliminate a weapon that makes such a significant contribution to deterrence. . . . We will not accept a solution that constrains important elements of U.S. conventional capability. The importance of antiship and conventional land-attack cruise missiles . . . is unquestioned.[52]

U.S. negotiators further objected that the Soviets' proposed restrictions were unverifiable. The Soviets suggested shipboard inspections, which the United States rejected as intrusive and unreliable. A widely publicized visit of the Soviet missile cruiser *Marshal Ustinov* and destroyer *Otli-*

chnyy to Norfolk in 1989 featured a charade inspection purporting to prove that the ships were not carrying nuclear weapons; it may have been the primary reason for their visit. Naval arms control was a major point of contention at the December 1989 Bush-Gorbachev summit in Malta.[53] START proponents feared that the sea-launched cruise missile issue would prevent any strategic arms agreement. The issue delayed the START I treaty from 1986 until 1991, after the Soviet regime had collapsed. The final START text authorized 880 nuclear sea-launched cruise missiles for each side, without a requirement for verification. The U.S. Navy reportedly planned a force of 637 nuclear Tomahawks. Reportedly 367 had been built when procurement ended after the demise of the Soviet Union. These weapons have been in storage ashore since 1991.

Strike Warfare during Operation Desert Storm

In 1991 Iraqi land-line telecommunications centers were destroyed at the start of Operation Desert Storm to force enemy commanders to use radios, which United Nations coalition communications intelligence units could locate and classify. Many Outboard-equipped ships were in the Mediterranean Sea, Red Sea, and Persian Gulf, where they could provide cross-fixes on Iraqi transmitters. By war's end, specialized earth-penetrator bombs built from 8-inch gun barrels were being used to destroy Iraqi underground command posts. The Aegis cruiser *San Jacinto* may have carried some nuclear-warhead Tomahawk missiles to deter Iraq from using nuclear, chemical, or biological mass-destruction weapons.[54]

All 288 Tomahawks fired during the 1991 Kuwait campaign were conventional-warhead land-attack versions (TLAM-C and TLAM-D). It was the first combat use of this weapon. The first Tomahawk missions took two weeks to plan. Fleet planners doubted that Tomahawk flight-testing had proved the missile's reliability and destructiveness under combat conditions. They feared that the missiles might collide ("clobber") with unknown terrain features, get lost in the featureless desert, fall to Iraqi air defenses, or fail to demolish the target. At first three or four missiles were fired at each target.[55]

Tomahawk attacks were coordinated with reconnaissance satellite passes so that the Iraqis would have no time to conceal, or to fake, damage.[56] Reconnaissance showed that reliability and

USS *Fife* sails through the plume of a Tomahawk fired from her vertical-launch system forward of the bridge. She fired 60 Tomahawk land-attack cruise missiles during Operation Desert Storm and earlier had used her Outboard system to uncover Iraqi blockade runners. (U.S. Navy photo)

accuracy were very high, and that the warhead did more damage than expected. Cable News Network transmissions from Baghdad showed Tomahawk entry holes in the sides of buildings, at least one of which was brought down in a follow-up attack. By the fifth day of the campaign one missile was being fired per target. Tomahawks kept pressure on Iraq when weather precluded manned aircraft strikes and were the only weapons used for daylight attacks against central Baghdad. When the fleet commanders requested follow-up attacks, strike planning staffs ashore planned flights to branch off earlier routes and transmitted the new flight routes to the ships by satellite within 48 hours.[57]

Despite some imaginative reports, no Tomahawk variant carried a "microwave warhead" or dispensed wires or carbon filaments to short-circuit electrical power stations.[58] Rumors had it that the idea began after a 1985 incident when chaff from an offshore naval exercise drifted onto a San Diego power line. It interrupted some commercial computers that took a few minutes to restart, but it did not cause an even momentary electrical blackout.[59] Such a science-fiction weapon would be operationally useless even if it did exist, which it does not.

Spruance-class destroyers and Aegis cruisers launched two Tomahawk strikes against Iraq in 1993. All missiles were unitary-warhead TERCOM-guided versions (block II-A). Since TERCOM fields were scarce, only a few approach routes to Baghdad were available. In the first attack, missiles arrived in waves from one direction, and Iraqi gunners shot down one. Others failed at launch or missed, but 34 of the 45 missiles designated for launch hit every building at a nuclear weapons laboratory. Contrary to rumors that damage was minor, destruction of equipment was total; the Iraqis tore down the buildings' shells. In the second attack, 16 of 24 missiles hit and destroyed a terrorist headquarters.[60]

Strike Weapons Modernization Programs

The success and limitations of the 1991–93 strikes against Iraq engendered upgrade programs for Tomahawk missiles, the ships' launch control systems, and targeting. The inventory will be larger (4,100 conventional missiles) and all will be far more capable than any of the earlier versions. The nuclear version (TLAM-N) is unchanged.

All antiship and TERCOM-guided conventional land-attack Tomahawks (blocks I and II) are being upgraded to GPS-guided configurations (blocks III and IV).[61] Block III missiles have time-of-arrival control, greater range, a GPS receiver, a stronger engine for better terrain avoidance, and a scene matching unit with less sensitivity to diurnal and seasonal conditions. The new blast warhead is lighter and equally powerful, and has better penetration. Most important, mission planning is much faster and can be done in theater with the Afloat Planning System aboard aircraft carriers and amphibious assault command ships. These missiles can converge on a target from many directions, synchronized with each other and with air strikes. Tomahawk multimission missiles (TMMM, block IV) have a penetrator warhead for both antiship and stringent land-attack missions. TMMM will not use a radar seeker but instead will relay high-resolution IR video images to an airborne controller, probably in an F-18 strike fighter, who will aim or retarget the missile over a new data link. (Russian long-range antiship cruise missiles worked similarly but used low-resolution radar.)

The final video will instantly confirm that the target was hit.

An outcome of misses in the 1993 strikes is that block III versions have a "precision strike Tomahawk" (PST) option. If this option is activated for a mission, a damaged or malfunctioning missile will divert to a programmed safe area to self-destruct. Although derided as a feature to let diplomats, who demand these strikes in the first place, feel better by risking fewer civilian casualties during future missile raids, a weapon that reduces collateral damage would be more defensible if used to attack urban targets. A subset of 100 missiles can relay their positions and health via a one-way data link. The Clinton administration gave no basis for the figure of 100, nor a policy as to whether these missiles should have a priority or a restriction when planning deployments and operations.

The Advanced Tomahawk Weapon Control System will support the new missiles aboard firing ships. It will replace the ships' present Harpoon and Tomahawk weapons control systems.[62] With a low-cost computer upgrade, *Spruance*-class destroyers could receive real-time video images from reconnaissance drones through the SRQ-4 Hawk Link. Interest is also being expressed in converting Harpoon missiles to SLAMs for ship-launched, SH-60B-guided strike missions.

Antisubmarine Warfare

ASW in Modern Strategy

Considerations of the armament and performance needed for antisubmarine warfare (ASW) at high speed with aircraft carrier battle groups were dominant influences on the DD 963 design.[1] Active sonar tactics dominated ASW in the 1950s and 1960s. Passive sonar dominated ASW tactics in the U.S. Navy from the time the *Spruance* class entered service until the collapse of the Soviet threat. Since then, interest is returning to active sonar for use in shallow waters, such as in straits or along the littoral of potentially hostile regional powers. *Spruance*-class ASW armament has progressed through three phases:

- As delivered: *Spruance*-class destroyers were equipped with the SQS-53A hull-mounted sonar, two triple-tube torpedo launchers, an ASW rocket launcher (ASROC) recycled from older destroyers, and a Mark 116 Mod 0 underwater fire control system.
- Interim modernization (1980s armament): Light Airborne Multipurpose System (LAMPS) Mark I equipment, a passive sonar analyzer, and torpedo decoys were added to *Spruance*- and *Kidd*-class destroyers after delivery. The first *Ticonderoga*-class Aegis cruisers were similarly armed. These systems constituted the ships' standard ASW system for almost all operations from the late 1970s through the late 1980s.
- Full modernization (1990s armament): In the late-1980s the *Spruance* class began receiving a major ASW weapons upgrade under the SQQ-89 modernization program, centered on the SQR-19 towed-array sonar and the LAMPS Mark

III helicopter system. At about the same time, on most *Spruance*-class destroyers vertical-launch cells replaced the ASROC launcher forward.

Ship features that boost ASW effectiveness include the fast-response gas turbines, silencing, endurance, reliability, and tactical intelligence systems. For ASW, distinguishing features of the *Spruance* class are speed in heavy seas, hull silencing, long-range sonar, endurance on station, ASW ordnance magazine capacity, and the ability to add the surrounding underwater picture to the battle group via computer data links. When built, these destroyers' large helicopter facility and their degree of silencing were unique. Their speed, endurance, reliability, and antimissile weapons reduce their demands on battle group logistics and defenses and let them operate far from the carrier. The fact that not one of these characteristics was an externally visible weapon launcher incensed critics for years.

Some insight into how ASW requirements influenced the *Spruance* design comes from contrasting the destroyers with other ASW forces. Land-based patrol aircraft cover oceanic straits to prosecute contacts detected by SOSUS and long-range shipboard sonars. Carrier-based ASW aircraft concentrate near the carrier, primarily to detect submarines in the sector ahead of the ship, which is where an enemy submarine would have the best chance to make a torpedo attack. American submarines are stealthy enough to patrol off hostile submarine bases, but they do not maintain contact data over radio data links. Frigates were bought to escort reinforcement convoys and amphibious assault groups, which are slower than carrier battle groups. They do not have the *Spru-*

ance class's speed, sensor diversity, or command-and-control capabilities. In the budget climate of the 1990s, frigates are on the U.S. Navy's endangered-species list.

Soviet strategy to defeat the West by warfare, if opportunity afforded, included a preoutbreak surge of their large force of torpedo-attack and cruise missile–attack submarines into the Atlantic to put them on station at the outbreak to split NATO.[2] Soviet warships would provide surface cover to disrupt NATO antisubmarine tactics in oceanic straits. As Chief of Naval Operations during 1970–74, Admiral Zumwalt continued former Secretary Nitze's policy of developing ASW systems for open-ocean sea control operations to protect shipping. The strategic mission was to support reinforcement convoys to the Army–Air Force garrison in Germany.

The Soviet strategy made it essential for the United States to position naval forces beyond the choke points, in particular in the Norwegian Sea, in a crisis before the Soviets opened hostilities. Destroyers, submarines, and ASW aircraft would maintain a moving ASW defense screen around the carriers and other high-value ships, in the same way that cruisers, fighters, and early-warning aircraft would concentrate on area air defense. ASW aircraft carrier task groups had been the vanguard for the attack carriers, but the Navy retired all ASW carriers (CVSs) by 1973, in part to pay for building the *Spruance* class. Half of a CVS air group—one squadron each of ASW patrol aircraft (first S-2G Trackers, then S-3A Vikings) and ASW helicopters (SH-3 Sea Kings)—was assigned to the large attack carriers. *Spruance*-class destroyers with LAMPS helicopters helped to make up for the reduced ASW defense of the attack carriers.[3]

Navy fighting power began to recover in the late 1970s, when the *Spruance*-class destroyers, *Los Angeles*–class attack submarines, F-14 fighters, *Nimitz*-class aircraft carriers, and *Tarawa*-class assault ships began arriving in the fleet. Finding crews for these advanced new weapons was a concern, as were inadequate plans for munitions and for sustaining military recovery. The Reagan administration fixed those problems. Under its Maritime Strategy, aircraft carrier task groups exercised inside Norwegian fjords and between Aleutian Islands, landforms that would obstruct antiship cruise missiles. *Spruance*-class destroyers would screen the carriers during open-ocean

transits and would guard underwater approaches to straits and fjords.

Sonar Operation

The ASW combat system supports the fundamental tactics of an open-ocean ASW operation:

- Detecting evidence that an unknown submarine may be present
- Classifying the contact as a possible, probable, or certain submarine, and identifying whether it is hostile
- Localizing the submarine sufficiently for attack
- Prosecution: getting to weapons range and attacking the submarine
- Countermeasures to defeat enemy submarine attacks

In open-ocean ASW the ship's crew evaluates intelligence reports, contact data from the ship's own sensors, and other tracks received over the computer data links. The ship and the LAMPS helicopter maneuver to improve sensor data and to bring the submarine, if that is what the contact proves to be, into weapons range. This detect-and-decide cycle continues throughout the ASW operation to refine the tactical picture and to attack discovered submarines. Shallow-water ASW is faster, a knife fight compared with an open-ocean shootout.

Sonar connects the ship to the environment of ocean acoustics and exploits how the ocean transmits sound. Light and radio-frequency energy propagate through seawater only in a few narrow-frequency bands that are not useful for detection (not yet, anyway). At low sonic frequencies sound is little absorbed by seawater, so can travel great distances. Speed of sound in water is about 0.8 nautical mile (1,600 yards) per second, over four times its speed in air.

Machinery on another ship or submarine produces continuous resonant vibrations, which are often too low in frequency to be audible in air but are detectable in water by sonar. Vibrations produce harmonic echoes that trail off into higher frequencies. In passive sonar operation the destroyer attempts to detect these resonant vibrations and harmonics. A passive sonar contact produces a line of bearing on the Naval Tactical Data System (NTDS) displays pointing toward the sound source. Any sector with significant ambient noise may produce a line of bearing, so false alarms are common. Destroyers use a submarine tactic to eliminate false alarms. The ship steers a sinusoidal course so that off-ship sound sources show changes in relative bearing.

Two factors complicate sonar operation. First, oceanographic effects distort acoustic signals so that they often are not detectable on a straight-line direct path from the source. A phenomenon known as the layer makes distant submarines below layer depth practically inaudible to hull-mounted sonar on a direct path (unless layer depth is shallower than the sonar transducer). Second, sonars always receive ambient noise interference. Ambient noise is the random noise received continuously over all acoustic frequencies from marine life, shipping, and ocean surface turbulence. The ocean is a surprisingly noisy acoustic environment. Ambient noises from sea life, surface wave crests, and other ships tend to mask direct-path sound waves from distant submarines above the layer. Quiet submarines may become audible on a direct path to a hull-mounted sonar in passive mode only at dangerously close range. The newest conventional submarines, built for export by Germany and Sweden, may not be detectable even then.

Destroyers detect submarines passively by sonar beyond surface direct-path range through convergence zones. Oceanographic research revealed that low-frequency sound waves from a submarine travel at long range deep below the surface layer, but eventually deepwater pressure refracts them upward. The path traced by the sound waves eventually converges on the ocean surface. Since a submarine's sounds radiate in all directions, the sound path creates an imaginary ring at the ocean surface. This ring is called a convergence zone (CZ). Convergence zones exist only in deep water (typically, over 1,000 fathoms) because the sound waves must dive deep to encounter the pressures that create the phenomenon. The submarine herself does not need to be operating at the depths where the sound waves penetrate to be detectable in a convergence zone. Only a few oceanographic research submarines can dive that deep at all.[4]

Surface range to a convergence zone varies from about 15 miles in deep warm water, such as the Mediterranean Sea, to 35 miles in deep cold water, such as the Norwegian Sea and the Northern Pacific. A convergence-zone ring is about 5 miles wide at the ocean surface because the sound waves disperse over distance. The SQS-53A sonar can detect targets in the first convergence zone in either active or passive mode, and under some conditions it can detect a loud submarine, such as the Russian Foxtrot class, in the second convergence zone. The ship will detect a submarine entering the convergence zone, but the contact will disappear after the submarine passes through it, even if the range is closer. The destroyer must localize the submarine along the line of bearing. By clever maneuvering, the destroyer can hold contact on a submarine at convergence-zone range. That buys time to launch a helicopter. In the contact area the helicopter can classify the contact as a submarine and localize it with sonobuoys.

The SQS-53 and the new SQR-19 TACTAS linear towed-array sonars provide two significant functions for sonar detection: low-frequency sensitivity and narrow beam width. Because low-frequency sounds are little absorbed by seawater, low-frequency sonar can detect a submarine at greater distance at these frequencies. A bearing that is relatively louder may include a submarine. This is spatial or broadband detection, so called because it checks a broad range of acoustic frequencies. Active sonar operates in broadband only. Narrow beam width enables the sonar technicians to focus the sonar more accurately to discern sounds received from a bearing of interest. This precision narrows the sector of ocean from which ambient background noise is collected, improving the signal-to-noise clarity and making detection of an actual signal more likely.[5]

Whether the ship succeeds in detecting a submarine depends on several factors, including distance, sonar sensitivity, effective tactics, ambient noise, self-noise, and oceanographic conditions of water depth and temperature. Convergence-zone detection takes advantage of the bending effect of very deep ocean pressures on sound paths. The depth of water needed to create a convergence zone depends primarily on ocean temperature. The destroyer receives predictions through satellite communications about oceanographic conditions and measures them directly with expendable bathythermographs (XBTs); each ship routinely drops several XBTs daily.[6]

A destroyer must determine that a sonar contact is a submarine, as opposed to a false echo. The destroyer might go active with sonar long enough to get a return echo, or in combat she might simply fire a homing torpedo toward the contact position and see what happens. Disadvantages are that active sonar reveals the destroyer's position and the supply of torpedoes is limited.

Submarine machinery generates sound patterns. The 50 Hz turbogenerators on Russian submarines, for example, generate a regular pattern of sharp lines when displayed over time (a Lofargram). These lines show that the signal and its harmonics are being received consistently over time. The destroyer can alter course to verify that the received signal is off-ship. Long-range low-frequency signals include blade tonals from imperfect machining of Russian submarine propellers. The initial sound generates harmonics. A particular initial sound might be indistinguishable against the ocean background noise, but the same signal and its harmonics create a mathematically consistent pattern over time. A computer program applying a mathematical analysis known as a Fourier transform can discern these patterns against ambient noise. Patterns show that the source is man-made. Different configurations of equipment among submarine classes (even among individual submarines within the same class) emit distinctive combinations of signals and harmonics.

The first generation of Soviet nuclear submarines (the Hotel, Echo, and November classes, or HENs) generated similar sounds and harmonic patterns because their propulsion plants were identical. The second generation (the Charlie, Victor I, and Yankee classes, or CVYs) shared another common plant. In 1985 *Hewitt* (DD 966) trailed a CVY-type submarine for 700 miles through the Indian Ocean until it surfaced.[7] Passive sonar is about the only technique to distinguish between submarine types. This has limits because, for example, a North Korean submarine made in China may not be readily distinguishable from Chinese Navy submarines of the same class.

Submarine crews hope to be detected only by a flaming datum, unambiguous evidence of submarine presence such as an otherwise inexplicable underwater explosion adjacent to a ship in company. This statement is not altogether facile, because it raises two significant points about ASW. First, submarines can be detected by means other than sonar. Second, convoys and battle group formations during war might lure ambitious submarines within range of escorting ASW forces, which may detect the submarines before they attack.

Initial ASW Armament

The initial sensors and weapons on the *Spruance* class supported ASW operations at the first

convergence zone. This armament is listed in table 11.1. Most of this armament was installed on the ships during construction, with the rest of it added during overhauls in the 1970s and early 1980s. Box 11.1 describes an ASW exercise with the armament as delivered; it is interesting in that it includes feedback from a submarine.

All ships in the *Spruance, Kidd,* and *Ticonderoga* classes mount a large, low-frequency SQS-53-series sonar. The sonar transducer array and its controlling electronics occupy almost the entire bow forward of the 5-inch gun. A 200-ton seawater-filled dome suspended beneath the hull at the bow holds the sonar transducer array. This is a 16-foot-diameter 30-ton cylinder with a vertical stave of eight transducers every 5 degrees (72 staves). Location at the bow secludes the transducer array from propeller and hull noise. No machinery that normally operates during an ASW operation is installed forward of the main engine rooms. The Prairie and Masker silencers and machinery isolation mountings further reduce interference from self-noise. The sharp forward rake of the stem is not for the sake of good appearance, although that is a commendable effect, but so that the anchors drop clear of the sonar dome. In a clear-water anchorage such as the Diego Garcia lagoon, the dome is visible from the bow. From that vantage point its size is spectacular.

The original SQS-53A sonar was based on the earlier SQS-26CX model fitted on the *Knox*-class destroyer escorts. Early references about the *Spruance* class in fact list the sonar as SQS-26CX.[8] The SQS-53A version differed in using a digital scanner, a solid-state power supply, and a digital interface from the operator consoles to the underwater fire control system.[9] These were major improvements. The solid-state power supply superseded a mechanical motor-generator in the SQS-26, a noisy device that degraded SQS-26 sonar sensitivity. The scanner is a computer processor that can be visualized as creating a sharply focused listening beam that sweeps around the ship. It listens for loud sectors (using digital multibeam steering, or DIMUS). The digital interface integrates sonar and fire control with the Command and Decision computer system to provide the ship's commanding officer and tactical action officer with immediate displays of the tactical situation underwater.

The SQS-53A has a passive receiver adjunct (Unit 31). In the initial ASW system the SQS-53A sonar was connected to an SQR-17 ("square 17,"

Spruance-class destroyers *Peterson*, shown here, and *Thorn* watched a Soviet Victor III nuclear attack submarine that surfaced after her propeller snagged a frigate's SQR-15 towed-array sonar in the Atlantic in October 1983. They trailed her back to Cuba. Autumn 1983 was busy for the *Spruance* class and its sisters. Some other operations then included the search for flight KE-007 wreckage, the Grenada invasion, and Lebanon. (U.S. Navy photo)

Table 11.1   Initial *Spruance*-class ASW armament, carried on most ships during the 1980s

Sonar
— SQS-53A active/passive sonar
— SQR-15 towed-array surveillance sonar (6 DD 963s)
— WQC-6 and WQC-2 underwater communication links
Tracking and fire control
— Command and Decision System (Naval Tactical Data System with enhancements for ASW)
— Mk 116 Mod 0 underwater fire control system
LAMPS Mark I
— ARR-75 sonobuoy receiver
— SKR-4 LAMPS data link receiver
— SQR-17 spectral analyzer (LAVA)
SH-2F Seasprite helicopter
— 25 sonobuoys
— ARR-75 sonobuoy receiver
— AKT-22 sonobuoy data link transmitter
— ASQ-81 magnetic anomaly detector (MAD)
— 1–2 Mk 46 ASW torpedoes
Ordnance
— 2 Mk 32 triple 12.75-inch torpedo tubes for Mk 46 ASW torpedoes (12–18 reloads)
— DD 963 class: ASROC 8-cell launcher for rocket-thrown torpedoes and rocket-thrown nuclear depth charges (16 reloads)
— DDG 993–996, CG 47–48: ASROC rocket-thrown torpedoes in Mk 26 guided missile launching system magazines
Countermeasures
— SLQ-25 Nixie acoustic decoys
— Prairie acoustic sound dampener on propellers and struts

or LAVA) passive sonar processor to analyze the passive sounds heard amid the ambient ocean noise. The SQR-17 and SQS-53A consoles were in Sonar Alley, a section of the combat information center closed off for the sake of silence. Over time the sounds of machinery showed up as frequency lines on SQR-17 video displays. SQR-17 also displayed signals received from sonobuoys dropped by the LAMPS helicopter.

By design, the SQS-53A was an active sonar to exploit bottom-bounce and convergence-zone phenomena with active sonar signals (called "pings," but heard as such only in movies). Bottom-bounce avoids the deep-ocean bending effect on sound paths by transmitting sonar signals down a steep angle (greater than 25 degrees) to penetrate to the ocean floor, as with a fathometer, in water depths from about 1,000 to 2,500 fathoms. Both the outgoing signal and the returning echo reflect off the seabed. The success of passive sonar and LAMPS, and the scarcity of flat, acoustically reflective seafloor, made bottom-bounce a rarely used tactic in blue-water ASW.[10]

Early reports on the *Spruance* class mention plans for the SQS-35 independent variable-depth sonar (IVDS), a medium-frequency, active-only towed sonar. For the *Spruance* class this was a fallback in case a suitable towed-array passive sonar did not appear. Earlier variable-depth sonars had shared the electronics of the ship's hull-mounted sonar, so that the ship could use either but not both at once. IVDS featured independent electronics, hence its name. The large towed "fish" was mechanically awkward and often broke from the towing cable. Some *Knox*-class frigates received IVDS. Eventually it was used as a sinker and signal line for a short towed-array passive sonar, like a big fishing weight attached to a lure.[11] Superiority of passive low-frequency sonar ended interest in IVDS. The SQR-19 tactical towed-array sonar is located in the space in the transom provided for the IVDS fallback.

Towed arrays are line-shaped passive sonars towed deep beneath the surface layer. Before they received the SQQ-89 modernization, several *Spruance*-class destroyers carried a deck-mounted

Box 11.1 Destroyer *Arthur W. Radford* (DD 968) did not yet have LAMPS or her complete passive sonar suite for this 1977 ASW exercise.

151

Antisubmarine Warfare

Feedback from the submarine is important to validate ASW tactics.

SPECIFIC COMMENTS FOR WEAPONS

1. (C) *General.* This section contains commentary on the performance of equipments and systems under the cognizance of the Weapons Department.

    a. (C) *ASW.* The potential for passive ASW of this ship class has become strikingly more apparent since [commissioning]. During refresher training, *Radford* received four days of dedicated ASW exercises. Throughout this period the ship was not detected beyond 4000 yards by a submarine below the layer. Under the same environmental situation, *Radford*'s Unit 31 [passive sonar receiver] consistently gained contact at 2500–3000 yards. Passive ranges above the layer for both *Radford* and the submarine were much greater, ranging out to 28,000 yards through the full speed range.

    During active attacks, contact was routinely gained at 4000–5000 yards with the submarine below the 80-foot layer. Contact was maintained within 5000 yards at speed in excess of 27 knots in almost all exercises. Quick acceleration, rapid deceleration, and a small turning radius allow this ship a great deal of flexibility while tracking actively. In addition, the constant shaft RPM at 12 knots and below presented a new dilemma to the submarine, who could not establish turn count [i.e., determine the destroyer's speed from propeller rpm] at any time.

    Often while trying to keep the submarine out of the baffles [i.e., astern, where the sonar was deaf]

at close quarters, hard rudder [and] flank speed turns of more than 360° were accomplished without losing contact. Though one would not normally engage in maneuvers of this type, the capability to track a submarine at close quarters is outstanding in this class ship. . . .

4.c. (C) *ASW Lessons Learned*

    (1) (C) We learned that our submarine adversary never detected any discrete frequencies during the week of exercises and detected broadband noises only "well within 4000 yards." Unit 31 was detecting the submarine at 2600 yards.

    (2) (C) Once sonar was active, the submarine was able to determine position accurately by ping stealing, but was unable to correlate the information when we were within 2000 yards.

    (3) (C) ASW was conducted successfully at 27 knots, but the small turning radius of the submarine made ship maneuvering interesting, particularly in close-in situations. The tactic of stopping the ship, securing Prairie air, and leaving the propellers turning, appeared to confuse the submarine and was useful in more than one scenario.

    The potential of this ship as a passive platform continues to be underscored. The value of passive ASW has infected the combat systems organizations on board; only the hardware [SQR-17, SQS-53 wide-band elements, etc.] remains to be seen. . . .

[Declassified 1983]

SQR-15 passive surveillance towed-array sonar. A large equipment van mounted athwartships between the Harpoon launchers and the mainmast housed its electronics. Array handling equipment was mounted on the fantail on the port side of the aft 5-inch gun. SQR-15 was primarily a long-range surveillance sonar that was towed at slow speed. It was not integrated with the ASW combat system, an operational disadvantage, but its ease of use made it popular. Plans in 1987 were for the Atlantic and Pacific Fleets each to have three SQR-15 systems, all for DD 963 use.[12] *Hewitt* (DD 966) carried SQR-15 starting in 1981 and used it for over-the-horizon tracking of surface ships.

Merrill (DD 976) and *Kinkaid* (DD 965) carried SQR-15 arrays in the early 1990s in the Pacific before the system was retired.

    The Kaman SH-2F Seasprite was the LAMPS Mark I helicopter. Small enough that the flight deck crew could move it around on deck by hand, one was embarked on most *Spruance*-class destroyers before the SQQ-89 modernization. It could dash to 30–50 miles' range for antiship surveillance and targeting, or to drop sonobuoys and an ASW torpedo. Fuel and payload limited the SH-2F effectively to about 35 miles' range from the ship for ASW patrol missions, or the first convergence zone.

ASROC and Nuclear ASW Weapons

As built, the *Spruance* class carried RUR-5A ASROC antisubmarine rockets. It included a Mark 16 eight-cell box launcher, a magazine beneath it, and a reload rammer.[13] Vertical launchers displaced the ASROC system on most *Spruance*-class destroyers. In 1993 ASROC box launchers were removed from all ships still mounting them. The ASROC payload was a standard Mark 46 12.75-inch torpedo or a 1-kiloton nuclear depth charge. ASROC fired a rocket-boosted Mark 46 homing torpedo from the box launcher out to 10,000 yards. The original (late 1950s) tactical plan for ASROC was to fire two torpedoes ahead of a targeted submarine, one to each side of her computed course. In theory such an attack would be lethal no matter how the submarine maneuvered. The ASROC box launcher had four parallel two-cell modules for that reason, although by the 1970s the tactic had been discarded, perhaps because two active torpedoes might interfere with each other.

In an attack exercise the ASROC launcher trained to the ordered bearing, and the selected cell elevated to 45 degrees. When within range, the ASW fire control officer fired it, and the ASROC rocket-thrown torpedo shot out of the launcher. At a preset time in flight the weapon broke free of the booster, parachuted into the water, and began searching for its target. For older Mark 46 torpedoes the ideal drop point was 800 yards 60 degrees off the submarine's port bow.[14] Sonar detected warhead detonation; target noises and movements would reveal the effects of the attack.

The launcher rotated backward, and modules elevated to the vertical for reloading. Two vertical rings in the magazine each held eight reload ASROCs. A transfer device rammed new rounds vertically into the launcher through deck hatches. Reload rocket-thrown torpedoes on the service rings sat on a plenum connected to a blow-out door, visible on the starboard side of the hull. If a booster ignited in the magazine, the service ring would restrain the weapon and the flame would be channeled out this door, so that it would not cook off other rounds and cause a magazine explosion. Novice sailors were told that the door covered a mine-watch lookout station.

Until 1989 *Spruance*-class destroyers sometimes carried ASROC nuclear depth charges. The W44 depth charge had a blasting power of about 1 kiloton of TNT, small for a nuclear weapon. A typical loadout was four rocket-thrown nuclear depth charge rounds, all stored in the magazine. Each rotating reload ring in the magazine held two depth charge rounds. An additional slot on each ring was left empty, since the launcher had to unload a torpedo round before the depth charge round could be loaded. One more slot held a dummy training round for the crew to practice handling nuclear weapons. This loadout reduced effective magazine capacity to only nine rocket-thrown torpedoes plus eight in the launcher.

ASROC depth charges were unpopular. The rocket-thrown torpedo had been regarded for years as having at least an equal kill radius to the nuclear depth charge. A submarine was as likely to be hit by a Mark 46 homing torpedo as to be destroyed by the shock wave of a nuclear depth charge dropped at the same aim point. A submarine surviving a torpedo attack might be quickly reacquired for follow-up attacks, whereas an underwater nuclear explosion deafens sonars with reverberations (blue-out) that echo through the ocean for minutes or hours. Surface ships would need to wash down radioactive contaminated water thrown up by the detonation and might suffer damage to electronics from electromagnetic energy transients.

Except during ship overhauls, when the ship carried no ammunition, destroyer crews had to guard the ASROC magazine all the time, since to change the security routine would reveal whether nuclear weapons were aboard. They frequently practiced firefighting for the unsettling problem of putting out a burning, explosive-laden nuclear bomb. Duty aboard the *Oliver Hazard Perry* (FFG 7)–class frigates became popular because without nuclear weapons capability, these drills were unnecessary. As an antinuclear protest, some countries, notably New Zealand, refused to let U.S. Navy warships visit.

ASROC nuclear depth charges were retired in 1989 and scrapped.[15] The ASROC box launchers and RUR-5A rocket-thrown torpedoes were retired in 1993, and the launchers were removed from all ships still mounting them. No other nuclear ASW weapons are, or ever were, carried. Surface ships could not fire the nuclear SUBROC missile, retired in 1989 along with nuclear ASROC. LAMPS helicopters are not equipped to carry the Navy's only remaining nuclear ASW weapon, a depth-fuzed B57 air-drop bomb.[16] The Sea Lance ASW missile, planned to supersede SUBROC and nuclear

ASROC, was canceled in 1990. Congress had canceled its nuclear depth charge warhead (W89) in 1986.[17] The new RUM-139A vertical-launch ASROC (VLA) carries a Mark 46 torpedo as its only payload.

Mark 26-equipped *Ticonderoga*-class cruisers and the *Kidd* class had almost identical ASW capability to the *Spruance* class. They fired ASROC missiles from their Mark 26 twin-arm guided missile launchers, controlled by a modified Mark 116 underwater fire control system. The long forward extensions of the Mark 26 launcher arms existed solely to aim ASROC missiles onto their unguided, ballistic trajectories. These ships cannot use any nuclear weapon remaining in the American arsenal, since Mark 26 launchers cannot fire Tomahawk cruise missiles.

SQQ-89 Modernization of ASW Weapons Systems

Development of new ASW systems began in the late 1960s to handle multiple submarines detected at long range by passive sonar. The DX/DXG program specified that the DD 963 class would have advanced silencing features and large reservations of space to support modernized systems, a very farsighted vision since in the late 1960s the necessary technology did not exist.

In the 1980s a major modernization program began equipping all *Spruance*-class destroyers with the SQQ-89 ASW combat system, providing the long-range ASW capabilities foreseen by the DX/DXG office. Table 11.2 lists the modernized ASW armament. The first SQQ-89 ASW combat system was installed aboard USS *Moosbrugger* (DD 980) in 1982. *Moosbrugger* earned a Meritorious Unit Commendation for surveillance operations against multiple Soviet submarines in the Mediterranean during 1986. USS *Spruance* (DD 963) reported that the new ASW system was "a tremendous advantage" on a prolonged exercise in the Norwegian Sea in 1988.[18]

This upgrade provides the LAMPS Mark III system with the SH-60B helicopter to attack distant targets, a very long-range towed-array passive sonar (SQR-19 TACTAS), and a large set of display consoles to handle the larger number of contacts that the ship can track. Other systems include vertical-launch ASROC and upgraded versions of the SQS-53 sonar and the Mark 116 underwater fire control system. The new SQS-53B

Table 11.2   ASW weapons systems after the SQQ-89 modernization program

Sonar
— SQR-19 tactical towed-array passive sonar (TACTAS)
— SQS-53B or -53C active/passive sonar
— WQC-6 and WQC-2 underwater communication links
Tracking and fire control
— SQQ-89 ASW tracking system
— Command and Decision System (Naval Tactical Data System with enhancements for ASW)
— Mk 116 Mod 5/9/10 (non-VLS ships) or Mod 6/7/8 (VLS) underwater fire control system
— UYQ-25 acoustic range predictor
— UYS-1 spectrum analyzers
LAMPS Mark III
— SRQ-4 data link receiver
— SQQ-28 sonobuoy analyzer
SH-60B Seahawk helicopter
— 25 sonobuoys
— APS-124 search radar
— ARQ-44 sonobuoy data link transmitter
— ASQ-81 magnetic anomaly detector (MAD)
— 1–2 Mk 46 or Mk 50 torpedoes
Ordnance
— 2 Mk 32 triple 12.75-inch torpedo tubes for Mk 46 and Mk 50 ASW torpedoes (12–18 reloads)
— VLS DD 963 class and CG 56–73: vertical-launch ASROC rocket-thrown Mk 46 torpedoes (VLA)
Countermeasures
— SLQ-25 Nixie acoustic decoys
— Prairie acoustic sound dampener on propellers and struts

and TACTAS sonars have built-in sonar signal processors that replace the SQR-17 analyzer.

The SQQ-89 system provides additional computer power to compute target motion of multiple submarine contacts. The combat information center (CIC) crew fights the ASW battle at four dual-screen consoles (UYQ-21). Each console can be switched to any of the SQQ-89 sensors: SQQ-28 LAMPS sonobuoy data analyzer; SQR-19 towed-array sonar; and SQS-53B hull-mounted sonar. Adjacent consoles might display the SQR-19 view of a contact and an SQQ-28 view of sonobuoys laid near the same contact. The CIC crew evaluates contact data to identify contacts and sends contact data to the Mark 116 underwater fire control system, which sends them to the Command and Decision System (CDS). The SQQ-89 system can replay track histories, a vital feature.

The SQR-19 tactical towed-array sonar (TACTAS) vastly increases passive sonar detection range over the SQS-53 hull-mounted sonar. Sonar transducers are contained in an 800-foot-

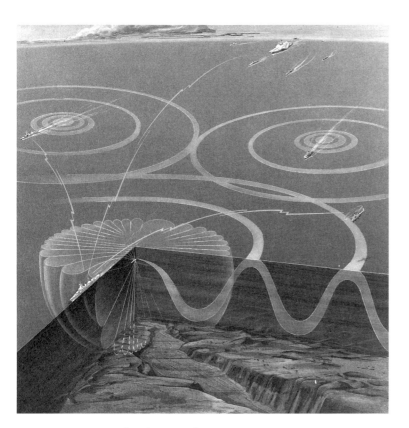

This diagram shows sonar convergence zones as rings traced by sinusoidal sound patterns. A destroyer can detect submarines and other ships in convergence zones at very long range using the SQR-19 tactical towed-array sonar. Seahawk helicopters or other ASW aircraft would localize the submarine with sonobuoys. (Courtesy Martin Marietta)

long flexible tube (the array), towed from a 5,600-foot cable down to 200 fathoms' depth. The ship's speed and the length of cable streamed out control the array's depth. Its linear shape gives the array minimum resistance, so that the destroyer can tow it at high speed and in heavy weather without risk of water resistance stripping the array from the cable. Distance isolates it from surface sounds, including those of the destroyer herself. Prairie air, Masker belts, machinery isolation mounts throughout the ship, and the gas turbines reduce the ship's self-noise interference with the towed array. Towed-array sonar handling equipment is mounted internally beneath the fantail. Only its circular deployment port in the transom is visible outside the ship.

Submarines cannot dissipate self-noise into the atmosphere, and their smaller hulls limit the size of their towed arrays and the isolation cable. The

destroyer's towed array is superior to a submarine's in some circumstances. Towed arrays perform well in heavy weather because hull slamming affects them less.[19] The SQR-19 TACTAS can detect and track submarines across several convergence zones. Open sources on ASW tactics discuss stationing SQR-19 ships 100 nautical miles from the main body of a formation to reduce acoustic interference.[20] This reveals two interesting capabilities. First, a range of 100 nautical miles equals three convergence zones in cold water. Second, the towed array can detect surface ships passively, providing targeting information for Harpoon and Tomahawk missiles.

The capability of TACTAS to detect targets passively at 100 nautical miles increases the number of contacts within sonar range, so that the ASW combat system must track multiple acoustic contacts simultaneously. The upgraded Mark 116 Mod 5 underwater fire control system tracks up to 99 underwater contacts. New operator consoles display more information, such as inputs from multiple sensors, or details about multiple contacts. Operators can enter synthetic data such as intelligence reports or contact data from radar detectors. SQQ-89 adds an on-board trainer to feed simulated contact data into the displays. Another SQQ-89 unit (SIMAS) models water conditions to help predict sonar range and sonobuoy field dispersion.

Under the SQQ-89 program, the hull-mounted sonar is upgraded to SQS-53B, which is easier to maintain and can track more contacts. An even more advanced version of the SQS-53 sonar, SQS-53C, was installed aboard USS *Stump* (DD 978). This sonar has a stealthy active mode in which a submarine cannot discern the pings until she is within detection range. SQS-53C can survey all sectors around the ship simultaneously for new contacts while tracking existing contacts.[21] *Stump* has routinely made second-CZ detections with her SQS-53C at high ship speeds. Current plans are to install the SQS-53C receiver section in all SQS-53A and -53B sonars, giving them near-53C performance, especially in shallow water.

The Mark 116 underwater fire control system is upgraded to track 99 contacts. It receives nonsonar data from CDS (NTDS), notably radar and ESM (electronic support measures) contacts, to identify which sonar contacts are not from submarines. By matching a sonar contact to a radar contact, operators can estimate range to other sonar contacts. This is especially important in

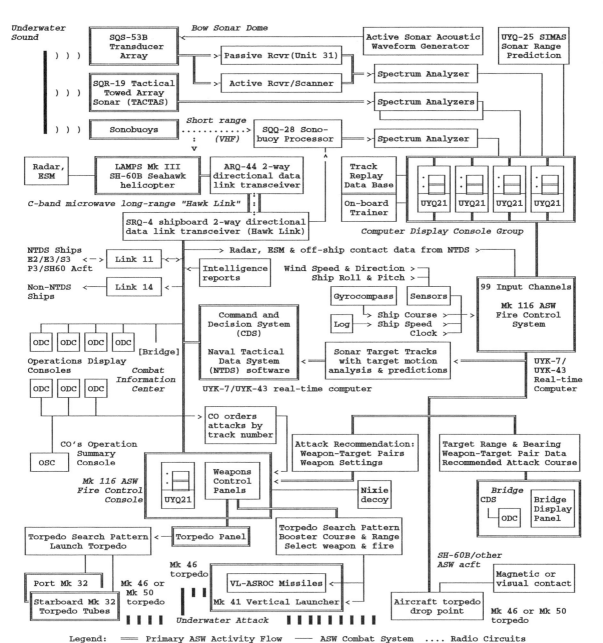

Underwater
Sound

SQS-53B Transducer Array — Bow Sonar Dome — Active Sonar Acoustic Waveform Generator

UYQ-25 SIMAS Sonar Range Prediction

Passive Rcvr (Unit 31)

Spectrum Analyzer

SQR-19 Tactical Towed Array Sonar (TACTAS)

Active Rcvr/Scanner

Spectrum Analyzers

Sonobuoys — Short range (VHF) — SQQ-28 Sono-buoy Processor

Spectrum Analyzer

Radar, ESM — LAMPS Mk III SH-60B Seahawk helicopter — ARQ-44 2-way directional data link transceiver

Track Replay Data Base

On-board Trainer

UYQ21 UYQ21 UYQ21 UYQ21

C-band microwave long-range "Hawk Link"

SRQ-4 shipboard 2-way directional data link transceiver (Hawk Link)

Computer Display Console Group

NTDS Ships E2/E3/S3 P3/SH60 Acft — Link 11 — Radar, ESM & off-ship contact data from NTDS

Intelligence reports

Wind Speed & Direction > Ship Roll & Pitch >

99 Input Channels Mk 116 ASW Fire Control System

Non-NTDS Ships — Link 14

Gyrocompass Sensors

Command and Decision System (CDS)

Log > Ship Course > Ship Speed > Clock >

ODC ODC ODC ODC [Bridge]

Naval Tactical Data System (NTDS) software

Sonar Target Tracks with target motion analysis & predictions

UYK-7/ UYK-43 Real-time Computer

Operations Display Consoles

Combat Information Center

UYK-7/UYK-43 real-time computer

ODC ODC ODC

CO orders attacks by track number

Attack Recommendation: Weapon-Target Pairs Weapon Settings

Target Range & Bearing Weapon-Target Pair Data Recommended Attack Course

CO's Operation Summary Console

OSC

Mk 116 ASW Fire Control Console

UYQ21

Weapons Control Panels

Nixie decoy

Bridge CDS ODC

Bridge Display Panel

Torpedo Search Pattern Launch Torpedo

Torpedo Panel

Torpedo Search Pattern Booster Course & Range Select weapon & fire

SH-60B/other ASW acft

Port Mk 32

Mk 46 torpedo

Mk 46 or Mk 50 torpedo

VL-ASROC Missiles

Magnetic or visual contact

Starboard Mk 32 Torpedo Tubes

Mk 41 Vertical Launcher

Aircraft torpedo drop point

Mk 46 or Mk 50 torpedo

Underwater Attack

Legend: ═══ Primary ASW Activity Flow ──── ASW Combat System .... Radio Circuits

This diagram shows the primary functions and subsystems of the SQQ-89 computer-based ASW system, which the destroyer can use to prosecute multiple submarine contacts.

shallow water, where sonar ranges change quickly. The system can compute a contact's track and speed with four active-sonar pings. This information is sufficient for launching an attack on the contact.[22]

Ship-Launched ASW Attacks

ASW ordnance now installed on the *Spruance* class includes vertical-launch ASROC and over-the-side torpedoes. Destroyers equipped with vertical-launch systems during modernization can fire vertical-launch ASROC (VL-ASROC or VLA, designation RUM-139A). The VL-ASROC missile fires a Mark 46 torpedo out to about 20,000 yards. Several hundred have been procured and are rationed to forward-deployed ships. It is undergoing extensive testing. Fire control uses the Mark 116 underwater fire control system.

Two Mark 32 over-the-side triple-tube torpedo launchers are mounted in internal torpedo rooms.

The Mark 32 torpedo tubes fire 12.75-inch ASW homing torpedoes through pneumatic doors recessed in the hull sides. Air-drop torpedoes are carried for the LAMPS helicopter; parachute kits can be attached to adapt the ship's torpedoes for LAMPS use. The reload capacity anticipates tactical reality, in which many torpedoes will be fired at false contacts or in urgent counterattacks, or will miss. The over-the-side torpedo tubes and the SH-60B Seahawk helicopter are being modified to fire the Mark 50 ASW torpedo for use against fast, deep-diving submarines.

In the original DD 963 configuration, two linked computer-based systems supported ASW tactics: the Command and Decision System and the Mark 116 Mod 0 underwater fire control system. Invisible to outside observers, these systems vastly increased the range of ASW tactics available to the ship. *Spruance*-class destroyers were the first surface warships that could prosecute multiple submarines simultaneously. The DD 963 Command and Decision System is a version of the antiaircraft-oriented Naval Tactical Data System, upgraded to display ASW-specific data. The *Spruance*-class destroyers were the first ASW ships designed for NTDS functions. The Command and Decision System is important in coordinating ships and aircraft to obtain cross-fixes and to attack targets. Its software runs on a UYK-7 real-time digital computer with three central processors (bays) accessing a set of common memory banks.

The Command and Decision System displays datum (last contact), sonobuoys, and farthest-on circles showing a submarine's possible location since last contact. Given a possible submarine contact, CDS displays torpedo danger zones: trapezoidal sectors where the submarine could make a torpedo attack, which are therefore areas to avoid. CDS expands the torpedo danger zone automatically over time for possible submarine courses and speeds since last contact. Another CDS function is to control helicopter missions, showing search patterns and sonobuoy locations and recommending helicopter courses to the ASW air controller. Two ships holding a contact can exchange passive cross-bearing data through the NTDS data links to fix the contact's location, course, and speed.[23]

The Mark 116 underwater fire control system tracks submarines using sonar contact data. Based on bearing change rate and other data, such as depth and range, if available, the underwater fire control system predicts a track showing the contact's speed and course. The prediction is displayed both on CDS consoles and on sonar display consoles, where the operators continue to enter updates based on the submarine's observed maneuvers. The computers can replay track histories. Early Mark 116 systems could track four contacts: two using sonar data, one track from CDS (such as track data relayed from another DD 963), and one entered by the ASW fire control officer (usually a senior sonar technician). The upgraded versions maintain 99 contact tracks, including track histories. Target tracks might be reported from other ships, from the LAMPS helicopter, from sonobuoys, from other aircraft, from intelligence reports, from land-based SOSUS operation control centers, and from mobile inshore undersea warfare units.[24] Sonar technicians and operations specialists correlate contact reports from different sources and times into tracks.

The Mark 116 underwater fire control system can set up simultaneous attacks on two target tracks, firing VL-ASROC at one and a torpedo from a Mark 32 launcher at another. Mark 116 recommends which target to attack with which weapon (weapon designation), recommends attack course and speed, and calculates aim points for each weapon. When Mark 116 achieves a fire control solution, the ASW fire control officer sets the course, depth, and search pattern into the torpedo and, for VL-ASROC, sets flight range. For over-the-side torpedo shots, the ASW fire control officer selects which torpedo to shoot and transmits the search pattern to the torpedo in the tube. Firing is controlled remotely from CIC. The torpedo tube door, mounted on vertical rails, is raised by pneumatic rams, and the Mark 32 trains out to point the muzzles over the side. A compressed-air charge shoves the torpedo out of the tube. It dives into the water to run out on the preset trajectory to search for its target. Its range of about a mile limits it to close-in targets. Detonation of its warhead gives the ship a perceptible jolt.

In open-ocean ASW, success of LAMPS reduced the likelihood of use of close-range ship-launched weapons for attacks against submarines penetrating the outer helicopter screen. In shallow water quiet conventional submarines encountered at close range would clearly be dangerous. The ship would fire torpedoes by VL-ASROC or Mark 32 toward the contact in urgent attacks to disrupt submarine attacks. Meanwhile, the ship would try to get enough time and contact data to compute a more accurate fire-control solution and to get a LAMPS helicopter airborne.

In close-range antisubmarine combat a DD 963 would fire ASW torpedoes through pneumatically raised doors in the hull. These are the muzzles of the port side Mk 32 torpedo tubes, trained out in firing position.

A question exists about whether a Mark 46 hit is certain to be lethal. The Soviets built their submarines stronger than American submarines to withstand combat damage. Large Russian-built submarines are designed to survive submerged flooding of a compartment, but frequent losses in accidents suggest that these designs remain inadequate.[25] No submarine operated by potentially hostile regional powers could outrun a Mark 46 or survive a hit by it.

LAMPS Mark III

LAMPS Mark III with the SH-60B Seahawk ASW helicopter is the weapon to attack convergence-zone contacts made by the SQR-19 TACTAS towed-array sonar. The TACTAS sonar detects contacts in the second and third convergence zones, and the Seahawk helicopter can localize and prosecute contacts at those ranges. The second convergence zone is a practical limit. Distance to a third-convergence-zone contact is

In the starboard internal torpedo room, the Mk 32 triple-tube launcher is trained out through the rising pneumatic door. Reload torpedoes, which can also be sent to the flight deck for helicopter use, are in racks at left. A deck-mounted monorail can shunt weapons between the port and starboard torpedo rooms and to the flight deck hoist. (U.S. Navy photo)

around 100 miles, so the tactical commander of an ASW force will prefer to send a faster fixed-wing ASW aircraft such as the S-3 Viking, the P-3 Orion, or an allied counterpart.[26] Table 11.3 describes the SII-60B helicopter.

The backbone of LAMPS Mark III is the "Hawk Link" microwave data link, between the shipboard SRQ-4 and the SH-60B's ARQ-44 data link radios. It supersedes the LAMPS Mark I one-way sonobuoy signal relay and eliminates the need to use TACAN (tactical air navigation) beacons and UHF (ultra-high-frequency) voice radios during LAMPS missions. The ship transmits contact data (from all sensors) over the Hawk Link directly into the Seahawk's mission computer for immediate display on the aircraft's consoles.[27] Similarly, the helicopter transmits contacts from its sensors to the ship. The helicopter has better surface-search radar coverage than the ship, provides an automatic cross-fix on ESM rackets, and does not reveal the location of the destroyer to enemy targeting systems. These are extremely useful services. Force ASW coordinators often find that SH-60B missions are more oriented toward surface patrol than ASW patrol.

Helicopters give destroyers a long-range, fast-response ASW capability. A plan for a helicopter ASW mission balances range, payload, and time on station. Sonobuoys transmit their signals over VHF radio channels. In high sea states waves quench sonobuoys and can block the radio to the ship. The SH-60B carries an acoustic-signal processor to analyze sonobuoy signals independently. A new acoustic-signal processor, SQQ-28, analyzes sonobuoy contact data aboard ship. Often the ship will monitor the sonobuoys so that the Hawk Link is free to transmit the helicopter's surface radar picture. The 99-channel capability lets the ship and aircraft monitor multiple sonobuoy fields simultaneously and switch attention to new buoys while old ones are still transmitting.

On ASW patrol a destroyer keeps a helicopter on deck on alert, loaded with sonobuoys, fuel, smoke floats, and a torpedo. Given a sonar contact or intelligence report of a submarine, the helicopter launches. The operations watch in the combat information center uses the Command and Decision System to vector the helicopter to the contact area and to plan sonobuoy drop patterns. A major value of LAMPS is that the helicopter can fly quickly to a contact area to start determining whether a contact may in fact be a target. The ship spends much less time tied down by false alarms

and is less likely to waste ASW ordnance on urgent shots against false contacts.

The combat information center estimates the target's position, course, and speed, and vectors the helicopter to drop sonobuoys around this location. Data are entered into the underwater fire control system, which computes a torpedo drop point. The helicopter drops patterns of sonobuoys until the target is redetected, or until the original contact is evaluated as a false alarm or as having escaped. The helicopter might detect the submarine's steel hull magnetically or make an above-water detection with radar, ESM equipment, or aircrew sighting. The aircraft commander would drop the torpedo when sensor data, or unambiguous detection, give confidence that the submarine is within the torpedo's acquisition range.[28]

The *Spruance*-class hangar was originally designed to house the big SH-3 Sea King ASW helicopter. *Elliot* operated an SH-3 for a week in 1979, storing it in the hangar and moving it to and from the flight deck by hand. Seas were calm.

Table 11.3   LAMPS Mark III SH-60B Seahawk helicopter (Sikorsky)

Crew	
3 (pilot, copilot/tactical officer, weapons operator)	
Engines	
2 turboshafts T700-GE-401 (1,690 hp) or -401C (1,900 hp)	
Weights	
Empty	13,648 lbs (6,191 kg)
ASW	19,500 lbs (8,845 kg)
Surface targeting	18,000 lbs (8,165 kg)
Utility	21,000 lbs (9,526 kg)
Dimensions	
Fuselage length	50 ft (15.26 m)
Overall length	64 ft 10 in (19.76 m)
Rotor diameter	53 ft 8 in (16.36 m)
Performance	
Speed	125 knots (233 km/hr)
Radius	50 nm (92.5 km) with 3-hour loiter
	150 nm (278 km) with 1-hour loiter
Electronics (partial)	
APS-124 surface search radar (X-band)	
ALQ-142 microwave ESM intercept sensor	
ARQ-44 directional data link to ship's SRQ-4 antenna	
AYK-14 mission computer	
Sensor operator display console	
ASQ-81 magnetic anomaly detector (MAD)	
Armament	
1–2 Mk 46 or Mk 50 torpedoes	
Penguin antiship missile	
Chaff and flares	
M-60 machine guns	

Helicopter takeoffs and landings are limited to sea states in which a DD 963 does not roll prohibitively. Acceptable conditions occur in the North Atlantic only 69 percent of the time during winter, or less considering precipitation, icing, and limited visibility.[29] Frigates are even more affected and can operate helicopters just 25 percent of the time in the same seas. Since oceanic storms are neutral and cover very large areas, this situation was acceptable if Soviet naval forces were themselves unable to attack with long-range antiship missiles in the same weather.

The Canadian RAST (recovery assist, secure and traversing) system is being installed for landing and flight deck handling of the heavier SH-60B helicopters. RAST supports helicopter operations during sea state 5, conditions that would ground helicopters flying from conventional shipboard flight decks. RAST is a descendant of the Beartrap recovery system on the Canadian *Iroquois*-class destroyers. It is another of the many links between the *Iroquois* and *Spruance* classes.[30]

The landing signal officer works in a station below the flight deck with bulletproof windows to give a view of the deck. To land, the SH-60B hovers about 15–20 feet above the deck and drops a messenger line from a probe on the bottom of the fuselage. Flight deck crewmen hook the line to a cable from the RAST deck trolley (recovery securing device, or RSD). RAST winches the helicopter down to the deck trolley and latches onto the probe. After the helicopter shuts down and folds its rotor blades, the trolley pulls the helicopter into the hangar. The pilots can release RAST and land without it. RAST is still necessary to traverse the aircraft into the hangar. Launch is similar. The trolley moves the SH-60B onto the flight deck, where the helicopter spreads its rotor blades and starts its engines. When the pilots signal ready for takeoff, the landing signal officer releases RAST's grip on the probe. The helicopter ascends, receives a visual check from the landing signal officer, turns 45–60 degrees off the centerline, and departs.[31]

The first 15 *Spruance*-class destroyers converted to LAMPS III have one RAST system. A large compartment has been attached to the starboard aside of the hangar structure to store sonobuoys and firefighting equipment and to house an aircrew ready room. Later conversions provide two RAST systems. On these ships, the port-side hangar wall has been moved outboard flush with the side of the hull, widening the hangar for two

The Sikorsky SH-60B Seahawk is the LAMPS Mk III helicopter, a far more capable aircraft than the smaller Seasprite. The Seahawk features an on-board display of contact data and a high-speed secure computer data link (Hawk Link) to the parent destroyer. This one is dropping a Mk 46 antisubmarine torpedo. They can also carry chaff, flares, machine guns, and forward-firing homing missiles. (U.S. Navy photo)

SH-60Bs. In addition, the hangar structure is extended to starboard as on the single-RAST ships.

Keel slamming still restricts *Spruance* ASW operations during heavy weather, sea state 6 (12-foot waves) or worse.[32] The ship must steer a parallel course to hold a submarine in a convergence zone or in direct pursuit, but frequent slamming can deter the ship from keeping that course. A dominant feature of the succeeding *Arleigh Burke* design is its broad waterplane hull to improve seakeeping in rough northern seas. *Arleigh Burke*'s designers sacrificed speed, range, and the helicopter facility, and were prepared to dispense with 5-inch gun armament, to obtain that improvement, which suggests its operational importance.

Outer-Zone ASW

A destroyer listening for a convergence-zone contact must station herself beyond convergence-zone distance from other ships, to avoid masking from their hull and machinery noises.[33] Sprinting to long range from the task group reduces noise interference from friendly ships. It extends the detection range against hostile surface ships, which can be targeted over the surface radar horizon when they pass through a convergence zone. Sonar sensitivity can improve when the ship stops and drifts in the ocean current in silence with even her own machinery shut down. A DD 963 might use sprint-and-drift tactics to reacquire a contact at closer range or from a different bearing. Fast shutdown and fast restart features of the gas turbine propulsion plant make sprint-and-drift passive listening tactics feasible. Silent-running discipline and the Prairie/Masker silencing systems make *Spruance*-class destroyers as efficient sonar platforms under way as with sprint-and-drift tactics.

Long endurance increases the time that a DD 963 can continue remote operations before having to approach noisy oilers to refuel and to replenish food and ammunition. A new class of replenishment oilers, the *Supply* (AOE-6) class, has gas turbine propulsion and other silencing features to reduce acoustic signature in fleet formations.

Destroyers and the SH-60B helicopters can be valuable in the inner zone, the area within 35–60 nautical miles, or about one to two sonar convergence zones, of the task group center. The SH-60B pencil-beam data links can connect remote forces by secure voice radio even during general radio silence. Battle group commanders leading a force on a raid or amphibious operation find this service obviously useful for final planning and coordination.

The cost is that the destroyers may not be allowed to sortie over 60 nautical miles from flagships, a restriction that reduces ASW efficiency.[34] Two of the *Spruance* class's main sensors (towed-array sonar and the Outboard communications intercept system) are most useful when the ship is in the outer zone of a task group, 100–200 nautical miles from the task group center. These systems collect combat intelligence about an area of operation. The long separation range reduces noise interference from friendly units. At 100–200 nautical miles' range, multiple destroyers cover a broader and deeper front, and they can use triangulation to get cross-fixes on contacts of interest. Tactics expert Captain Wayne Hughes calls this scouting.[35]

A ship scouting in the outer zone risks attack from hostile ships and aircraft. The Soviets intended to deploy much of their surface fleet to attack scouts. Outboard, the radar plot, and the Link-11 radio data links inform supporting aircraft of the tactical situation surrounding a destroyer. Harpoon, Sea Sparrow, chaff, and Phalanx weapons are obviously vital defenses for outer-zone operations. One assumption about outer-zone ASW is that submarines will hunt the aircraft carriers and other high-value units in the inner zone, or that they will attempt to pass through the destroyer screen en route to another operating area. A submarine on such a mission might in theory not risk attacking a lowly destroyer, because an attack (flaming datum) would reveal the submarine's certain presence. More likely, however, a submarine commander would consider the destroyer a suitably valuable target, or would simply attack the first target of any kind encountered. Outer-zone ASW might well be as intense as ASW in the inner zone, with multiple submarines making repeated attacks. To paraphrase Winston Churchill, destroyers might be safer in the inner zone, but that is not what destroyers are for.

Active Sonar Operations

In active mode the SQS-53 transmits pings on all bearings to search for a contact, or it can focus active search signals toward a sector. The SQS-53 forms sound patterns for best penetration in the water conditions. The active sonar ping starts in tone at a low sonic frequency, rises quickly to a higher pitch in about a second, and continues at that pitch for about another second. Pings repeat at an interval of several seconds. The SQS-53 listens for echoes from the ping on all sectors simultaneously and can detect an echo returning on any bearing. Time for the echo to return indicates the distance to the submarine. Echoes from sequential pings are used to compute target motion. The SQS-53 centers its active tones at 3.5 kHz; this is lower than other sonars and is the reason that the SQS-53 is termed a low-frequency sonar. A squadron of DD 963s can go active with their SQS-53s simultaneously on different frequencies around 3.5 kHz.

The advantages of active sonar operation are that it can classify a contact as a submarine, that it can find quiet submarines, and that it can more readily find the target's range, course, and speed. Of course, a submarine may be altering course, speed, and depth drastically during combat. Disadvantages of active sonar include its shorter range in open-ocean ASW and the fact that it alerts a submarine that the destroyer is present and hunting.

Active sonar echoes are broadband and do not identify the specific submarine type. The submarine may counter with a decoy or by firing a torpedo or a pop-up antiship missile toward the source. Such an attack is, of course, both a risk during ASW and the reason for it. Notice, however, that a fired weapon is a flaming datum that immediately answers the critical questions in ASW of whether a submarine is actually present, whether she is hostile, and possibly even where she is.

The DD 963 WQC-6 Probe Alert system uses the SQS-53 sonar in active mode as a one-way underwater data link, to transmit short encrypted digital signals at long range to a friendly submarine. It is not a tactical data communications link, but it can alert a friendly submarine that radio traffic awaits it.[36] Range and modulation require high sound intensity. The well-known WQC-2 Gertrude underwater telephone supports close-range two-way communication. Underwater communication generally requires that the submarine be expecting the call. The destroyer can alert a submarine by dropping a Sound Underwater Signal, a type of depth charge that detonates explosive charges in a coded sequence. This method and others notify a friendly submarine to take a prearranged course of action, such as rising to receive urgent radio traffic or to identify herself.[37]

Shallow-Water ASW

Shallow coastal water requires ASW tactics very different from those used in the open ocean. Low-frequency sound does not propagate far through shallow water. Convergence zones rarely occur, but sometimes bottom-bounce is useful. Ambient noise from marine life and coastal industry is high. Since no regional power has nuclear submarines, defending submarines will be conventional, running slowly on batteries or chemical systems without the noisy gears and pumps of nuclear submarines. These conditions make passive sonar much less useful than active sonar. Active sonar will often encounter wrecks, seamounts, and man-made dumps where defending submarines might, or might not, be hiding. Water conditions, thus sonar effectiveness, change rap-

John Hancock (DD 981) streams her SQR-19 towed-array sonar. The mile-long towing cable is visible extending aft from the circular deployment port on her transom. Notice that her name is in script, replicating John Hancock's signature on the Declaration of Independence. (Larry Marchlewski)

idly.[38] Shallow-water ASW ranges will often be surface direct-path ranges, typically less than 3,000 yards. False contacts will be frequent. Urgent attacks on nearby underwater contacts will be common.

Submerged submarines raise periscopes, snorkels, search radars, and radio communication antennas, which provide detection opportunities for a patrolling destroyer. DD 963 systems to detect radio and radar signals are Outboard, SLQ-32, and sensors aboard the LAMPS helicopters. At close range submarine periscopes, snorkels, and electronics masts can be detected by lookouts, high-RPM SPQ-9 radar, and helicopters. If the radar contact disappears, then the helicopter is still in a better position to drop sonobuoys or a torpedo. The opposite problem is that the submarine might surface to fire hand-held antiaircraft missiles at the helicopter.

During the Falkland Islands War in 1982 Argentina's sole active modern submarine, the German-built Type 209 *San Luis*, operated in shallow water for almost a month. Shallow water, many targets, and knowledge that no friendly ships were in her area made conditions ideal for a small submarine. The ASW-oriented Royal Navy expended 239 rounds of ASW ordnance, including

44 ASW torpedoes, during the campaign but never obtained a solid contact on the submarine.[39]

This ordnance was not wasted, since the active British ASW defenses so intimidated the Argentines that *San Luis* accomplished nothing. She made only two attacks and declined to risk exposing her periscope during either. On her first attack, about May 1, *San Luis* fired a torpedo at excessive range toward a warship. It missed, and the submarine dived to the seabed while helicopters dropped depth charges and ASW torpedoes in the area for 20 hours. The British then gave up and left, and *San Luis* escaped. The attack indirectly affected British operations on May 4 when, under an illusion that the submarine was in the area, the task force abandoned attempts to save the missile-damaged destroyer *Sheffield.* North of Falkland Sound on May 11, *San Luis* fired an acoustic homing torpedo at the frigate HMS *Arrow* at close range and at a favorable angle ahead of the ship. *Arrow*'s towed antitorpedo decoy lured away the torpedo, which rammed it without detonating. *Arrow* hauled in her successful but pulverized decoy, attributed the damage to clobbering on the seabed, and sailed on. *San Luis* was slow and could not pursue for another attack.[40] In combat, non-nuclear submarines are intelligent minefields.

Two other ASW incidents occurred. In April British naval helicopters found the old Argentine submarine *Santa Fe* (ex–USS *Catfish* [SS 339]) on the surface off South Georgia Island. They severely damaged her with rockets and depth charges. The Argentines ran her into the harbor and abandoned her. After *San Luis*'s unsuccessful first attack on the British, the nuclear submarine HMS *Conqueror* pursued the Argentine cruiser *General Belgrano.* Two old ex-American destroyers were in close company, but their sonars were not active, and they probably had no passive sonar capability. *Conqueror* torpedoed and sank *General Belgrano* at a range of 1,400 yards. Before Argentina bought her, the cruiser was USS *Phoenix* (CL 46). The city of Phoenix, Arizona, had donated a presentation silver collection to this ship in the 1930s. The U.S. Navy retained the silver and assigned it to the *Spruance*-class destroyer *Elliot.*

The Falkland Islands War was the most intense naval action since World War II. Other works cover it extensively and only a brief narration is given here. Given no warning, the Royal Navy assembled an amphibious task force and deployed it to the South Atlantic. Equally important, a large logistics squadron of supply ships, tenders, tank-ers, hospital ships, and minesweepers supported the combat ships and transports. The single notable category of British numerical superiority over the Argentines was in ships. Surface warships screened the amphibious assault. Argentine troops under naval bombardment retreated from prepared positions into infantry killing zones. Four modern destroyers and frigates and two transports were lost to land-based air attacks (two to Exocet ASCMs, four to bombing), but Argentine aircraft losses were so heavy that these raids became ineffective. The newest air-defense missile systems were the most lethal. The decisive elements in Britain's victory were Prime Minister Margaret Thatcher's leadership, the outstanding quality of personnel, and the availability of a versatile navy. If U.S. forces of the same quality and size had attempted to fight the campaign under the political interference practiced throughout the 1960s and 1970s, they would have lost. Afterward the U.S. Navy adopted British-style protective clothing and increased warship magazine capacity with vertical launch systems.

Hostile coastal waters are likely to feature not just quiet submarines but minefields, too. Attacks from aircraft, fast missile boats, and shore-based cruise missiles like China's Silkworm are additional threats. The *Spruance*-class destroyers' multiple radar, sonar, and optical sensors improve their chances.

Torpedo Defenses and Mine Countermeasures

Submarines detect and identify a ship from her noise signature. The noise signature results from the wave-forming propulsive resistance of the hull, from machinery transmitting vibrations to the hull and thence to the water, and from the propellers. DD 963s were built to be quiet with specific silencing features, described previously. Quiet modern submarines remain lethal foes. One officer has described how his nuclear submarine exploited oceanic conditions in an exercise:

Starting at test [near-maximum] depth and greater than 20 knots, it went to periscope depth for one seven-second look, then returned below layer while firing exercise Mark-48 [torpedoes] in an undetected combined attack against an active *Spruance* (DD 963)–class destroyer and the "high-value unit" she was escorting. The total elapsed time of the evolution was five minutes, less than two of which were above layer. Maritime patrol aircraft and ASW helicopters were supporting the surface units.[41]

During updates in the 1980s the ships were equipped with SLQ-25 Nixie decoys. The visible evidence of Nixie is a pair of small vertical doors in the transom. To attract active/passive acoustic homing torpedoes, Nixie generates the noises of a more valuable alternative target. It returns active-sonar echoes.[42] Nixie's decoy signals are advertised as not interfering with sonar. Typically two Nixie fish are towed. If collision with a successfully seduced homing torpedo destroys the first Nixie fish, the second decoy remains active.[43] The Mark 116 underwater fire control system has an interface with Nixie, evidently to set the ship's torpedoes so that they do not home on friendly decoys.[44]

Deficiencies of Nixie include threats from more than two torpedoes and from nonacoustic weapons, such as Russian wake-homing torpedoes, mines, and old-fashioned periscope-aimed torpedoes.[45] The successor to Nixie is a joint United States–United Kingdom project that will feature "defense in depth," which suggests both long-range and close-in torpedo-defense weapons.[46] Many nations own short-range, optically aimed torpedoes of the sort that sank *General Belgrano*. Often they are Russian copies of captured straight- or pattern-running U-boat weapons.[47] Deaf, they cannot be decoyed, only evaded. Short range means little warning. Destroyers are maneuverable, so they can elude such an attack. Their sonars could offer the only warning to other ships to turn away from a rapidly closing salvo.

Developments have been under way since the late 1980s to add mine detection capability to the *Spruance* class's SQS-53 active sonar. The newest version, SQS-53C, generates sonar wave-forms precise enough for shallow-water ASW, including mine avoidance.[48]

Research Directions

The SQQ-89 modernization program equipped the *Spruance* class for operations against submarines of 1970s technology, a category that still covers almost all submarines of potentially hostile nations. SQQ-89 is a passive system, and it relies on submarines making detectable noise. In the late 1970s, however, the Soviet Union introduced the Victor III–class nuclear submarines, which were much quieter than previous Soviet boats. Their decreased noise emissions made them much harder to detect. The likelihood increased that they might not be detected during an operation at all.

Soviet attention to submarine silencing stemmed from insight into ASW gained after 1968 from the Walker-Whitworth spy ring.[49] After 1985 Soviet submarines became quieter again by using new propellers made on precision machinery exported illegally by Norwegian and Japanese firms. The latest Russian nuclear submarines are detectable passively within only 5 percent of the range of earlier designs.[50] Some Russian submarines have plastic coatings on their hulls to absorb high-frequency active sonar pings. The independent Russian regime is completing nuclear submarines already on the building ways, but according to President Boris Yeltsin, it plans no new ones. The U.S. Navy began research for an improved SQQ-89 system, SQY-1, to use a very-low-frequency active sonar for detection. The plan was for one unit to broadcast low-frequency pings; destroyers and submarines would detect echoes from contacts through their towed-array sonars. Range in deep water would presumably span multiple convergence zones. The plan was canceled in 1991 after the collapse of the Soviet Union shifted naval attention to littoral theaters.

The immediate ASW problem today is the threat of modern non-nuclear, air-independent submarines operating in shallow coastal waters. Russia, Sweden, and Germany sell small, air-independent submarines for export. Multiple enemy submarines might decimate an amphibious force off their coast or block a task force from passing through a strait such as the Straits of Hormuz. Without the noisy pumps of nuclear reactors or large hull size to generate flow noise, these submarines may be undetectable by passive sonar, at least in shallow water. As noted, during the Falkland Islands War the Royal Navy never made a good sonar contact on a modern Argentine submarine of this type. However, submarines must still locate targets, and bad sonar conditions, notably the false-target echo problem, affect them, too. Attacks on their command posts ashore may deprive them of orders and intelligence updates while they are under way. Countermeasures might include dense sonobuoy barriers and offensive minefields. Competition continues.

The *Kidd* (DDG 993) Class

The *Spruance* Class Replaces DXG

In the original plans, the DX and DXG classes were to have been sisters, sharing as many common components as possible. Work proceeded first on DX (DD 963) because it was simpler. Total Package Procurement (TPP) for the DXG class would occur after the DD 963 design was in hand. Admiral Elmo Zumwalt's Major Fleet Escort Study in 1967 suggested designing DX for conversion to DXG armament. Admiral Thomas Weschler, DX/ DXG program coordinator, adopted the suggestion, a change that reduced the urgency to design DXG.

Meanwhile, a third class of new warships, DXGN, was planned, to be nuclear-powered and mount DXG armament. Since only the Navy had design experience with nuclear warships, the Navy would design DXGN. Admiral Weschler argued that the DXG class should be designed to support the Aegis air defense system once it was ready, so that an Aegis ship would be a simple modification of the DXG design. As with nuclear propulsion, Aegis was a Navy development, so that the Navy would have to design DXG. Industry could not provide a fixed-price TPP contract for an Aegis ship. If the Navy would be designing the Aegis-capable DXG class anyway, there would be no TPP contractual barrier to nuclear propulsion. Admiral Weschler concluded that DXG should be Navy-designed and nuclear-powered, not an industry-designed product of a TPP contract competition.[1] That was as far as plans for DXG ever developed.

A conventionally fueled DXG class, without Aegis but whose basic design might include design margins to accommodate it, remained a theoretical possibility for a TPP competition to follow the DD 963 class. The DD 963 design competition convinced Admiral Weschler that TPP could not be used for a guided missile destroyer class, even without Aegis and nuclear propulsion. Summarizing the use of Total Package Procurement for the DD 963 class, he reasoned that TPP was impractical for a warship as complex as DXG:

> A major combatant ASW ship is so complex that it is the upper limit of what can be competed [for] wisely by [TPP]. A missile ship is beyond such competition [because of] the long time required to generate and evaluate the competing proposals, the large government team and government investment required to evaluate the proposals[,] and the dollar cost and impact on industry in waiting through this competitive approach. The results for DD 963 are considered outstanding, but the size of the investments demanded such results if the investments were to be justified. Outstanding results cannot be guaranteed.[2]

Navy interest in DXG ended in 1970 for another reason: Litton's design for a converted DD 963 would be more powerful than DXG would have been. The DD 963 design included margins so that the ship could be converted to add DXG armament without sacrificing the helicopter facility or gun armament. Converted DD 963s would be more heavily armed than new-design DXGs. In any event, the Office of the Secretary of Defense (OSD) canceled Total Package Procurement after approving the DD 963 contract in 1970. That ended the DXG class as a separate shipbuilding project. DXGN became the *Virginia* (CGN 38) class of guided missile cruisers. To replace the FRAM fleet, Chief of Naval Operations Admiral Zumwalt initiated the *Oliver Hazard Perry* (PF 109) class, later redesignated the FFG 7 class. *Oliver Hazard Perry* (FFG 7) was his idea of what DXG always should have been, a missile-armed destroyer escort that he could present to Congress as affordable in large numbers.

The Shah's Cruisers

With the Suez Canal blocked after the 1967 Six-Day War, all tanker traffic from the Persian Gulf to Europe and the Americas had to pass around Africa. The United States was more distracted than ever by Vietnam, first by the 1968 Tet Offensive and then by attempts to disengage from the war. After the debacle in Vietnam it was difficult to develop national support for expanded military operations anywhere, and the security of the Persian Gulf was left to the British. But in 1968 Britain began withdrawing its permanent forces from the gulf, and by 1971 they were gone. Two ambitious powers intended to replace the British: the Soviet Union and Iran.

From 1968 on, Soviet warships patrolled the Indian Ocean more often and in greater numbers than ships of the U.S. Navy. The closing of the Suez Canal required the Soviets to support their ships from their Pacific Fleet bases instead of from the Black Sea. Their willingness to bear that expense showed the importance to them of an Indian Ocean naval presence. The Soviet Union signed an alliance with anti-Western Iraq and flooded it with Soviet arms. Soviet naval presence was influential. Small nations around the Indian Ocean could afford to maintain only a few embassies, and Moscow often got one of them.

Shah Mohammed Reza Pahlavi of Iran built up military strength to defend his nation, its impoverished eastern neighbors, and the Persian Gulf.[3] He bought small new warships abroad and in 1971 acquired two old but relatively large *Allen M. Sumner*–class destroyers from the U.S. Navy.[4] Oil-rich Iran had the old ships rearmed and modernized extensively, and both were still active in 1988. Presidents Richard Nixon, Gerald Ford, and Jimmy Carter all saw Iran's buildup into a regional power as very useful in offsetting Soviet influence and in easing strategic pressure on the United States to move into the gulf. This was convenient for the shah, who regarded potential American military presence as unwanted competition.[5] Iranian forces seized several disputed fron-

Heavily armed, handsome, and inexpensive, *Kidd* was a very well regarded addition to the fleet in 1981. She went on to prove reliable and highly suc- cessful in operation, notably in the Persian Gulf in 1987 and 1991. This is her original configuration. (U.S. Navy photo)

The *Kidd* class was not a DXG design but the *Spruance* design converted to a guided missile destroyer. Barely two months before launching, *Callaghan* (DDG 994) officially had still been the Iranian cruiser *Daryush* (Darius the Great). Her gas turbine air intakes, twice as high as on the *Spruance* class, house additional dust separators for Persian Gulf service. (U.S. Naval Institute collection)

tier areas from Arab neighbors, including the eastern bank of the Shatt-al-Arab estuary and Abu Musa and two smaller islands inside the Strait of Hormuz, the entrance to the Persian Gulf.[6]

The U.S. government thought that the shah considered himself the protector of the Persian Gulf. It is more accurate to say that he considered himself the owner of the Persian Gulf, and of the Indian Ocean, too. Iraq and Saudi Arabia had land pipelines to the Red Sea and the Mediterranean, but Iran relied entirely on tankers for exporting oil. Every nation on the Indian Ocean littoral, except for Australia at the southeast and South Africa at the southwest, looked vulnerable to takeover by anti-Western forces.

The shah began looking for state-of-the-art warships with good endurance. Britain hoped to sell him an *Invincible*-class light aircraft carrier, then in design for ASW and amphibious operations. Instead he chose a version of the *Spruance* class, built to Litton's DD 963 conversion design. Contract kickbacks and bribery were rampant in Iran, but by the late 1970s the shah had made visible

reforms. He fired and imprisoned the commander of the Iranian Navy for corruption and forbade imperial relatives to make any government business deals that might benefit them. Several American arms firms were involved in such activities, but I have found no published report that mentions Litton.[7]

The DD 963 conversion design that the shah chose was the same as the *Spruance*-class destroyers that Ingalls was delivering to the U.S. Navy, plus the Tartar-D antiaircraft missile system. It was not a DXG design. With helicopters and the Naval Tactical Data System (NTDS) these ships could keep a large area of ocean under continuous surveillance. As a concentrated force, they would have far better antisubmarine and air defense armament than one small carrier. On December 15, 1973, Iran announced that it would order two, and on August 27, 1974, it announced an increase to a total of six. Iran transferred funds for the ships to a U.S. government account for foreign military sales. The U.S. Navy would order them, supervise their construction, and pay Litton for progress. In its plans, the Navy assigned these ships destroyer hull sequence numbers DD 993 through DD 998 (see table 12.1).

The shah designated his planned new ships as cruisers. Seeking to legitimize his self-appointed position as virtual emperor, he named them for kings from ancient eras of Persian imperial greatness: Kouroosh, Daryush, Ardeshir, Shapour, Nader, and Anoushirvan. Kouroosh and Daryush are known to the West as Cyrus the Great and Darius the Great. Darius was the Persian king who menaced the Greeks at Marathon in 490 B.C. The modern marathon foot race repeats the cross-country run of the exhausted battlefield messenger who brought Athens the news of Darius's defeat and then, as every runner knows, dropped dead. Ardeshir and Shapour, who were father and son, founded the Sasanian empire; Shapour defeated the Romans in the third century A.D. Shah Nader, the last great Persian emperor, conquered Afghanistan and northwestern India as far as Delhi in the eighteenth century and kept the British out until his assassination.

The most interesting name was Anoushirvan. Anoushirvan Sharaf al-Maali was the son of a capable Persian ruler in the eleventh century A.D., an era when the Islamic world was more advanced than Europe. Anoushirvan became king while still a boy, upon the death of his father, whereupon a powerful chieftain excluded him from rule. After eight years of growing and thinking, he arrested the chieftain, probably had him put to death, and assumed the throne himself.[8] It was an occupational hazard for medieval princes. His rise was similar to that of Ivan the Terrible and Peter the Great in Russia centuries later. To the shah, Anoushirvan's precedent legitimized his own CIA-sponsored countercoup to recapture power from Iran's demagogic Prime Minister Mohammed Mossadegh in 1953. By naming a cruiser for the obscure young monarch, the shah set Anoushirvan, and indirectly himself, as the equal of the five conquerors whom Iranian schoolchildren knew as ancient Persia's greatest heroes.[9]

Knowing about the delays at Ingalls, Congress insisted that construction of the Iranian ships must follow completion of the 30 *Spruance*-class destroyers. No doubt for reasons of expense, on February 3, 1976, Iran reduced its order to four ships: *Kouroosh, Daryush, Nader,* and *Anoushirvan.* Since the ships were to be identical, the shah's choice of those four for construction was probably based on their names. Western navies often name ships for heroes and victories, in part to evoke fortitude among the crews, but mostly to allow the ships to serve as monuments. To Iranians the cruisers' names symbolized a reincarnation of the ancient kings' souls. Kouroosh and Daryush were conquerors who had led ancient Persia to its zenith. Nader was a relatively recent imperial hero, and the shah had a special tie to Anoushirvan. The latter pair had the added advantage of being Muslims, so naming ships for them might placate Iran's reactionary mullahs. On March 23, 1978, the Naval Sea Systems Command contracted with Litton to build the class of four ships. What no one could know was that the shah had less than a year to reign. In less than two years the United States and Iran would virtually be at war.

The *Kouroosh*-Class Guided Missile Cruisers

With the design work already complete, Ingalls laid down *Kouroosh* on June 23, 1978, only three months after contract award. *Daryush* followed on October 23. Recall that the first 17 *Spruance*-class destroyers were built in two stages: contract delivery at Ingalls, and installation of new weapons later at a naval shipyard. As ordered for Iran, the *Kouroosh*-class guided missile cruisers were similar to Litton's contract-delivery configuration of

A sheep is butchered for King Ibn Saud of Saudi Arabia, embarked aboard the World War II destroyer USS *Murphy* (DD 603) in the Red Sea, 1945. Other sheep stand nearby. Published in 1976, this photograph spawned the story that the *Kouroosh* design included a sheep pen, too, by incorrect projection of a Bedouin tribal custom onto non-Arab Iran. (U.S. Naval Institute collection)

the *Spruance* class (see table 12.2). The major difference was the addition to the *Kouroosh* class of the Tartar-D missile system with two guided missile launching systems. Litton's original design proposal for missile conversion of the DD 963 class had one guided missile director radar, which was on the mainmast. For the *Kouroosh* class Litton added a second missile director radar on a platform on the signal bridge. A missile guidance channel (a continuous-wave illuminator) was added to the SPG-60 gunfire control radar to guide missiles to intercept aircraft or ships. The *Kouroosh* class thus had three missile control channels. A large SPS-48 height-finding air search radar replaced the SPS-40 air search radar on the mainmast.

Iranian officers at Ingalls were adamant about maintaining U.S. Navy standards for damage control equipment, but otherwise they did not pay much attention to the military aspects of these ships. Despite all the weapons, the shah had planned his military buildup inexpertly.[10] Iran did not specify the new equipment that the Navy was adding to the *Spruance* class after delivery from Ingalls. These systems included Harpoon, LAMPS,

Table 12.1 Iranian *Kouroosh*-class guided missile cruisers

Iranian Pennant Number and Name	FMS Hull Number		Became in U.S. Navy
	Plan	Actual	
11 *Kouroosh*	DD 993	F-DD 993	*Kidd* (DDG 993)
12 *Daryush*	DD 994	F-DD 994	*Callaghan* (DDG 994)
13 *Ardeshir*	DD 995	(never ordered)	
14 *Nader*	DD 996	F-DD 995	*Scott* (DDG 995)
15 *Shapour*	DD 997	(never ordered)	
16 *Anoushirvan*	DD 998	F-DD 996	*Chandler* (DDG 996)

Table 12.2 Comparison of the *Spruance* class in the late 1970s with the *Kouroosh* class

Weapons Systems	*Spruance* (DD 963)	*Kouroosh* (F-DD 993)
Search radar	—SPS-40B 2-D air search —SPS-55 surface search	—SPS-48C 3-D air search —SPS-55 surface search
Helicopter	—Hangar for 1 SH-3 or 1–2 SH-2 LAMPS	—Hangar but no LAMPS support equipment
Guns	—2 Mk 45 5-inch/54, 1,200 rounds —Mk 86 Mod 3 gunfire control system —Radars: SPG-60 and SPQ-9 —2 remote optic sights (ROSs)	—Same guns and ROSs —Mk 86 Mod 5 gunfire control system —Continuous-wave illumination channel added to SPG-60 for missile guidance
ASW weapons	—1 8-cell ASROC with 24 rocket-thrown torpedoes and nuclear depth charges —2 triple Mk 32 12.75-inch TT —Mk 116 Mod 0 underwater fire control —Sonar: SQS-53A	—Mk 26 fires rocket-thrown torpedoes —No nuclear depth charge capability —Same TT —Mk 116 Mod 1 fire control —Same sonar
AAW weapons	—1 NATO Sea Sparrow: 1 8-cell launcher —1 Mk 91 fire control system —1 Mk 95 illuminator	—2 Mk 26 launchers: Mod 0 forward (24 rounds), Mod 1 aft (44 rounds) —1 Mk 74 Tartar-D fire control system —2 SPG-51D illuminators
Other weapons	—Naval Tactical Data System (NTDS/CDS) —8 Harpoon antiship missiles —WLR-1 or SLQ-32(V)2 radar detector —SRBOC chaff decoy rockets —WSC-3 satellite communications	—Naval Tactical Data System (some functions deleted)

satellite communications, chaff, and electronic warfare sensors. Without chaff decoys and radar-warning sensors, the *Kouroosh* class was highly vulnerable to missile attack. It is anyone's guess whether Iran would have had these vital systems installed after receiving the ships.

The Iranians did attend closely to shipboard habitability. Their planned crew composition included more officers, probably to make up for an expected scarcity of skilled enlisted men, so staterooms had more bunks. A fourth air-conditioning plant was added for the main engine rooms, which were uncooled on the *Spruance* class. Additional dust separators were installed on the gas turbine air intakes, raising them several feet so that they were level with the top of the funnel casings.

Water distillation capacity was increased to 20,000 gallons per day, up from 16,000 gallons per day on the *Spruance* class.

A U.S. Navy officer in 1982 told me a rumor that Iran had specified installation of a pen on the broad fantail aft, to keep goats or sheep for banquets. I thought the rumor was nonsense, but I heard the same story from others.[11] These ships' U.S. Navy crews seem so convinced of the truth of this legend that this book would be incomplete without exploring it. For the record, I found nobody actually associated with the design of this class who has heard of such pens, nor any report that mentions them. Lamb could be stored in the refrigerator spaces like any other meat. Livestock need a lot of perishable forage that the destroyers could not store out of the weather. Keeping and butchering sheep at sea would be unsanitary, to say the least, and the ships could hardly embark very many. Besides, banquets of the sort imagined are a custom of the Bedouin Arabs, not of the entirely different Iranian culture. One wonders what dignitaries boarding the Third World's most powerful surface warship would think upon encountering a flock of sheep at the aft quarterdeck. It would look and smell like a stockyard.

The exotic legend probably arose from a 1976 Naval Institute *Proceedings* article that described the Red Sea cruise of King Ibn Saud of Saudi Arabia aboard the destroyer *Murphy* (DD 603) to meet President Franklin D. Roosevelt in 1945.[12] The king arrived quayside at Jidda with his entire court, including odalisques of the harem. After diplomatic discussion by liaison officers who pointed out the size limitations of a World War II destroyer, *Murphy* embarked the king, several sons, a few other members of his entourage (none from the harem), a tent for the fo'c's'le, and indeed some sheep, which were penned by the depth charge racks for the king's meals during the short voyage. A photograph of this no doubt spawned the legend mentioned above.

The Americans and Arabs got along very well at sea. It was the first time Ibn Saud had been outside Saudi Arabia, which he had named for himself. The crew fired the destroyer's guns for the old warrior, kept radio contact with Mecca, and issued him a set of binoculars to observe the Hejaz coast. Later the Americans gave him the binoculars and two submachine guns from the ship's armory as gifts. Officers joined the king under the tent to dine on fresh mutton from the fantail while his sons sneaked out to watch the crew's movie. Upon arrival in the Great Bitter Lake, *Murphy* transferred Ibn Saud to the heavy cruiser *Quincy* (CA 71). He and President Roosevelt discussed oil and Palestine amicably.[13] No doubt to the relief of *Murphy*'s supply officer, the President approved the gift of the guns and binoculars. Forty-five years later, when the king's descendants let American troops land for the Desert Shield buildup, they cited this cruise as proof enough that Saudi Arabia could trust the United States.[14]

The U.S. Navy Acquires the "Ayatollah Class"

In January 1979 the shah fled Iran, and a moderate antiroyalist government took over. On February 3 the Iranians told the United States to cancel *Nader* and *Anoushirvan,* and on March 31 they canceled the remaining pair. Iran had already paid about $800 million of the total cost of $2.16 billion for progress on the class. The Iranian trust fund still held $327 million, which might pay Ingalls to scrap the ships, with the balance to be returned to Iran. As an alternative, the Navy might complete all four ships for only $1.353 billion, a stunning 40 percent discount during an era of severe inflation and cost overruns. Commonality with the *Spruance* and *Virginia* classes meant that logistical support already existed. The Navy needed presidential and congressional authorization to get the money, which would take months.

Even a slow rate of progress at Ingalls, just enough to keep the buy-out option alive, cost over $7 million a day. If Iran would let its trust fund cover that cost, the Navy would have about two months to act before the trust fund was depleted. The new Iranian regime agreed that the Navy could continue to pay Ingalls from the trust fund to continue construction of the ships while the Navy sought funds from Congress to take over the ships.[15] This saved Iran from having to pay contract termination costs. Once funds were approved, the Navy would reimburse Iran for funds used, less deductions for Iranian-ordered equipment that the U.S. Navy did not need. President Carter, an Annapolis graduate, authorized the Navy to ask Congress for a supplemental defense appropriation. Following the lowest-cost schedule, Ingalls laid down *Nader* and *Anoushirvan* with trust fund money.

Congressional debate was fierce. Opponents from across the political spectrum argued that the Tartar-D system was obsolescent since Aegis was already in production, that the Navy was rewarding Iranian revolutionaries and should simply take the ships, and that the scheme was a pork-barrel project for Mississippi. Perhaps mindful of the strategic disasters that befell Britain following the Royal Navy's seizure of two Turkish battleships from shipyards in 1914, the Navy defended its agreement with Iran.[16] The winning argument was the need for more warships. With the strategic balance shifting badly against the United States, any new warships would be welcome. These four would be the most modern and heavily armed surface warships in the fleet. The usual lead time for warships was five years; these could be delivered in half that time. For the price of four frigates of half their power, they were a stunning bargain.

Congress approved the supplemental appropriation, and on July 25, 1979, the Navy acquired the *Kouroosh* class. The ships were renamed for four admirals killed in the Pacific during World War II: Kidd, Callaghan, Scott, and Chandler. They were redesignated from destroyers (DD, or F-DD under the Foreign Military Sales program) to guided missile destroyers (DDG), but they kept their destroyer-series hull numbers. That put them out of sequence with all other guided missile destroyers, but avoided inserting their hull numbers into the already-authorized Aegis DDG 47 class sequence.

In Iran a series of regimes rose and fell at the whim of Ayatollah Ruhollah Khomeini, the Muslim cleric who had inspired the revolt against the shah. Khomeini's followers revered him as holy and incorruptible; Westerners thought he was psychotic. Officially the *Kidd* class, the new ships are perhaps better known as the "Ayatollah class" in gallows humor at how they came to the Navy. Just 99 days after the Navy bought the *Kidd* class from Khomeini's regime, his radical followers seized the American embassy and took its staff hostage to trade for the shah. American anger at this treachery has barely abated 15 years later. On Christmas Day 1979 the Soviet Union took advantage of the turmoil in Iran to invade Afghanistan, a nation the shah had been willing to defend with American weapons. In September 1980 Iraq invaded Iran and started a war that eventually would involve the *Spruance, Kidd,* and *Ticonderoga* classes. They would have a major role in ending it.

The *Kidd*-class DDGs received the accolades that the original *Spruance* class had missed. They quickly earned the happy reputation of being the best destroyers the Navy had ever built. Observers regarded them as an ideal design, the type of warship that the Navy should have been building for 20 years. *Combat Fleets of the World* called them "superb" (and still does).[17] Criticism of the *Spruances* often accompanied praise for the *Kidds*: "The *Kidd* is the best shipbuilding buy the Navy has made in recent times. . . . The *Spruance* destroyers and *Oliver Hazard Perry* frigates are the kinds of ships we have been able to afford in numbers, but are they the kinds of ships we need to do the job? I do not think they are."[18]

Visible weaponry remained an incomplete indicator of total capability. The *Kidds'* heavy armament precluded installing the long-range surveillance systems planned for the *Spruance* class, in particular towed-array sonars and the Outboard communications intelligence system. This limited the *Kidds'* value as scouts for aircraft carrier battle groups. As early as 1984 Navy planners assigned all *Spruance-* and *Ticonderoga*-class ships to carrier battle groups for war, but assigned the *Kidd* class to escort amphibious convoys.[19] Previously that had been something of a secondary mission, since amphibious forces would follow into an assault area only after attacks by the Air Force and carrier battle groups had secured air and sea control. The aggressive Maritime Strategy required stronger escorts, since it reportedly planned amphibious operations against the Soviet Union itself.

DDG 993 Combat Systems

Upon taking over the *Kidd* class, the Navy completed the ships with the same new systems the early *Spruance*-class destroyers had received

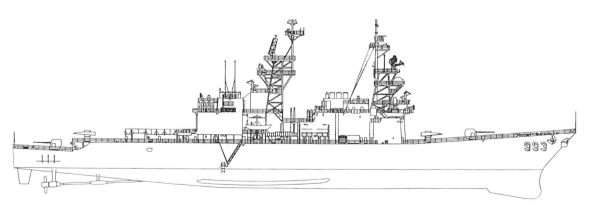

This drawing depicts the DDG 993 class after New Threat Upgrade modernization. (Lt. Cdr. Robert Lenson, USNR)

after delivery from Ingalls. New weapons included two quadruple Harpoon antiship missile launchers and two Vulcan/Phalanx 20 mm CIWS (close-in weapons system) guns. The *Kidd*s received Phalanx before any of the *Spruance* class. Electronic warfare systems included the SLQ-32(V)2 radar warning detector and SRBOC chaff decoy rockets. Iran would not have received the full set of Naval Tactical Data System functions. All sections were restored. WSC-3 ultrahigh-frequency (UHF) satellite communications transceivers were added. Kevlar ceramic laminate armor was added around the combat information center. The armor would protect CIC from ballistic fragments of antiship missiles shredded by the Phalanx guns at very close range.

Additions brought full-load displacement to 9,200 tons for initial service, compared with 8,040 tons for the *Spruance* class in its interim modernization (1980s) configuration. Combat system upgrades accounted for perhaps half this increase, revealing that the ships carried ballast. Ballast is undesirable in a weight-critical ship design, since the weight of the ballast would be carried more profitably as payload or in improvements to hull and machinery. But the *Spruance, Kidd,* and *Ticonderoga* classes, like most modern warships, are volume-critical designs. In a volume-critical design, ballast costs almost nothing in terms of ship performance, and it supports a larger payload higher in the ship. Submarines, the most volume-critical ships, carry ballast by design.[20] (See chapter 2 for a discussion of volume-criticality.)

Two Mark 26 twin-arm guided missile launching systems were the most prominent change from the *Spruance* class. Forward, one Mark 26 launcher replaced the ASROC launcher. The second Mark 26 was installed in the large spaces reserved for it aft of the helicopter deck. Below decks a conveyor holding the missiles runs in a loop fore and aft beneath each launcher arm. The forward Mark 26 Mod 0 launcher holds 12 missiles on each conveyor (5 on each parallel track and 1 at each end), for a total of 24. The aft Mark 26 Mod 1 launcher has twice the track length on its conveyors and holds 44 missiles. These missile launchers can fire two missiles every 10 seconds. The missiles are Standard medium-range radar-intercept missiles (RIM-66) and, until the weapon was withdrawn in 1993, RUR-5A ASROC rocket-thrown torpedoes. For ASROC the rail on the lower part of the arm switched to align with the upper extension, forming a long track to aim the ASROC on a ballistic

Visible additions to create the *Kouroosh/Kidd* class from the *Spruance* design included two Mk 26 guided missile launchers, two SPG-51 targeting radars, an SPS-48 height-finding air-search radar, and a large radar electronics room at the base of the mainmast. This is *Callaghan* during the search for the flight data recorders from Korean Airlines flight KE-007 off the Siberian coast in 1983. (U.S. Navy photo)

trajectory. Conversion of the ships for U.S. Navy service apparently did not restore a capability to launch ASROC nuclear depth charges, which would have been deleted from the Iranian design. The *Kidd*s probably never carried nuclear weapons.[21]

The DDG 993 Command and Decision System UYK-7 computers were enlarged to run seven real-time programs simultaneously, compared with three on the *Spruance* class. The same computers and consoles support Naval Tactical Data System functions and missile engagements. The distinction is that the NTDS software builds track files on contacts from the search radars and data links. Weapon control software uses the NTDS track files to designate targets for the missile fire control radars.

The *Kidd*s originally carried an SPS-48C long-range air search radar. The SPS-48C was a height-finding or three-dimensional radar, in that it determined target aircraft altitude in addition to range and bearing. This is important for pointing a fire control radar toward a target quickly to start a missile engagement. At long range the ship can direct fighter intercepts. Link-4A/4C NTDS data links exchange target data with F-14 Tomcat fighters. This is a point-to-point link normally transmitted over a UHF radio channel.

The main antiaircraft fire control system is a Mark 74 Digital Tartar (Tartar-D) system with two SPG-51D missile directors. A large pyramid-shaped deckhouse at the base of the mainmast contains transmitters and other equipment for the SPS-48 search radar and the aft SPG-51D missile illuminator. The forward SPG-51D is atop the bridge forward of the foremast. The SPG-51D has servomotors, not mechanical gears, for steering. Gears are susceptible to jamming from the whipping effect of explosive hits on the ship during combat.

The SPG-51D transmits signals in two radar bands. A C-band radar signal searches for the target and upon finding it tracks it using mon-opulse signals. Radar C-band burns through weather clutter. Aboard the target aircraft a radar detector will show that the SPG-51D has locked on. The SPG-51D uses monopulse, frequency agility, and Doppler to overcome jamming and chaff. It computes predicted target motion so that it can maintain a track even if the aircraft has a range jammer. The shipboard SPG-51D antenna receives and acts on the C-band signal. A second set of microwave horns on the SPG-51D emits a high-precision X-band continuous-wave illumination radar beam. The ships first carried Standard SM-1 medium-range missiles. After launch, the Standard SM-1 homes on the target's reflection of the X-band illumination beam. This is semiactive homing. The reflection is sharp enough for the missile to intercept the target closely. The missile's proximity fuze detonates the warhead to disable the target, which will break up under aerodynamic stresses.

On the DDG 993 class the SPG-60 gunfire control radar has an X-band continuous-wave (CW) emitter so that it too can illuminate targets for Standard antiaircraft missiles. The SPG-60 works similarly to the SPG-51D except that both the tracking and illumination beams are X-band. The *Spruance*-class SPG-60 radar does not have the CW illuminator. As an aside, stealth technology attempts to defeat precise targeting radars such as SPG-60. Longer-wavelength radars such as the S-band SPS-48 and SPY-1 might acquire a stealth aircraft target, but such aircraft reportedly do not reflect intelligible short-wavelength echoes for a radar-intercept missile to intercept it. As on the DD 963 class, the SPG-60 has a bore-sighted TV camera (a remote optic sight, or ROS) for range safety checking and damage assessment.

The New Threat Upgrade

A ship firing homing missiles must illuminate the target throughout the missile's flight from launch to intercept. This is acceptable for a point-defense system such as Sea Sparrow, which is intended only to intercept a target closing on the launching ship. Its short range means short time of flight for the missile, so that a second engagement can quickly follow the first. Defeating a saturation attack of high-speed, high-altitude antiship cruise missiles is more difficult. The defenders must begin shooting down the "vampires" at long range to prevent saturation of point defenses.

The Standard SM-1 missile had limitations. With its three missile guidance radars, the *Kidd* class could engage only three targets simultaneously. The ship cannot fire until the target returns a radar echo strong enough for the missile's homing sensor to detect, which reduces engagement range. Range falls off as the one-fourth power of intensity, so that doubling radar illumination power would extend range to the target about 20 percent. Once launched, the missile flies an inefficient path in pursuit of the target. If the target is headed toward the launch ship at medium altitude, the pursuit trajectory is relatively simple and a hit is likely. If the target's course is crossing the ship, so that angular bearing or elevation changes rapidly, interception is difficult.

Among the antiship missile defense developments stimulated by the Charlie-class submarines and the *Elath* incident was a project to upgrade the Standard SM-1 missile systems to defend against them. This project was called simply the New Threat Upgrade. It provides advanced radar for automatic detection and tracking of targets, in particular new sea-skimming antiship cruise missiles. It also uses the Standard SM-2 radar-intercept missile, designated RIM-66C or SM-2(MR), to engage high-speed, high-altitude targets, in particular Soviet anticarrier cruise missiles. Soviet bomber-launched AS-4/6 (Kh-22/KSR-5) missiles and the SSN-N-3/12/19 (P-25/35/500) missiles, launched from ships and submarines, could fly above the intercept altitude of American naval missiles earlier than the SM-2.[22]

The *Kidd*-class destroyers received the New Threat Upgrade during 1989–90 at the Philadelphia Naval Shipyard. This was a major modernization of the class and changed the ships' appearance and arrangement significantly. The new SPS-49 long-range air search radar and four SYR-1 missile telemetry receivers were installed. The ships were out of internal space, so a new deckhouse was built beneath the foremast behind the forward SPG-51D pedestal for SPS-49 equipment, which weighs 7 tons. The SPS-49 antenna was located on the platform originally occupied by the SPG-60 gunfire control radar. The SPG-60 was relocated to a new platform on the mainmast above the aft SPG-51D.

A new SPS-48E radar has replaced the original SPS-48C. The new SPS-48E radar uses the radar S-band for long range and good weather rejection. S-band is higher-frequency than L-band, so its pulses carry more information. The SPS-48 scans with different frequencies on different elevations, so that the frequency of a returned echo indicates target altitude. If jamming distorts echo frequency, then the radar can switch to a kind of monopulse multiple-frequency signal, transmitted at one elevation to burn through the jamming signal. Range is

230 miles, and maximum beam elevation is 65 degrees. Accuracy is reportedly 450 yards in range and 0.6 degree in bearing.[23] From the front the rectangular SPS-48E antenna looks like the original SPS-48C, but it is completely new.

In the original design, the SPS-48C sat between the masts and was blocked on fore and aft bearings. Electromagnetic interference from the masts increased the SPS-48C's side lobes, making the radar more susceptible to jamming or attack by antiradiation missiles.[24] The new SPS-48E antenna was set higher on the mainmast, and the topmast was stepped forward of it. This improved radar performance, but at the expense of the ships' original handsome appearance. The original masts were balanced and blended well with the lines of the ship. They appeared to thrust forward and imparted a handsome, aggressive, military profile. Since the New Threat Upgrade, the mainmast is visually much more massive than the foremast, ruining the balanced profile. Now the *Kidd*s look like they have an oil derrick amidships. An alterna-

Installed during 1988–90, the New Threat Upgrade greatly increased the *Kidd* class's air defense capability. An SPS-49 long-range radar is on the foremast, and a new SPS-48E radar is mounted on the mainmast, higher and on an aft-facing platform. The SPG-60 radar has been moved from the foremast to a new mainmast platform above the aft SPG-51 missile fire control radar. (U.S. Navy photo)

tive would have been to cut the foremast down to a stump for just the SPQ-9 and SPG-60 gunfire control radars. Other antennas from the foremast would have been relocated to an enlarged mainmast. The SPS-48E and SPS-49 antennas would have been mounted on a single turntable on the mainmast that would have rotated around an enlarged pole topmast.[25] That might have eliminated the electromagnetic-interference problem, but the ships would have looked grotesque.

Other new systems included an SPS-64(V)9 navigation radar, two SLQ-34 cover-and-deception units,[26] and the SRQ-4 data link for SH-60B helicopters. The SRQ-4 directional data link supports exchange of electronic support measures (ESM) intelligence and surface radar data with the helicopter. No other components of the SQQ-89 ASW modernization program were fitted, nor were RAST recovery systems for landing SH-60Bs. *Callaghan* and *Chandler,* the Pacific Fleet ships, were repainted with all-gray top hamper to reduce their visual detectability. The ship's service gas turbine–powered generators (SSGTGs) were upgraded to 2,500 kW each. These changes and probably more ballast increased full-load displacement to 9,950 tons. A *Kidd*-class destroyer displaces more than an Aegis cruiser.

The SPS-49(V)5 long-range two-dimensional air search radar has a detection range of over 200 miles. It uses L-band for long-range search. Operation at a slightly higher frequency than the SPS-40 radar of the *Spruance* class provides more information in the pulse. Accuracy is very high: range within 60 yards and bearing within 0.5 degree. It has features to counter clutter, chaff, and enemy jamming, and it displays jammers' bearings. A pulse-Doppler moving-target detector automatically feeds target data to NTDS. The radar antenna is stabilized against the ship's rolling.

The radars and the new SYS-2 Integrated Automatic Detection and Tracking (IADT) System automatically feed digital target data to NTDS for prompt engagement of sea-skimming missiles. The New Threat Upgrade uses the Standard SM-2 medium-range radar-intercept missile, designated RIM-66C or SM-2(MR), the same weapon used by Aegis. Externally almost identical to the Standard SM-1 missile, its internal systems are more advanced. Its fundamental feature is that the ship no longer needs to illuminate the target before missile launch. Instead, the Weapons Direction System tracks each target and computes the time and point in the sky where the Standard SM-2 missile

should arrive before intercept. The missile has an autopilot that the ship sets before launch. The missile flies to the point in space without guidance from the illuminator radars. Missiles in flight report their locations during flight to shipboard SYR-1 telemetry receivers.[27] The ship can transmit a new intercept position to the missile's autopilot via a radar uplink. When the missile and target close and a shipboard SPG-51D becomes available, the ship tracks the target and illuminates it with the continuous-wave beam. The missile homes on the reflection to intercept.

This lets the ship fire missiles at more than three targets, so that new salvos can be launched while the ship is still engaging targets with an earlier salvo. Since the missile is not then pursuing the target echo, it flies on a least-energy trajectory. That consumes much less propellant, so the missile's fuel load can take it about 40 miles, nearly twice the range of the Standard SM-1.[28]

During 1989 *Kidd* and *Scott* tested cooperative engagement, a capability shared with Aegis. Radars off, *Scott* remained electronically silent, so that an adversary would have no clue to her identity and might not notice her presence. *Kidd* provided radar coverage and relayed targeting data to *Scott* via Link-11. On *Kidd*'s command, *Scott* fired a Standard SM-2 missile that *Kidd* guided to intercept. In combat the destroyers might be widely separated, say by 30 miles, with *Scott* lying in ambush closer toward the enemy. Using a new data link, the operational system correlates sensor data from multiple ships and aircraft into targetable tracks. It can target a Scud launch site in seconds.[29]

The DDG 993 Warfighting Improvement Program

The next modernization planned for the *Kidd* class will outfit the ships to carry two SH-60B Seahawk LAMPS Mark III helicopters. The flight deck will be widened to the full width of the hull. Two RAST recovery tracks will be installed along with necessary internal hangar modifications. Electronics additions will include USC-38 extremely-high-frequency- (EHF) band MILSTAR satellite communications and SLQ-32(V)3 active electronic countermeasures jammers. ASW enhancements will include a torpedo defense weapon and possibly the new Mark 50 lightweight ASW torpedoes. The *Kidd*s will not receive the SQQ-89 ASW upgrade, the SQR-19 towed-array sonar, or the

SQS-53C upgrade to their existing SQS-53A sonar. Plans do not include Tomahawk.

Aviation-Capable *Spruance* Designs

Needing more work for the Ingalls yard, during the late 1970s Litton researched a series of futuristic ship classes based on the *Spruance* design. The DD 963 design was highly modular, with large growth margins, so designing variations was relatively simple. New systems could be added without trade-offs elsewhere. Litton seemed to have had two markets in mind: low-cost destroyers with modular weapons stations, and aviation-capable ships to support vertical/short-runway takeoff and landing (V/STOL) aircraft.

Two of Ingalls's designs were for surface combatants.[30] One, the "DD 963 SEAMOD" version,

was a standard DD 963 except that it showed cutouts in the deck and superstructure to insert pallet-mounted gun mounts, missile launchers, and electronics systems. This design may have been intended to compete with the German firm of Blohm & Voss, builders of the famous *Bismarck*, who in the 1970s were selling MEKO destroyers and frigates to foreign navies. MEKO designs have mounting cutouts for weapons and electronics pallets with standard mechanical and electrical connectors. A Navy buying a MEKO ship could specify the systems it preferred and still get the ship rapidly and within budget.

One Litton design showed an 8,650-ton "DDX 963 Enhanced Combatant." The Litton design advertised vertical launchers and armor over magazines and other vital spaces. It was unusual in having a secondary general-purpose gun battery:

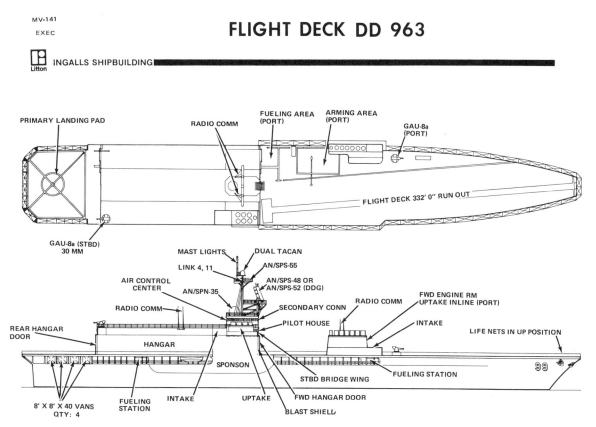

Litton proposed many possible variants of the DD 963 design. The most radical was this 12,000-ton flight-deck version for vertical/short takeoff and landing (V/STOL) aircraft. The concept was that her light scout V/STOL aircraft would find targets for long-range ship-launched missiles. The Navy never requested funds for it. In 1980 Grumman Aircraft designed a V/STOL aircraft for the flight-deck "DD

963F." With wings folded for storage it would be only 16 feet wide and could also fit in a LAMPS helicopter hangar. It used the nose and tail of the Mitsubishi MU-2M transport with a new wing and engine mount. The engines pivoted for takeoff and landing. Missions were foreseen as surveillance, early warning, targeting, and radar jamming.

Another *Spruance* design variant was for this "DDM 997" that Litton hoped the Navy would build instead of the *Arleigh Burke* (DDG 51)–class destroyers. The Navy proceeded with the *Arleigh Burke* class instead. (U.S. Naval Institute collection)

two Italian 3-inch Mark 75 guns were located amidships in place of Harpoon canisters. A sketch showed a flush-deck hull with the aft 5-inch gun mount aft of the flight deck. The fantail had a vertical missile launcher and a 30 mm CIWS gun on each quarter. Propulsion would be two 50,000 shp gas turbines with a cross-connection so that either one could drive both shafts. A later design was an Aegis-armed missile destroyer Litton called a "DDM."

Other Litton designs were for aviation ships. Success of the AV-8A Harrier jet made V/STOL a popular research topic in the 1970s. Admiral

James Holloway, Chief of Naval Operations during 1974–78 after Admiral Zumwalt, foresaw a Navy need for two types of V/STOL aircraft and commissioned studies concerning them. The attraction was that aircraft carriers would not need catapults or arresting gear, greatly reducing those ships' cost. "Type A V/STOL" would be a transport-type design that could be equipped for ASW, airborne surveillance and targeting, and cargo. The current MV-22A Osprey tilt-rotor assault transport is an aircraft of this type, although it resulted from a different research project (JVX, joint-service V/STOL aircraft). "Type B V/STOL" would be a high-performance fighter-bomber, much larger than a Harrier so that it could carry sufficient engine power, fuel, avionics, and ordnance. The Type A/Type B V/STOL program ended in 1978, but interest continued. Destroyer-based aviation would cost more per aircraft than a full-deck aircraft carrier, but it would be cheaper per hull.

With several V/STOL aircraft, such a ship might replace the small, aging aircraft carriers (mostly ex-British from World War II) that many foreign navies operated.[31] The Royal Navy found that the most efficient flight deck operation for the Harrier was a short takeoff, where the wings helped to lift aircraft and payload, and vertical landing, which was safer aboard a ship without arresting gear and required minimal deck space.

Litton designed a 12,020-ton "DD 963F" with a

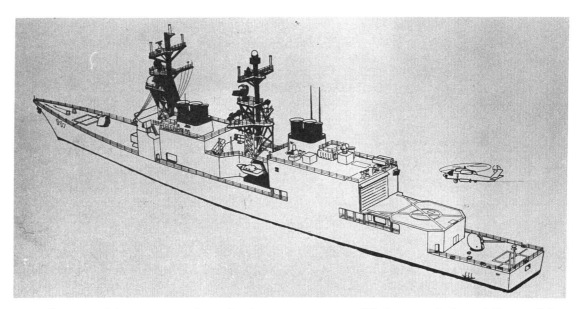

A partially successful Litton proposal was this design for "DDH 997" to carry four medium helicopters such as the SH-60B Seahawk. In 1978 Congress appropriated funds for a 31st *Spruance*-class destroyer, modified to carry "substantially more" than one to two aircraft, as long as it would not cost more than a standard *Spruance*. (U.S. Naval Institute collection)

USS *Hayler* (DD 997) was built as a straight DD 963. She has a larger SPS-49 air-search radar on the mainmast instead of the SPS-40 carried by all other DD 963s. Notice the SH-3 Sea King helicopter on deck. *Spruance*-class destroyers have often embarked Sea Kings for search and rescue missions, commando assaults, and other duties. (L&L Van Ginderen collection)

332-foot flight deck at the 02 level forward for short takeoff. The hull was widened to 68 feet beam, and draft was 34 feet. A small port-side island held the forward gas turbine intakes and uptakes. An area to arm and fuel the aircraft was abaft the island. A large hangar amidships extended the full width of the ship. The bridge was above the hangar. A landing pad was on the fantail, built up level with the hangar and forward flight deck. With all aircraft handling on one deck, the ship did not need aircraft elevators. It might operate eight Harriers and four LAMPS Mark III Seahawks. That was too few to guarantee tactical security of open-ocean operations against opposition. The ship was feasible but impractical.

Grumman had designed a small Type A V/STOL aircraft based on the Mitsubishi MU-2M STOL transport. Two TF34 turbofans mounted on the fuselage on a pivoting box would provide VTOL capability. Research continued until 1980 with NASA and Navy funding. Grumman envisioned it as a small version of the firm's E-2 Hawkeye carrier-based early-warning-type aircraft, with a "conformal" long-range radar wrapped around the fuselage. The VTOL aircraft would guide long-range ship-launched air-targeted (SLAT) missiles against Soviet missile-carrying naval bombers. In this concept the flight-deck-equipped DD 963 would carry the aircraft and 100–120 antiaircraft missiles, type not specified.[32] Weaknesses included the vulnerability of the aircraft; reliance on light sea states for flight deck operations; and no

explanation of how the missiles would actually be targeted. Serious research interest in these ideas persisted until 1982.[33]

Litton advanced a "DDH 997" design with an extended hangar for four LAMPS helicopters. Two such ships could keep a helicopter on station almost around the clock, very useful for surveillance. This design came closer to reality than the other DD 963 air-capable variants. Representative Charles Bennett obtained the House Armed Services Seapower Subcommittee's support for two ships for FY 1978. Agreeing, the Senate Armed Services Committee stated, "The committee does not intend for these funds to be used for acquisition of two standard 963-class destroyers; rather, it is the committee's intention that these ships be the first element in a new-technology approach to the problems of designing surface escorts. The standard 963-class design should be modified to substantially increase the number of helicopter aircraft carried."[34] Congress funded one destroyer of this type, provided she would not exceed the cost of a standard *Spruance*-class destroyer.

Litton went on with design of "DDH 997." The flight deck was moved aft to just ahead of Mount 52. The hangar was lengthened and extended aft by almost the same distance, about 40 feet. The hangar deckhouse was widened to 55 feet, the beam of the hull, and formed a continuous slab amidships from the waterline to the 04 level. A rising, segmented hangar door assembly replaced the horizontal sliding doors, giving access to a full-width hangar. The NATO Sea Sparrow launcher was relocated to the hangar roof. A new elevator inside the hangar carried Sea Sparrow missiles to the hangar roof. Two large boundary-layer infrared suppression uptakes, as in the *Ticonderoga* design, replaced the six smaller exhaust-cooling cylinders of the *Spruance* class. Both WSC-3 UHF antennas were relocated to the mainmast.

Litton described the aircraft complement as "two SH-3 Sea King, or four SH-2F, or four SH-60B LAMPS helicopters, or one Type A V/STOL."[35] Flight deck width would limit operations by multiple helicopters. Drawings did not show RAST equipment, which in practice would limit the ships to SH-2F Seasprite LAMPS Mark I helicopters, a significant weakness given the short combat radius of the SH-2F. Officer complement grew to 36 from 23, and enlisted complement from about to 333 from 295. Aviation jet fuel stowage was increased.

The aft retracting cargo kingpost and aft ammunition elevator apparently did not change location and opened onto the flight deck. Open space and wind protection by the superstructure made this the favored station for underway replenishment on *Spruance*-class destroyers. The DD 963 flight deck is usable during underway replenishment. Helicopters routinely land to transfer mail and passengers. LAMPS helicopter operations, such as emergency recoveries or quick-alert takeoffs to investigate a contact, are unimpaired. Since the "DDH 997" flight deck was co-located with this equipment, use of the aft replenishment station would close the flight deck. Replenishment leaves an inevitable residue of pallet splinters and broken cargo bands. This would increase the risk of foreign object damage (FOD) to aircraft at night, when the flight deck crew might not find all the debris.

The cost of designing the *Kidd* class was very low, $6.5 million, compared with over $17 million for the *Oliver Hazard Perry* (FFG 7) class at about the same time. Detail design and production engineering cost for the *Kidd* class was $110.8 million. That was less than for the FFG 7 class and a little higher than for an oil tanker.[36] For "DDH 997" the cost at Ingalls would be much lower than that, since little of the DD 963 design would change. However, for a production run of one or two, that might still be a significant additional cost. Cost estimates predicted that "DDH 997" would cost more to build than a standard DD 963. Referring to the restriction in the congressional appropriation, the Carter administration's OSD told the Navy not to build the ship as a "DDH."[37]

Government agencies are famous for opposing initiatives that are "not invented here." The original *Spruance* class exemplified that tendency. Litton and the congressional committees on the armed services, not OSD and the Navy, were the inventors of the "DDH 997" plan. Naval analyst Norman Polmar reported that the carrier aviation community strongly opposed further development of V/STOL aircraft, and thus of a V/STOL-oriented destroyer.[38] The additional aviation capability that the ship would really bring to an operation was marginal. *Kidd*, with an unmodified hangar, operated four helicopters in the Persian Gulf in 1991: two Navy SH-2F Seasprites and two Army OH-58 attack helicopters.

For whatever reason, on September 29, 1979, the Navy ordered the ship as DD 997, a straight repeat of the DD 963 design. The new destroyer was named *Hayler* (DD 997) in March 1981. The sole notable difference between *Hayler* and the rest of the class is that she carries the SPS-49 long-range air search radar instead of the SPS-40B. No trace remains of the original intent to make her a new kind of aviation ship, unless one considers the trace to be the ship herself.

The Shield of the Fleet: The Aegis Cruisers

Failure of the Typhon Air Defense System

In the 1960s the threat of Soviet antiship cruise missiles had jeopardized the future of the Navy. The Office of the Secretary of Defense wondered whether long-range land-based nuclear bombers like the F-111 could replace aircraft carrier task groups. Fortified air bases, many more bombers, and fleets of air-refueling tankers would be costly but might reduce needs for naval fighters and fleet air defense missile systems. It would also reduce the opportunity to divert large numbers of enemy aircraft from other fronts and to destroy them. "The entire question of the cost and capability of the fleet in relation to the cost of defending it against air attack is still in need of a most thorough analysis," Secretary McNamara told Congress in 1963.[1] The Navy's difficulty in answering that challenge helps to explain cutbacks in construction of high-end surface combat warships in the 1960s.

The Navy initially attempted to build the prototype Typhon weapon for fleet air defense against mass air raids and pop-up antiship missiles.[2] Large Typhon ramjet-propelled missiles would replace the long-range Talos missile. Smaller Typhon missiles would replace Terrier and Tartar. Both models of Typhon missiles would arm a new class of cruisers. The medium-range missile and a smaller derivative of the radar would arm new destroyers.

Instead of the homing-all-the-way guidance system that Tartar used, Typhon would use command guidance, in which the ship would send course corrections during flight, and the missile would home on the illuminated target only in the last few seconds. A Typhon ship could in theory fire missiles at many distant targets and command

them into position almost simultaneously. A new electronically scanned shipboard radar (SPG-59) provided all radar functions: air search, target tracking, target illumination, and missile command channels. With no need for dedicated target-tracking antennas, the Typhon radar system could in theory track 120 targets and control 30 missiles.[3] The advantages included longer missile range, the ability to have more missiles in flight, and the elimination of the hand-off delay for a target-tracking antenna to locate its designated target.

Prototype Typhon missiles flew well enough in tests, but engineers could not surmount problems with the radar. The Typhon radar was physically gigantic. Even the smaller version became too heavy and expensive for any destroyer. All Typhon radar signals were pulsed, including target illumination beams. Missiles could not carry that era's bulky electronics to turn pulsed-beam echoes into target interception data. To get around that, a Typhon missile in flight received echoes from the target and relayed these echoes back to the Typhon radar aboard the ship.

The shipboard radar processor updated the intercept point and transmitted the new flight path orders back to the missile. That simplified missile design, but it consumed radar processing time and signal power and reduced the Typhon radar's actual capacity for target detection and tracking. Signal quality eroded inside the radar much more severely than the designers had predicted. Loss of internal signal strength degraded target detection performance to unacceptable levels. Signal amplifiers added to the cost and weight of Typhon but did not fix the performance problem.

The Typhon radar system was in effect a gigantic analog computer, probably the largest and most complex analog system ever built. In that era analog systems were faster than digital technology. In any radar, strength and other physical attributes of the radar microwave signals carry information. Typhon attempted to maintain these characteristics in internal processing. However, since consistency in analog signal-processing requires an unobtrusive background, the Typhon radar processor was highly vulnerable to enemy jamming.

All Typhon microwave signals were in the radar C-band. These signals focused well for missile accuracy, but over long distances they dissipated in weather and atmospheric effects. At long range only a target with a large radar cross-section, generally implying a physically large aircraft, returned a detectable echo. At close range the echo returned before the Typhon radar had finished transmitting the outgoing pulse, so that the echo was not recognized. Caught between these limits, the prototype Typhon radar aboard the test ship *Norton Sound* could not detect small aircraft under any conditions.[4] Typhon had failed as a practical system.

In 1980 the Soviet Navy deployed a track-via-missile air defense system (SA-N-6) on the *Kirov*-class cruisers. The Soviet Navy thus had a multiple-target antiaircraft weapon years before the U.S. Navy deployed the Aegis and New Threat Upgrade systems.

The Origin of the Aegis Fleet Air Defense System

Late in 1963 Secretary McNamara canceled the Typhon project and authorized a successor project, the Advanced Surface Missile System (ASMS) program. Rear Admiral Eli Reich, the first ASMS program officer, solicited the defense industry to propose a new missile system. The contractors presented systems very much like Typhon but with simpler planar phased-array radars.

Not pleased with this conservatism, Admiral Reich convened a committee of Navy, Army, and industry experts to design an ASMS configuration and to assess its cost and effectiveness. Rear Admiral Frederick Withington, retired Chief of the Bureau of Ordnance, chaired this committee. Several defense contractors sought to place their salesmen on the committee instead of the invited technical experts. Admiral Withington told them to send the engineers he named or not to show up. He required the subcommittees to make majority reports so that industry engineers would go beyond merely defending their parent firms' earlier proposals. Minority reports could forward individual ideas.

Meeting early in 1965, the ASMS committee chose four top-level measures of performance for the new missile system:[5]

- Short reaction time to defeat low-flying missiles
- Firepower against saturation raids
- Immunity to weather clutter and enemy electronic countermeasures
- Reliability

Later, missile range was added to this list. Admiral Withington decided to measure effectiveness for the conceptual ASMS as minimum probability of failure against any realistic threat in the spectrum of threats that ASMS was to address. Presciently, he saw the primary threat as cruise missiles.[6]

Developments brought forward from the Typhon project included multifunction phased-array radar and midcourse command guidance of missiles. Missiles would use semiactive guidance only for terminal homing. During the missile's final seconds of flight, a shipboard radar would illuminate the target. The missile would home on the reflection of this beam. An issue was whether to use a pulsed beam for terminal illumination, as in Typhon, or a continuous-wave beam.

The primary argument against a continuous-wave beam was that, lacking range resolution, it might illuminate two targets, one near the other as seen from an approaching missile. A pulsed radar beam, such as a tracking beam, would distinguish between the targets, but a missile homing on continuous-wave reflections from both would fly between them. If the targets were separated by more than twice the lethal radius of a conventional warhead, detonation of the warhead would damage neither target. The ASMS panelists doubted that two aircraft could maintain the separation and relative orientation necessary for this, and for that matter doubted that it had ever occurred.[7] The ASMS committee chose continuous-wave illumination for terminal homing.

To jump ahead, during the Falkland Islands War in 1982 the Royal Navy guided missile destroyer *Coventry* (D 118) proved the ASMS planners correct for practical purposes. Her Sea Dart missile system, which used continuous-wave semiactive homing, illuminated two A-4 Skyhawk bombers at long range over West Falkland Island. One Sea Dart missile flew between the Skyhawks, detonated, and shot down both. Proving the ASMS planners correct too on needs for fast reaction and antisaturation capacity, later that month Argentine naval Skyhawks ambushed *Coventry* faster than her old Sea Dart target-tracking radars could lock on to them, and sank her.[8]

The most important issue for the multifunction radar was selection of the radar signal frequency, or wavelength. A phased-array radar with a frequency band lower than S-band would be too large for ships to carry. The choice was between S-band and C-band (electronic countermeasures bands F and G). Long-wavelength S-band radar signals penetrate atmospheric effects and weather well and give long detection range. A C-band phased-array search radar would weigh less than an S-band array. Its antenna could be mounted higher on the ship to give longer detection range against small, low-flying targets such as antiship cruise missiles. Both bands offered advantages to counter jamming: S-band had lower side lobes, and C-band had a wider signal band width for frequency agility.

After debating the issue for two months, the ASMS committee settled on C-band, as Typhon had used. A C-band radar could track low-altitude targets more easily. The Navy had more fleet experience with C-band radar installations. The failure of the SPS-33 radar systems on *Enterprise* and *Long Beach* tainted S-band. However, the crippling problem in the SPS-33 system was its computer processor. Its phased-array radar worked about as reliably as other electronics of the era.

Captain Bryce Inman documented the ASMS committee's decisions. He had supported C-band but became convinced that S-band would be better for ASMS. Microwave characteristics of the S-band provide the highest likelihood of detecting small and distant targets despite sea clutter, bad weather, and enemy electronic countermeasures. For a multifunction radar (search, tracking, and fire control) no other radar capability had meaning until the target was detected. "For two cents," Captain Inman ruefully told a Bell Laboratories scientist sitting nearby, "I'd change my vote." The scientist produced two pennies. They went to Admiral Withington and within half an hour had persuaded him to approve a change to S-band.[9]

The ASMS search radar was required to be at least as good as the SPS-48 radar for air intercept control, whereby the ship identifies air targets beyond her missile range and directs fighters to intercept them. Air intercept control requires a long-range radar with a height-finding capability so that controllers on the ship can inform the fighters of the range, bearing, and altitude of the distant targets. It was a frequent mission off North Vietnam. Many U.S. Navy fighter crews scored kills by following interception vectors relayed from radar console operators aboard cruisers in the Tonkin Gulf.

The use of programmable digital computers to control ASMS was the key to its success. The primary computer was the UYK-7 computer, in development for the DD 963 class. The ASMS radar would use electronic scanning from flat phased-array antennas, replacing Typhon's hemispherical dome. The system would send guidance commands to missiles in flight and illuminate the target for the missile's homing seeker only in the terminal phase, the final seconds of the missile's trajectory. Search radars use pulses of microwave energy to find range to a target (determined by the amount of time an echo from the target takes to return). The illuminator radar points a continuous beam at the target, and the missile follows the target's reflection of this beam to home in for the kill. Selection of continuous-wave illumination for terminal homing ruled out use of the multifunction radar for illumination. It was a technologically reliable solution that met the paramount need for minimum chance of failure.

In 1969 ASMS was renamed Aegis. Its multifunction radar was designated SPY-1.

Missiles, Launchers, and the Aegis Modular Deckhouse

Originally the ASMS program office expected that Aegis would require a new launcher and missile. The Standard medium-range antiaircraft missile family was being developed separately to replace the Terrier and Tartar missiles. The ASMS office decided to use Standard missiles, the new Mark 26 twin-arm launching system, and Tartar-D missile illuminators. Early Aegis designs showed six X-band continuous-wave illuminator radars. Two of these illuminators included C-band target-tracking channels, making them functionally identical to SPG-51D Tartar missile guidance radars (see chapter 12). These might lock on to targets that the Aegis multifunction radar might be unable to track, such as sea-skimming missiles. Later, once tests had established with some certainty that the Aegis radar could track anything, the pair of tracking illuminators was dropped.

During the mid-1960s the U.S. Army was developing the Patriot antiaircraft missile system, then called SAM-D for digital surface-to-air missile, to replace the unreliable Nike-Hercules missile batteries. The Office of the Secretary of Defense (OSD) decided to merge Patriot and Aegis into one development effort, using Patriot, because their targets were similar. OSD ordered a joint Army-Navy commonality study.

The study group recognized that Army and Navy operational requirements were very different. Patriot would cover a 90-degree sector of the front against high-altitude targets at short horizontal range. Aegis had to cover the entire 360-degree horizon simultaneously at all altitudes and ranges. Patriot's entire guidance system had to fit on board one compact, air-transportable, tracked vehicle that could move and set up to shoot quickly. The Army chose a compact single-face C-band phased-array radar for all functions, including illumination. The Navy would realize definite combat advantages with a larger S-band search-and-track radar and separate continuous-wave illuminators. Patriot needed a large new interceptor missile, but Aegis could use existing Standard missiles. Secretary of the Navy Paul Nitze backed the Navy, and OSD dropped the common-system development plan when the cost for a common system seemed likely to exceed the costs of developing Aegis and Patriot separately.[10]

By 1969 the ASMS program office had answered all the fundamental scientific questions associated with designing Aegis. RCA was selected as the contractor to build the Aegis system. RCA and the Aegis team began to design the actual physical and electronic components for their new system. The decision makers in the Navy, OSD, and Congress had only to choose the ships that would mount Aegis. Making that decision would take six years.

The Navy needed to sell projects to OSD to get funds to continue to work on them. A group in OSD was pushing broader use of the construction approach planned for the DX project of assembling warships from modules. To obtain support there, the Navy added a requirement for a modular deckhouse, called Deckhouse Mark 20.[11] The entire Aegis system, except for the missile-launching systems and the ship's combat information center, would be built in a pair of two-level deckhouses.

This approach had three theoretical advantages. First, the naval architects could design a ship separately from Aegis. It would be sufficient for the ship design to provide structural weight and stability to carry the deckhouses, accommodations, and the supporting ship's services of electrical power, air conditioning, compressed air, and cooling water. Second, the Aegis electronics and radar could be updated or replaced with less need to modify the ship herself, which would correspondingly reduce the time and expense of modernization. Third, and probably most important to OSD, was the assumption that the deckhouses could be assembled and tested in an electronics factory more cheaply than in a shipyard.

The modular Aegis deckhouse was actually a pair of 200-ton structures, fore and aft. Each deckhouse had two radar faces mounted at 45-degree angles from the centerline. The deckhouse structure had to be strong enough to support itself and the heavy components, such as power supplies, inside it. Functionally, the power supplies and other heavy components could have been located below the deckhouses. Two modular deckhouses with all components would weigh the same as three armored triple 6-inch gun turrets. The Aegis deckhouses were to be carried above the bridge in a ship that would not have barbettes, side armor, or boilers to lower her center of gravity.

Carrying this massive weight at such a height posed a formidable challenge for naval architects. An Aegis ship had to support the deckhouses and to remain stable despite flooding from firefighting or battle damage. That would make her a large ship, probably displacing around 12,000 tons. Admiral Rickover's congressional clients backed his demand for nuclear propulsion for any warship over 8,000 tons. Getting Aegis, the most vital single weapon in development for the Navy, to sea threatened to be prohibitively expensive. The *Spruance*-class destroyer project brought Aegis to the fleet.

Aegis and the *Spruance* Design

In 1970 Admiral Mark Woods, commander of the Naval Ordnance Systems Command (NavOrd), noticed a particularly talented officer at the Naval Ship Weapon System Engineering Station at Port Hueneme, California. Admiral Woods ordered this officer, Captain Wayne Meyer, to join his staff to take over the Aegis project. Meanwhile, in June 1970 the Navy had signed the Total Package Procurement contract for the *Spruance* class.

Captain Meyer, who quickly became by far the

most important figure in the development of Aegis, knew that the DD 963 design had space and weight margins reserved for the same Mark 26 launchers that Aegis would use. The DD 963's large margins of volume, power, stability, and reserve buoyancy for addition of the Tartar-D system interested the Aegis project office. Aegis project office and Naval Ship Engineering Center (NavSEC) staff personnel met after the DD 963 contract award to study modifying the DD 963 design to mount the entire Aegis system. The Aegis deckhouses were too large to incorporate as a payload during conversion of a DD 963. However, it did appear feasible to modify the *Spruance* design to support Aegis as installed equipment, with the ship and the radar system sharing common strength members.

Captain Meyer, Admiral Woods, and Admiral Sonenshein, commander of the Naval Ship Systems Command, reviewed these results in Sonenshein's office. All agreed that the Aegis/DD 963 study and other studies into the feasibility of installing Aegis on other ship designs showed that the DD 963 was the best Aegis platform. They further agreed to support the DD 963 as the best platform during Shipbuilding and Conversion, Navy (SCN) budget planning at the Naval Material Command and the Office of the Chief of Naval Operations. The DD 963 contract design was still evolving into the detail design. Admiral Sonenshein ordered that any changes to the DD 963 design had to preserve the stability margin, internal volume, and electrical power necessary for the Aegis system.[12]

The Aegis project office in NavOrd quietly began a major redesign effort to fit the large Aegis components within the available DD 963 design margins. The unobjectionable official explanation was that the DD 963 was simply a convenient sample platform to guide the simplification effort for a physically smaller and less expensive Aegis system.[13] As noted in chapter 6, this did not arouse Admiral Rickover and the nuclear power lobby into pressing Congress to prohibit use of Aegis project funds for design of a large non-nuclear Aegis ship, the true objective of the Aegis redesign.

The Aegis deckhouses would not fit over the off-center uptakes and intakes for the gas turbines. Arranging the after deckhouse atop the hangar was particularly difficult because the starboard gas turbine intakes were already flush with the after end of the hangar. That left no room for the starboard aft SPY-1 radar face. The bridge would

The Aegis system generates over 1,000 times more information than a conventional mechanically scanning radar, making it a valuable battle-management system. Four large-screen graphic displays in the combat information center show the status of operations. Two displays are for the ship's use, and two are for an embarked task group commander and staff.

need to be strengthened. The solution was to shift the SPY-1 radar faces 45 degrees to put them in plane with the DD 963 superstructure sides.

Captain Meyer, who had risen from seaman, faced a formidable task in turning the plans for Aegis into a deployed weapon. The challenges included generating confidence and enthusiasm in the project, showing the feasibility of Aegis, building prototypes, testing, budgeting, planning logistics, leading the organization and contractors, maintaining visible excellence, and creating the largest real-time computer program ever written. Along the way, the opposing technological threats of antiship cruise missiles and electronic jamming increased in capability.

Captain Meyer led the project very well and was promoted to rear admiral in 1974. He drummed into each Aegis staff officer, "You're only one man-year, so if you're going to get anything done, you've got to find a way to amplify, and the only way you can amplify is through people."[14] He enlisted other Navy offices to take on "double-hatted" responsibility for parts of Aegis, so that they reported to the Aegis project office for Aegis work and to their regular chain of command for other work. He kept unusually close contact with contractors and subcontractors, visiting their plants to "fire up the crowd" and to hand out Aegis Excellence awards.[15] Eventually he had several thousand people working on Aegis. Despite all this, political disputes over the best Aegis ship nearly killed the project. The criticism of the *Spruance* class influenced the decisions that led to an Aegis-armed *Spruance* design, but not in a way that any critic could have imagined.

The DG/Aegis Destroyer

Admiral Elmo Zumwalt disliked the *Spruance* class as "far too expensive" but assented to the contract with Litton a week before he took command as Chief of Naval Operations (CNO).[16] Aegis would be ready to order in 1973. If the plan for an Aegis-armed DD 963 was presented to Admiral Zumwalt, he did not approve it. He knew that to get enough ships in the fleet, they had to be cheap. In August 1971 he restricted the Aegis-armed escort to a maximum displacement of 5,000 tons and a cost of $100 million in 1973 prices. This design was known only as DG/Aegis. For comparison, OSD requested $612 million for FY 1973 for seven *Spruance*-class destroyers, about $87 million each and obviously without Aegis or initial design cost.[17] To get a budget for ships of even that cost, Admiral Zumwalt retired all the remaining ASW aircraft carriers and many other older ships. The long-foreseen problem of the block obsolescence of the World War II–built ships had arrived.

DG/Aegis was a single-purpose escort that never received a name, hull number, or even class type. It would mount a reduced-size Aegis system of the same type that the Aegis project office was redesigning for the *Spruance* class. This Aegis system would have the original layout of four faces at 45-degree angles from the centerline. The Aegis project office continued redesign work to integrate Aegis into the *Spruance* platform. Norman Friedman described the DG/Aegis ship that the Naval Ship Engineering Center could design on the CNO's cost-and-displacement budget: "Of 139 computer runs of the first series, only one alternative fell within the allowed bounds, and it was too austere even for Admiral Zumwalt: it would have had a single Mark 22 [single-arm] launcher (with only 16 missiles, as in a *Brooke*-class DEG), no

2-dimensional radar, no ASW capability at all, no Harpoon fire control system, no digital data links (and, therefore, no ability to make use of task force data resources)."[18]

Admiral Zumwalt now considered canceling Aegis. Charles DiBona, who had left active duty to head the Navy-sponsored Center for Naval Analyses, supported killing it. Against Aegis: Other new air defense systems were coming on line. The new Grumman F-14A Tomcat fighter could attack four, possibly six, bombers at long range. New, cheap point-defense weapons such as NATO Sea Sparrow and Phalanx could defend ships against individual missiles and aircraft that leaked past the Tomcats. For Aegis: Fighters required warning for enough aircraft to arm, launch, and take station, and were restricted by severe weather. The Navy's new Harpoon missile first flew in July 1972 and reminded Aegis skeptics of the power of antiship cruise missiles. Medium-range saturation attacks from Soviet Charlie-class missile submarines and coastal forces remained threats.

Captain Meyer promoted Aegis as a command post for coordinating all battle group air defense actions. The Aegis SPY-1 radar was far better than any other radar system for detecting, tracking, and displaying targets. An Aegis ship could link its displays with aircraft carriers, other air defense ships equipped with the Naval Tactical Data System, E-2C Hawkeye airborne early-warning aircraft, F-14 fighters, and Air Force E-3A Sentry AWACS aircraft. Captain Meyer cultivated Congress, OSD, and Admiral Zumwalt's staff at the Office of the Chief of Naval Operations to keep Aegis independent of any particular ship type.[19]

In November 1972 Admiral Zumwalt approved production of Aegis, if the Navy and Congress could agree on a ship design to carry it. He ordered development of Aegis to proceed at the pace of development of a conventionally powered ship design, smaller and cheaper than the DD 963 class.[20] He raised the limits for DG/Aegis to 6,000 tons and $125 million.

Reuven Leopold, the young designer of the *Spruance* class, had left Litton to become technical director of the Naval Ship Engineering Center. Under him NavSEC drew up a new DG/Aegis design for two Mark 13 single-arm launchers with 40-missile magazines. It would carry the Aegis SPY-1 radars, an SPS-49 radar, and a LAMPS helicopter. Design full-load displacement was 6,161 tons and cost was $136.1 million, above the CNO's limits.

NavSEC next recast the DG/Aegis design to replace the single-arm launchers with one Mark 26 twin-arm launcher and a 64-round missile magazine. Other sensors and weapons included a short-range sonar (probably the SQS-56 installed in the FFG 7–class frigates, or the similar Canadian SQS-502), one LAMPS helicopter, the SPS-49 air search radar, two 20 mm Phalanx guns, and Aegis SPY-1 radars, but only two illuminator radars for missile terminal homing.[21] Dr. Leopold personally thought this ship was "too puny" for Aegis.[22] Performance requirements were less stringent than for the DD 963 design. It had a single main engine room, a vulnerable point. Propulsion would be two FT9 marine gas turbines with 35,000 shp per engine driving two propellers. The FT9 was a marinized derivative of the new Pratt & Whitney JT9 turbofan. The next cheapest alternative in terms of acquisition cost was a steam plant with 850 psi boilers for 63,000 shp.[23] With gas turbines DG/Aegis displaced 5,884 tons, but its cost exceeded $200 million. Work on this design went on until 1974, but the Navy never requested funds to build it.

Redesign of Aegis for the DD 963 platform continued in the Aegis program office. Perhaps the main importance of the Aegis/DD 963 design was to keep up confidence among Aegis proponents that the platform disputes would be overcome. Aegis was very much a high-end project, but Admiral Zumwalt demanded low-end ships that he could get in numbers. Admiral Rickover's clients in Congress demanded nuclear-powered escorts, no matter how few were affordable, and did not care whether they mounted Aegis. The disputes were getting worse.

The Strike Cruiser

In 1973 the first two *Virginia*-class nuclear-powered guided missile cruisers were under construction, and the third had been authorized. In May 1973 Secretary of the Navy John Warner asked about designing a new Aegis-armed nuclear cruiser to replace the *Charles F. Adams* (DDG 2)–class destroyers. This design, initially designated as DG(N), was the origin of the strike cruiser (CSGN). The first DG(N) might be ordered in 1979.[24] Late in 1973 Admiral Rickover sought to divert $244 million from the FY 1975 shipbuilding budget, which OSD was preparing to submit to Congress, from Admiral Zumwalt's Sea Control Ship toward two more *Virginia*-class cruisers with-

out Aegis. Admiral Zumwalt stopped that diversion, but Admiral Rickover did get $100 million for long-lead-time material for a fourth *Virginia*-class cruiser.[25]

The *Virginia*'s armament was two Mark 26 twin-arm missile launchers with 68 missiles between them, two 5-inch guns, and ASW weapons. Some of those 68 missiles would be ASROC antisubmarine weapons, and others might be Harpoons and training rounds, reducing the antiaircraft missile load to perhaps 40. The *Virginia*-class design had originally included provisions for integration with the Aegis deckhouse, but now this ship design was alleged to be incompatible with Aegis.

More likely the problem was that Admiral Rickover's Naval Reactors Division needed a new ship design as a make-work project.[26] The nuclear-propulsion community had evaded the ship-design procedural changes of the 1960s. The Naval Reactors Division still designed nuclear warships independently from the operational requirements, if any, for such ships. The reduced-size Aegis system for DG/Aegis or the DD 963 class would, of course, be available for a similarly modified *Virginia* design. However, the Naval Reactors Division might well have preferred to start on a new ship design anyway, to keep its designers employed on a new destroyer-series reactor plant. Concern may have existed that the Secretary of Defense and Congress might regard a load of 40, or even 68, missiles on a repeat *Virginia* as too few to justify the very high cost of an Aegis-armed nuclear cruiser, especially since a cheaper Aegis-armed DD 963 could carry at least that many missiles herself.

When Navy planners began budgeting to build non-nuclear Aegis ships, Admiral Rickover tried to win on a technicality a case that he was losing on merit. Early in 1974 he lobbied Congress to prohibit the Navy by law from requesting non-nuclear escort ships for aircraft carrier battle groups, so that only nuclear ships could be designed for the fast task force escort mission. Congress obediently passed this as Title VIII of the FY 1975 defense appropriation bill.

A provision let the President, not the Secretary of Defense, waive this restriction by certifying that nuclear ship construction was not in the national interest. President Richard Nixon signed the bill with Title VIII into law in August 1974 but said that he would recommend nuclear propulsion only when the national interest justified the cost.[27]

He resigned four days later, his defense in the Watergate scandal demolished by leaks from Alexander Haig and Alexander Butterfield of the White House staff. In uniform a decade earlier, they had been McNamara's shadow Joint Chiefs of Staff, with then-Captain Ray Peet as Navy representative before he started the DX/DXG project. In direct opposition to the Chief of Naval Operations, Congress killed the Sea Control Ship and authorized a fourth *Virginia*-class cruiser.[28]

On July 1, 1974, Admiral James Holloway relieved Admiral Zumwalt as Chief of Naval Operations. An aviator, Holloway had been through the nuclear power program to qualify for command of the nuclear aircraft carrier *Enterprise*. Admiral Isaac Kidd was the alternate contender for CNO, but he was in disfavor in Congress after having attempted, on White House orders (probably Haig's), to discipline a bureaucratic critic of President Nixon's appointment of Roy Ash as director of the Office of Management and Budget. Sensitive to the misinformed criticism within the Navy that the *Spruance* class was inadequately armed, Admiral Holloway announced that his number one priority as CNO was to increase the firepower of the fleet. Appeasing critics was easier than explaining the procurement strategy for the *Spruance* class. He promptly canceled DG/Aegis and demanded a nuclear-powered version, DG(N).

In April 1974 the DG(N) design had displaced 11,900 tons and featured the same dual-reactor plant as the smaller *California* and *Virginia* classes. DG(N) would have been slow. The design was similar to DG/Aegis as described above, but with two 64-missile Mark 26 launching systems. More important, the Aegis system would have four illumination channels. Other new weapons included two LAMPS Mark III helicopters. Following Admiral Holloway's orders, the designers added two quadruple Tomahawk launchers, two quadruple Harpoon launchers, and an SQS-53A hull-mounted sonar. Gun armament was two 20 mm Phalanx mounts. The design now was for a ship 580 feet long, displacing 12,000 tons.[29]

The challenge was to show that the DG(N) was worth its higher cost over the Aegis-armed DD 963. Both designs were similar in capability. The gas turbine–powered DD 963–based ship would be better at ASW than the DG(N) design. With Aegis it would be equal at air defense. The DG(N) design was further enlarged to improve its survivability and antiship armament. It was redesignated a "strike cruiser" with a new ship type

designation of CSGN. The strike cruiser design increased to 666 feet waterline length and 17,172 tons full load. Its armament now included the DG(N) armament and one 8-inch gun but no 5-inch guns. New, more powerful nuclear reactors would be needed to increase its speed to 30 knots. An innovative compound plant, such as combined nuclear and gas turbine, would have been useful, but Admiral Rickover's Naval Reactors Division never considered it.

Survivability was the major combat improvement in the strike cruiser. The strike cruiser was designed to withstand five hits by cruise missiles or 1,000-pound bombs with a reasonable probability of retaining combat capability. The Aegis-armed DD 963 could be disabled by five 5-inch shell hits.[30] A task group command post, the tactical flag support center, was added to the strike cruiser design. The never-named strike cruiser went through more design changes. One version had a full-length flight deck for V/STOL aircraft (24,648 tons).[31] All designs showed Mark 26 twin-arm missile launchers instead of the new faster-firing vertical-launch system, because the twin-arm systems looked more like weapons, and so might increase political support for the ship.[32] Admiral Kidd quietly observed that the strike cruiser was an updated pocket battleship, a supposedly powerful all-purpose warship that proved to be an operational failure.

In 1975 the House Armed Services Committee demanded a nuclear-powered Aegis ship under Title VIII, but the House Appropriations Committee refused to fund it. OSD objected to the cost of the proposed nuclear-powered strike cruisers. Admiral Holloway favored the nuclear-powered DG(N), but Vice Admiral James Doyle, Holloway's deputy for surface warfare, persuaded him to support the Aegis/DD 963 design.[33]

Admiral Holloway warned the Secretary of Defense that unless OSD backed some Aegis ship, Congress would kill Aegis. With no Aegis ship design in sight, the Research and Development Subcommittee of the Senate Armed Services Committee voted to cancel Aegis. He proposed 8 nuclear-powered Aegis strike cruisers and 16 Aegis-armed *Spruance* destroyers. In June 1975 Secretary of Defense Donald Rumsfeld and the Defense Systems Acquisition Review Council (DSARC) directed development of both classes.[34] An indication of how much work the Aegis program office had already done on integrating Aegis with the DD 963 design is that the destroyer version would be ready for ordering in 1977. That was a year earlier than the strike cruiser, even though the destroyer design had never even been an approved project.[35] In 1976 the Aegis program office took official responsibility for all Aegis ship designs.[36]

Congress agreed to keep Aegis alive, but the dispute over ship propulsion continued. Congress rejected the costly strike cruiser but approved funds to install Aegis on the nuclear cruiser *Long Beach* (CGN 9), funds that the Ford administration then refused to spend. Reports that Admiral Rickover lobbied to kill the *Long Beach* conversion are consistent with a preference for projects that would keep his Naval Reactors staff in work, which rearmament of an existing ship would not do.

Instead the Ford administration approved an item for the 1978 budget plan to build the first Aegis-armed DD 963. This ship drew the next guided missile destroyer hull sequence number: DDG 47. In 1977 the new Carter administration approved the $930 million requested for DDG 47, and Congress authorized the ship in the FY 1978 budget.

The Navy abandoned the strike cruiser designs after Admiral Holloway retired as CNO in 1978. Work resumed on design for an Aegis-equipped version of *Virginia*. This ship was tentatively designated CGN 42 in budget plans. It would mount the same armament as DDG 47, and four armored box launchers for Tomahawks and Harpoons. The Navy never requested funds for it.

Critics have described DDG 47 as a "lowering of sights compared with the nuclear-powered strike cruiser (CSGN) and the CGN-42 class."[37] However, the distinctive features of the strike cruiser design (side armor, an 8-inch gun, even V/STOL aircraft) had been added to justify the cost of the nuclear-powered ship, not to meet additional military mission requirements. The CGN 42 cruiser would not have had even those capabilities. No additional nuclear-powered cruisers were authorized after the *Spruance*-class destroyers showed they were by far the best ASW ships in the fleet. Their tactical cruising range was good enough, and they were probably faster. Until then the nuclear power lobby had maintained that its ships were the best, despite their higher cost. The *Spruance* class showed that for the critical ASW mission, nuclear ships were not even as good as the new, computerized, gas turbine–powered destroyers.

DSARC authorized the Navy to change the DD 963 design and its mechanical and electrical systems only as necessary to install Aegis.[38] It authorized routine fixing of minor DD 963 class deficiencies. This directive bound the design team at the Naval Sea Systems Command (NavSea) very tightly.

Adding the Aegis SPY-1 radars to the DD 963 design caused two immediate and drastic problems. First, the center of gravity of the Aegis ship would be 23.35 feet above the keel, but the center of gravity of the DD 963 design could not go higher than 22.85 feet. Second, full-load displacement of the new ship would exceed 9,000 tons, but the DD 963 limit was 8,800 tons. An otherwise unchanged Aegis DD 963 would float so deep in the water that if battle damage opened less than 15 percent of her hull to the sea, she would risk sinking. To install Aegis the naval architects had to lower the ship's displacement and center of gravity, provided they could do that without violating OSD's restriction that they make only "necessary" changes.

The DD 963 hull is divided into watertight compartments by unbroken bulkheads that extend from the bottom of the ship to the main deck. Most of the main deck is inside the hull, and the main weather deck is actually the 01 level, one deck above it. Internal changes, such as increasing the strength of the internal bulkheads on the main deck up to the 01 level, raised the V-lines. This let the ship float deeper and withstand battle damage to Navy specifications. Converting four voids to fuel tanks provided additional weight at the bottom. A cross-flooding duct between two side-by-side compartments aft eliminated a possible location for dangerous asymmetric flooding. These changes increased the limits for displacement to 9,700 tons and for center of gravity to 23.35 feet above the keel.

The first concept design for DDG 47 took this modified DD 963 hull design and added four SPY-1 radar arrays in two deckhouse expansions, two Mark 26 guided missile launchers with 44-round magazines, four missile target illuminators, and related Aegis systems such as a new identification-friend-or-foe (IFF) antenna and four missile telemetry receivers. This used almost all of the design margin normally allowed for the modifications that inevitably occur during a ship's construction and operations. The SPG-60 antiaircraft tracking radar for the 5-inch guns was deleted and the foremast cut down to an ugly stump. Displacement was 8,843 tons, but 200 tons of lead had to be added along the keel to lower the center of gravity to 23.35 feet, which was only right at the limit.

The next stage was the initial preliminary design, which was more detailed and added 100 tons to displacement. Of this weight 55 tons was for Aegis equipment, spare parts, and upgraded (2,500 kW) electrical generators, and 45 tons was more lead ballast to keep the center of gravity from rising past 23.35 feet again.

The need for so much ballast was surprising, and the design team began to look for a way to decrease it. A margin of 580 tons was the normal allowance for a new ship design (9.2 percent of the light-ship displacement). Deciding that DDG 47 was not a new ship design but a modified DD 963 design, NavSea changed the margin policy for her. NavSea reasoned that both the DD 963 design and Aegis itself were mature enough that fewer unforeseen weight increases would occur during construction than might occur with a new design. The first *Spruances* were already in the fleet. The prototype Aegis system had been built while waiting for a ship to mount it. A much lower margin of 163.5 tons was approved. Ballast dropped to 180 tons on a displacement of 9,089 tons. In this form DDG 47 went on to the contract design stage.

Several interesting changes now occurred. The sonar room in the bow was armored with antifragmentation protection. The four LM-2500 gas turbines were upgraded to 25,000 hp each, and stronger and heavier propeller shafts were added to handle this load. Most noticeably, the gas turbine exhaust stacks were changed to use the boundary-layer infrared suppression system (BLISS). Two tall uptakes replaced the six low funnel caps of the DD 963 design. The heavy exhaust silencers were eliminated. These changes reduced displacement to 8,910 tons. With no ballast, the center of gravity was at 23.35 feet.

Ceramic fire insulation was added to the aluminum bulkheads in the deckhouse that served as fire-zone boundaries or that formed the side of the deckhouse against the weather. This insulation would provide 30 minutes of protection from fire before the aluminum melted. The designers reasoned that if a fire were still burning after 30 minutes, then the ship must be in dire trouble anyway. In 1987 fires burned aboard USS *Stark* for days after two Exocet missiles hit her. *Stark's* aluminum bulkheads and other fittings disintegrated, but her steel hull kept her afloat.

In March 1978 NavSea ordered the *Kouroosh* (F-DD 993) class of four guided missile destroyers for Iran from Litton. They incorporated many improvements learned from the early DD 963s and were heavier. In September 1978 the Navy awarded Litton a contract for the detail design and construction of DDG 47. With the bicentennial celebration of the American Revolution under way, the Navy named the ship *Ticonderoga* (DDG 47).

In February 1979 Litton submitted its detail design for the Aegis destroyer. DDG 47 would displace 9,270 tons, and her center of gravity would be an excessive 23.57 feet above the keel. The Navy now discovered the costs of OSD intrusion into ship design and of the Navy designers' neglect of the DD 963 project. Litton had based its DDG 47 detail design on the shah's ships instead of the DD 963 design that NavSea had used, following the 1975 DSARC directive. NavSea and Litton wrestled with the detail design for the next three years.

Ticonderoga was redesignated a guided missile cruiser on January 1, 1980, and was laid down as CG 47 on January 21, 1980. She was launched on April 25, 1981, and was christened by the new first lady, Nancy Reagan, an indication of the attention the new warship drew. Mrs. Reagan had christened another ship previously, but the champagne bottle had not broken. *Ticonderoga's* prow successfully smashed the bottle on her first swing, splashing champagne on the first lady. She patted the cruiser's stem in approval, the first "bravo zulu" (well done) the *Ticonderoga* class earned. By December 1981 displacement had grown to 9,287 tons, but the center of gravity was down to 23.25 feet. This good news was shattered in April 1982 when the almost complete ship was inclined, a routine shipbuilding test. It showed that *Ticonderoga* was 200 tons overweight and that her center of gravity had risen again.

One hundred ten tons of lead ballast was packed into the skeg, increasing the ship's displacement to nearly 9,500 tons. The Navy was now concerned about how close *Ticonderoga* was approaching the hull's full-load displacement limit of 9,700 tons. The problem was at her stern, where the transom would have a large aperture for the new SQR-19 tactical towed-array sonar. If flooding in the after part of the ship submerged this opening, then the additional mass of water entering through it might sink the ship by plunging. NavSea required the transom opening to be watertight except when the SQR-19 array was launched or recovered. Those are quick operations that a ship coping with severe flooding probably would not be performing anyway.

This increased the CG 47 design's limiting displacement to 10,200 tons, but that revealed still another problem: the structural strength limit for the hull was 9,950 tons. Ingalls checked the ship as built to determine how to strengthen the hull for a 10,200-ton displacement. This time the news was good: weight was well distributed throughout the hull. In 1979 the Navy had approved an Ingalls request to use metric-measurement hull plates. In yet another case of neglect of the DD 963 design, which some in the Navy seemed to regard as an illegitimate child, this change had been forgotten. The metric plates were thicker, thus stronger, so the hull was already stronger than necessary for 10,200 tons.

The Navy took delivery of the ship in December 1982 and commissioned her as USS *Ticonderoga* (CG 47) on January 22, 1983, on time and within budget. *Ticonderoga* displaced 9,589 tons at full load with a center of gravity 23.29 feet above her keel. For the first time a warship with a major increase in her capabilities was smaller than preceding warships of the same type. This was more of political interest than of engineering interest. Unlike the *Spruance* and *Kidd* classes, *Ticonderoga* was delivered with her full armament installed.

The second Aegis cruiser, *Yorktown*, was already in the water and would be a twin of *Ticonderoga*. The next three ships, *Vincennes*, *Valley Forge*, and *Thomas S. Gates*, were built to operate the new SH-60B Seahawk LAMPS III helicopters with the Canadian RAST equipment to recover them. The ships' armories were expanded for more small arms and ammunition. These changes added 40 tons high in the ship and raised the center of gravity again. Higher-strength-alloy steel plates reduced hull weight but kept strength within the required limit. The four-legged masts were replaced with tripod masts, which saved 8.9 tons high in the ship. The Phalanx magazine deckhouse between the Phalanx guns was moved down one level, replacing Aegis equipment rooms, which were moved into the hull. Over 160 miles of lightweight fire-resistant cable was installed.

Marilyn Quayle, wife of Indiana's maladroit but ambitious Senator Dan Quayle, christened *Vincennes* (CG 49). The ship was delivered in 1985 at a displacement of 9,407 tons. Her center of gravity was 23.34 feet above the keel. These figures apparently do not include ballast, if she carried any.

Even before being laid down, the *Ticonderoga* design was criticized as "ponderous and overweight."[39] But the ships are in fact lighter than their highly regarded sisters in the *Kidd* class and perform well at sea: "The engines can achieve no-load idle speed from cold iron in 60 seconds or less, and can accelerate from no-load to full speed and maximum power in 30 seconds or less. As a result, the *Ticonderogas*, like the *Spruances*, accelerate almost instantaneously, and will stop from all-ahead Flank III (top speed) to dead-in-the-water in two ship lengths or 60 seconds."[40]

On the *Kidd* class the weight of the weapons added to the DD 963 design was concentrated amidships and aft, where the hull is full. This left *Kidd*'s bow relatively light and buoyant. The *Ticonderoga* class, however, had a greatly enlarged forward deckhouse and a higher-capacity forward missile magazine for 44 (Mark 26 Mod 1) or 61 (vertical-launch system, VLS) missiles, compared with 24 missiles in the *Spruance* (pre-VLS) and *Kidd* classes. An Aegis cruiser pitches hard in heavy weather, especially with a full missile load in the magazines. During design of the CG 47 (DDG 47), a steel bulwark was added at the bow to keep the fo'c's'le dry. It was switched to aluminum to reduce top weight on later ships. It helps in moderate seas. If the ship takes green water over the bow, it acts as a scoop and throws water back toward the bridge. Water intrusion has damaged the windlasses. On several ships the bulwark has separated in heavy seas. One lost hers entirely while rounding Cape Horn. Steel bulwarks are being reinstalled.[41] Structural modifications made to ships in service include steel stiffeners in the deckhouses.

The *Ticonderoga* class retains its original destroyer maneuverability. At 25 knots an Aegis cruiser can turn 90 degrees in two ship lengths, but the ship heels steeply from the weight of the SPY-1 radar and deckhouses high up:

> A good rule of thumb is to never use a combination of rudder and speed that totals more than 30. In other words, at 25 knots, about 5° of rudder is appropriate. Much more will generally put the next meal squarely on the deck in the center of the galley. . . . In seas in excess of 20 feet with winds gusting more than 50 knots, she will roll 25°–35°, and occasionally will roll in excess of 40° [on a steady course].[42]

A naval architect would not willingly design a new warship with these seakeeping characteris-

Aegis cruisers use the DD 963 design with the Aegis electronically scanning radar system and two rail or vertical missile launchers. Taking the base DD 963 design to its limits, Aegis cruisers ride deeply and roll even in calm seas, as *Antietam* demonstrates here. These ships turned the surface fleet into an offensive fighting force.

tics, but external budget constraints and internal Navy bickering over nuclear propulsion had blocked the development of other designs. During the early 1980s the Navy designed the *Arleigh Burke*–class destroyers for seakeeping, lower cost, and better survivability on less displacement. The trade-offs were six years of design time, a broader and much shorter hull, and reduced armament, speed, and range.

Baseline 2 Aegis Cruisers

Beginning with the sixth ship, *Bunker Hill* (CG 52), Mark 41 vertical-launch systems replaced the twin-arm Mark 26 missile-launching systems. Magazine capacity increased from 88 to 122 Standard-2 missiles and vertical-launch ASROC rocket-thrown torpedoes. Design began in 1980. In 1981 the new Secretary of the Navy, John Lehman, ordered the depth of the vertical-launch system increased to carry Tomahawk cruise missiles. Tomahawks supported the Maritime Strategy. Given the already tight ship design, NavSea developed a notional load of 96 Standards and 26 Tomahawks on a displacement of 10,100 tons and a center of gravity at 23.35 feet. The first two ships, with twin-arm Mark 26 launchers, were known as

the Baseline, and CG 49 was Baseline 1983. The new VLS design was Baseline 1984. Later, the CG 47–CG 51 series became Baseline 1, and the VLS-equipped CG 52 design became Baseline 2.

NavSea began a campaign, "Take off Tons Sensibly" (TOTS), to reduce the weight of the VLS-equipped cruisers (Baseline 2). The use of high-strength HY-80 steel alloy, already in use in submarine pressure hulls, saved hull weight. The aft Aegis deckhouse was lowered 2 feet. The major change was to move heavy auxiliary equipment from the second deck down into the auxiliary machinery room (Aux One). Aux One was still roomy as a holdover from the early Litton design for the DD 963 class, which had located a large motor-generator there as a combining gear for the LM-2500 turbines. With that, many systems that had been located higher in the ship were shifted progressively downward, each moving into lower spaces vacated by systems previously moved down. This reduced the need for ballast and lowered the ship's profile amidships. Final CG 52 design (Baseline 2) was 9,410 tons, 7 percent below the initial estimate.

Baseline 2 included seven cruisers: *Bunker Hill* (CG 52) through *Philippine Sea* (CG 58). Again, a more powerful ship was smaller than her predecessor. The Navy invited Bath Iron Works to compete with Ingalls to create a second Aegis-capable yard. Ingalls had built the first 4 cruisers and won contracts for 15 more after competition began. Bath won contracts for eight. Since Bath had to get design changes from Ingalls, Bath's ships were behind the Ingalls ships in design updates. This led to rumors in the fleet that Bath-built cruisers were inferior.

ASW armament changes for Baseline 2 cruisers (CG 52 through CG 58) included addition of the SQR-19 towed-array sonar aboard *Antietam* (CG 54) and *Leyte Gulf* (CG 55). Their SQQ-89 ASW display system is almost identical to the SQQ-89 configuration on SQR-19–equipped FFG 7–class frigates. Their Mark 116 underwater fire control system uses only the SQS-53A hull-mounted so-

nar for targeting. All Baseline 2 ships carry two SH-60B Seahawk LAMPS III helicopters, RAST tracks, SRQ-4 data link transceivers, and SQQ-28 sonobuoy processors. The complete SQQ-89 ASW system was added to all Aegis cruisers starting with *San Jacinto* (CG 56), including the SQR-19 towed-array sonar and SQS-53B hull-mounted sonar.

Political Controversy over the *Ticonderoga* Class

After 1982 Congress approved new Aegis cruisers at the rate of 3 a year and closed the program with 5 in 1988, for a total of 27. Political controversy dogged the program throughout this period. This did not have the profound influence on the

The first Baseline 2 Aegis cruiser, *Bunker Hill* (CG 52), introduced vertical missile launchers. These are visible here as eight-by-eight matrices fore and aft. Several hatches are scorched from missile plumes. A reloading crane occupies three cells in each matrix.

The vertical launch system increases magazine capacity and firing rate, fires Tomahawk cruise missiles, and is less expensive than rail launchers. (U.S. Navy photo)

Navy that the *Spruance* controversy had had a decade earlier, or indeed any influence at all. The debate was a measure not of the ship but of misunderstanding in Congress about the military.

Congressman Joseph Addabbo of New York City wanted the Navy to issue a second-source contract to build Aegis systems in a plant in his district. Admiral Meyer rejected the proposal. Addabbo, a powerful member of the House Appropriations Committee, attempted to cut 10 percent of the Navy's entire budget for research and development and to divert it to designing a phased-array radar in this plant for use on the small *Oliver Hazard Perry*–class frigates. Navy officials ridiculed the idea. Taking a lesson from the infamous Senator Joe McCarthy, in 1982 the House Appropriations Committee sent its staff to extort the Aegis project with an investigation. The Navy helped the investigators but classified their report. It immediately showed up in a private newsletter. A spokesman for the committee proudly declared that the findings would be "devastating."[43] Addabbo claimed that the leaked report was accurate.[44] It is difficult to avoid suspicion that the leak was from Congress.

The report predicted that the *Ticonderoga* class would be "ineffectual."[45] The basis for this charge was that the ships risked becoming "initially unstable" because the design was tight and the center of gravity was higher than planned, so that the margin of stability available for future growth was proportionately lower than usual for a new design. TV news and the general press repeated the assertion of instability. This completely misrepresented the concept of stability margins, which measured not seaworthiness but the amount of equipment that could be added to the design. The ship would be stable even in a badly damaged condition in heavy seas, specifically with an 80-foot hole in the hull and a 35-knot beam wind.[46]

The report further criticized the class as unable to keep up with an aircraft carrier steaming at 34 knots, which was not a military requirement and which ignored sea conditions. The report also claimed that the SPS-49 air search radar, necessary for operational deception, was unnecessary.[47] These charges revealed the ignorance of the investigators. Their destructive approach and their leak raised questions about their honesty. Despite the House investigators' attempt to advance his hoped-for abuse of the procurement system, Congressman Addabbo went to his grave without getting the contracts he wanted. The report nearly did

have one technically damaging effect. In 1982 Congress voted to delete the SPS-49 radar and the LAMPS Mark III data links from the FY 1983 ships (CG 54–CG 56) but later restored funding, implicitly acknowledging that the staff criticisms had been invalid.

Misunderstandings about the term *margin of stability* continued in Congress for two years. A new issue was the testing and evaluation of Aegis. During 1983–84 several congressional committees were baffled about the purpose of developmental weapons testing. One purpose of these tests was to find possible weak points in the Aegis system. Congress focused on the test procedure and was visibly nervous that just 5 of 16 targets had been hit. The Navy pointed out that it was testing new systems and doctrines, not retesting what it already knew worked, and that the development tests were not a representative sample of combat engagements. It did not help that the Navy had announced that the tests were a success.

A top-level Navy explanation was reportedly accurate but was classified, and so was useless for answering press inquiries. The tests did show that Aegis had trouble hitting multiple supersonic sea-skimming antiship cruise missiles, such as the new Soviet SS-N-22 (P-80/3M-80). Aegis was built to hit hostile missiles at greater range early in their flight, when they cruised at higher altitude for speed and fuel economy, and possibly to search for their target. Several congressmen and senators doubted that anything short of full-scale worst-case combat simulation was a valid test case.

Concerned that Aegis was in danger, in 1984 the CNO, Admiral James Watkins, ordered USS *Ticonderoga* home early from deployment and had her conduct an expensive set of new tests that showed that Aegis could shoot down targets. It was a politically wise if costly ($30 million) and technically unnecessary exercise. This time 10 of 11 targets were hit.[48] Admiral Watkins announced that this showed that the Navy had fixed problems found earlier.[49] By 1985 *Ticonderoga* had fired over 100 missiles in tests. The testing controversy died.

Small controversies have cropped up. In 1989 Representative John Dingell, hoping for support from industrial polluters, attacked Aegis because Bath Iron Works was building Aegis ships in Maine, whose congressional delegation was trying to restrict pollution in hopes of reducing acid rain. Discovering that antiradiation missiles (ARMs) existed, Dingell announced, "The situation is far grimmer than anyone without very special security clearances would have guessed. The only way the CG 47 cruisers or DDG 51 destroyers can avoid being seriously damaged or sunk is to turn off the radar."[50] An attack with antiradiation missiles would be no safer or more reliable, from the attacker's viewpoint, than any other type of missile attack. A fundamental mission for Aegis was to force attackers to come out and fight. The ships carry the SPS-49 long-range radar to lure attackers within Aegis range, and would switch on the SPY-1 transmitter when they were too close to escape. Aegis cruisers can set up "remote track/launch on search" cooperative engagements where a radar-silent cruiser can ambush aircraft approaching another radiating cruiser. Additional classified capabilities await discovery by an aggressor. The Navy dismissed Dingell's statement as ignorant: "The ARM was wrongly popularized as a unique anti-shipping missile threat in the early 1980s. Those who hold this view are constrained by their 'lone citadel' perspective of fleet anti-air warfare."[51] Dingell doubted that measures to reduce radar cross-section would work, but the Navy corrected his error on that, too. A hostile aircraft that must approach a warship more closely to detect it and to distinguish it from decoys is in greater danger of being shot down while its missiles are still on its wings. Dingell abandoned his attacks on Aegis and the *Ticonderoga* class after that.

Baseline 3 and 4 Aegis Cruisers

Baseline 3 includes nine Aegis cruisers, *Princeton* (CG 59) through *Shiloh* (CG 67). The primary change in Baseline 3 is that the lighter SPY-1B phased-array radar replaces the SPY-1A. SPY-1B is far more capable for antimissile engagements; Aegis receives missile telemetry through the SPY-1B instead of through separate receivers. The last six cruisers, *Anzio* (CG 68) through *Port Royal* (CG 73), are Baseline 4. Baseline 4 changed the computers to the faster UYK-43 (32-bit) and UYK-44 (16-bit) models from the older UYK-7 and UYK-20. UYK-43 is the latest descendant of Seymour Cray's original design for the Naval Tactical Data System (NTDS) computer in the 1950s.

The *Ticonderoga* class totals 27. That number was justified as 1.5 cruisers per aircraft carrier for 15 aircraft carriers, and 1 per battleship for 4 battleships. In 1995 the battleships have all been stricken, and the carrier force seems likely to hover around 12, but all 27 Aegis cruisers are in service.

The Aegis System

SPY-1, the multifunction phased-array radar, is the heart of Aegis. Aegis did not provide a visibly new weapon, since the missiles and launchers looked and performed like any others. Instead Aegis provided new radars, computer programs, and displays: electronic weapons. The original SPY-1 radar was a prototype. The first production version was the SPY-1A. Advances in microelectronics led to a more compact version, SPY-1B. (SPY-1C was a never-built version for aircraft carriers. SPY-1D is on the DDG 51–class destroyers and has fewer displays.) For convenience, in the following discussion SPY-1 refers to both the SPY-1A and SPY-1B variants.

A conventional mechanical-scanning radar focuses a narrow beam of radar pulses in one direction relative to the rotating radar antenna. A reflected echo gives the target's range, and the antenna's azimuth shows bearing. The SPY-1 radar is motionless and steers the microwave beams electronically to scan the sky from horizon to zenith. Each octagonal SPY-1 radar face has over 4,000 radar elements, grouped into 140 arrays. The radar steers coherent beams by timing the phases of radar pulses emitted from the radiating elements, hence the term *phased array*.[52]

Certain features of the SPY-1 phased-array radar provide valuable military benefits. First, the radar can instantly steer a beam in any direction in space. Second, it can transmit from different elements on different frequencies simultaneously and steer them in different directions. Third, SPY-1 can create coded signals to send midcourse guidance commands to the Standard SM-2 missiles in flight.

Aegis generates over 1,000 times more information than a conventional mechanically scanning radar. SPY-1 sends out multiple beams to scan for sea-skimming targets out to 45 miles and for air targets out to at least 175 miles.[53] It can locate higher-altitude targets at 250 miles.[54] When a target is detected, SPY-1 tracks it with a beam many times per second. For comparison, a rotating search radar antenna makes about 6–10 scans per minute. This introduces ambiguity, since an echo will often show a contact in a location different from that shown on a previous scan. Has one contact disappeared from detection and a second

USS *Princeton* (CG 59) was the first Baseline 3 Aegis cruiser, introducing the lighter-weight SPY-1B radar. Internal rearrangement lowered her profile amidships compared with earlier Aegis cruisers.

With SPY-1B, Baseline 3 and 4 Aegis cruisers are the most lethal air defense and missile defense units in the U.S. military. (USS *Princeton* photo)

Mobile Bay (CG 53), a Baseline 2 Aegis cruiser, undergoes an explosive shock test. This test showed that missiles in the vertical launch system (VLS) needed better holders. Equipped with fixes, *Princeton* lost no missiles in her VLS magazines when she triggered two mines, one underneath her, in 1991.

Notice the compressed-air bubbles from her Masker sound-suppression belts amidships and from her Prairie anticavitation silencers on her propellers. These are features of all DD 963-type ships. (U.S. Navy photo)

Sent off as a decoy during the Grenada operation and then used as a distant radar picket off Lebanon, *Ticonderoga* had a chilly introduction to the fleet. But when photographed here off Beirut in December 1983, she was proving herself to be a very valuable asset. She is starting one of her aft gas turbines, hence the white plume. She first fired her guns in earnest less than a week later. (U.S. Navy photo)

been discovered? Did the original contact simply move? Or are one or both contacts spurious (false contacts)? Is the contact of interest, or is it clutter? Through almost continuous contact, SPY-1 maintains precise information about a target's identity and movement.

SPY-1 tracks over 200 targets as much as 250 miles from the Aegis ship. For missile fire control, the Aegis computer system can maintain 128 tracks. SPY-1 can track targets at any elevation including directly overhead (90 degrees elevation). A mechanical-scanning radar cannot search for targets at an angle much higher than about 45 degrees above the horizon, so the ship must use a fire control radar to track a target at higher elevation.

As a computer-based system, Aegis can be programmed. The crew codes doctrine statements into the computer to be executed automatically. Doctrine statements prevent the system from tracking contacts of no interest; one account hints that the SPY-1 can track objects as slight as a flock of birds. Doctrine statements can launch missiles automatically after the commanding officer has released the missile battery to fire. Because an Aegis cruiser is a logical choice for a command post, the *Ticonderoga*-class cruisers are equipped

with an independent Aegis display system for use by the force antiaircraft warfare (AAW) coordinator. Each ship has four large-screen displays: two for the battle group air defense coordinator, and two for the ship's officers. Five smaller monitors provide additional data for each pair of large-screen displays.

Standard SM-1 missiles, lacking autopilots, can only pursue the shifting echoes from the targets. Recall that with its original armament of two SPG-51 radars and one SPG-60 radar to illuminate targets throughout the Standard SM-1 missiles' flight, a *Kidd*-class destroyer could engage three aircraft at a time. Even with Standard SM-1 missiles, Aegis provides faster reaction, since the target track is developed quickly and there is no delay for a separate tracking radar to acquire the target.[55]

With Standard SM-2 missiles, an Aegis cruiser can attack 16 aircraft and missiles ("vampires") almost simultaneously and hit them at twice the range of less efficient systems.[56] Aegis forecasts intercept points, the volume of space ahead of each Standard SM-2 where it can intercept a target, depending on remaining fuel and glide range, and on the target's ability to maneuver relative to the Standard SM-2. Since Aegis can control four terminal interceptions simultaneously, it schedules launches so that Standard SM-2 missiles will arrive at points in space where the Aegis system can sequentially intercept the vampires.

An Aegis ship fires Standard SM-2 missiles on energy-efficient trajectories so that they fly to the intercept points independently. If a target changes course, Aegis sends midcourse guidance commands to the missile in flight to update the intercept point. During the last seconds before intercept, Aegis points an SPG-62 dish antenna to illuminate the target with a high-precision (X-band) continuous-wave illumination beam. The Standard SM-2 follows this beam's reflection and closes in on the target until its proximity fuze detonates the warhead. Standard SM-2 missiles use blast-fragmentation warheads. The ship would probably follow a shoot-look-shoot doctrine against distant targets, since Aegis could assess whether a kill had been scored ("look") and fire another missile if necessary, and a shoot-shoot-look doctrine against high-priority closer targets.

Since Aegis has no tracking antenna emitting a characteristic C-band or X-band signal, a manned enemy aircraft never knows whether it is under attack. Few military aircraft even carry radar-detection (ESM) equipment instrumented for S-band radar. In the last seconds before intercept, a missile-warning radar aboard an aircraft under attack would detect the SPG-62 X-band continuous-wave illumination beam.

Target defenses might include decoys or stealth (skin surfaces that would not reflect enough X-band energy for the missile to detect). Nuclear, biological, or chemical area-attack warheads might still be lethal even if the platform carrying them were hit. The ships are hardened against transient-radiation electromagnetic effects from nuclear bursts.[57] The Navy has publicly said that the SPY-1 radar can detect stealth aircraft. Shooting them down, however, might be more difficult. In practice, no adversary could afford reliable defenses against Aegis.

Revolution at Sea

Ticonderoga deployed late in 1983 with an aircraft carrier battle group. The fleet commanders at first regarded her as simply an advanced antiaircraft ship. Her first combat mission was to lead part of her task group across the Atlantic as a decoy force for Soviet ocean-surveillance satel-

lites, while the other ships sneaked off to Grenada. Arriving off Lebanon shortly after the Marine barracks bombing, she was at first stationed so far offshore that Beirut was beyond the SPY-1A radar's surface range. Her stock began to rise when other ships, aircraft, and unit commanders noticed the volume and quality of the information she was reporting to them over the data links. *Ticonderoga* was brought further in and soon became responsible for detecting and tracking all surface and air traffic in the area. She patrolled 30–50 nautical miles off the coast. Her combat systems officer wrote this account of her contribution:

> [The *Ticonderoga*'s patrols] permitted Aegis/ SM-2 coverage of the American embassy, Beirut airport, and the amphibious group. Amphibious warfare ships routinely carried out night steaming operations close to the *Ticonderoga* before moving shoreward for daylight cargo transfer operations [in other words, the lightly armed amphibious ships thought they were safer near her]. The *Ticonderoga* routinely performed flight-following of dozens of helicopters in the area (U.S., French, British, Israeli, and commercial). She became an unofficial, *de facto* sea-based air traffic control center. In addition to air and surface tracking, the ship maintained an active sonar search at all times off Beirut to counter the genuine threat of Libyan and Syrian diesel submarines. . . .
>
> During the Lebanon crisis, the *Ticonderoga* controlled 2,550 intercepts made by combat air patrol (CAP) aircraft against radar targets. The ability of the Aegis/SPY-1A system to engage targets anywhere within the battle group's warning area, many of which were not held by any other sensor in the battle group, was demonstrated repeatedly. . . . This was accomplished under electromagnetic wave propagation conditions that were as bad as might be found anywhere. Proximity to land, severe ducting, and dust storms contributed to the problems of tracking and target identification. . . .
>
> Most fire control radars used in AAW applications are pulse-doppler type, and have been optimized for targets with high closing speeds. Such radars are relatively ineffective with slow targets such as light aircraft and helicopters. The SPY-1A radar is designed to track surface and air targets with equal facility. Because of

the unique ability of the *Ticonderoga* to provide tracking and fire control data on any target in the radar horizon, surface or air, the ship usually was assigned a position where radar coverage included all amphibious forces stationed inshore from the battle force, as well as the Beirut airport, thus protecting the most likely targets of air attack.[58]

An old DDG 2–class destroyer in the harbor reported seeing a small unmarked single-engine aircraft that *Ticonderoga* had not detected, a report that the general press repeated as criticism of Aegis. Intelligence reports warned that a kamikaze mission against U.S. warships was possible; on the other hand, it might have been a press aircraft or just a student pilot on a practice flight. *Ticonderoga* was far offshore at the time, and the aircraft was probably below the SPY-1A's radar horizon. The aircraft flew near the *Spruance*-class destroyer *Moosbrugger* (DD 980), which was prepared to shoot it down with 5-inch guns if it turned toward the ship. The DDG ran missiles out on the launcher, but *Moosbrugger* told her to hold her fire. The aircraft flew around for a while without provocative maneuvers and left the area without incident.[59]

Ticonderoga brought other capabilities to the force off Lebanon. Her Aegis displays overlaid the radar plot with geographical maps of the Levant and could replay track histories. The Aegis SPY-1A radar detected and tracked aircraft against the clutter of the land background, increasing the warning time available against air attack. This reduced the need for naval aircraft to stay aloft on combat air patrol, thereby conserving fuel and repair parts aboard the aircraft carriers. Aegis both improved the endurance and readiness of the aircraft carrier for combat operations and reduced the cost of support. "For the first time in my 30+ years in the Navy, I'm able to manage the antiaircraft warfare problem, not just react to it," reported the battle group commander. Aegis proved completely reliable and had no failures off Lebanon.[60]

Other notable successes followed. *Ticonderoga* (CG 48) was involved in the interception of the *Achille Lauro* hijackers in 1985 (see chapter 14). Both *Ticonderoga* and *Yorktown* were in the force that operated against Libya in 1986. *Yorktown* detected unprovoked antiaircraft missile launches from Libya and tracked them from the moment they left the launch rails. Warned by her, no

aircraft were hit. Her Aegis system's tracks of missile trajectories revealed the launchers' locations. Further, her track history file was unambiguous legal evidence justifying the retaliatory strikes that knocked them out.[61] Congress approved eight more Aegis cruisers in the next two years.

Together Aegis, Tomahawk, and the vertical-launch system gave the surface Navy powerful new offensive tactical capabilities, a "Revolution at Sea" as one enthusiastic admiral put it. He could have added gas turbines, integrated all-digital weapons systems, advanced silencing, and room to grow, and noted that the *Spruance*-class destroyers had introduced every one of the new capabilities. Aegis and the SQQ-89 ASW systems, he admitted, worked better than the fleet had anticipated.[62] He could have said that about the *Spruance*-class destroyers, too.

With Aegis, a ship other than an aircraft carrier can perform all warfare missions against modern threats. An Aegis cruiser can go on assignments that would not justify the cost of, or the wartime risk to, an aircraft carrier. Such an assignment frees the aircraft carrier to perform her primary mission of launching decisive air strikes from a mobile base. Work is under way to use Aegis ships and new Standard missiles (SM-2 Block 4) to intercept Scud-type ballistic missiles that an enemy might use to attack a beachhead or allied cities. The cruiser, fully equipped, can be in position promptly without demanding scarce and expensive airlift support or diplomatic clearance. By comparison, moving a single Army Patriot battalion by air requires over 90 C-5 sorties, airfields at the destination, and security against enemy attack during the buildup. The SM-2 Block 4 Standard has new sensors to distinguish an incoming warhead and to detonate to kill the warhead. The missile retains all its existing antiaircraft/anti-ASCM capabilities.[63]

With her tremendous radio communications capability, a *Ticonderoga*-class cruiser can handle all messages for an aircraft carrier battle group and can pick up the guard (identify and record all addressed messages) for all units in the force in minutes. The Ship Signal Exploitation Space receives tracking and targeting information from tactical intelligence sources and provides it to the ship's Command and Decision System and the battle group NTDS links.[64] The surveillance capability of the SPY-1 radar system allows an Aegis cruiser to serve in place of the aircraft carrier as the command post for the battle group. It is common for an Aegis cruiser to be the antiaircraft warfare commander for the battle group.

The *Ticonderoga* class and the Maritime Strategy complicated the Soviets' problem of seaward defense against naval strikes. Soviet submarines might make simultaneous torpedo attacks against air defense escorts, especially the *Ticonderoga*-class Aegis cruisers, to reduce the defenses facing their first-strike missile attacks on aircraft carriers. So the capability to detect and destroy ambushing submarines had the strategic deterrent value of depriving the Soviets, should they have chosen to become military antagonists, of the opening they would have needed to attack the aircraft carriers.

Soviet leaders feared a technical breakthrough in antimissile defense under the Strategic Defense Initiative, an indicator less of SDI's prospects than of the Soviets' dependence on their own military power. The Maritime Strategy, backed by naval forces adequate to carry it out, was an analogous competitive strategy. The Soviets' reaction was to make repeated nuclear threats during 1983–86, which led nowhere either in diplomacy or in raising domestic confidence. If Soviet leaders contemplated threatening a theater conventional war to mobilize their population and to do something more successful than Afghanistan, the Maritime Strategy was an obstacle they could not overcome.

Loss of the facade of military invulnerability shook the Soviet Union to its foundations, already weakened by the American economic siege. In 1991 Soviet military-aligned politicians attempted to seize power but found they had neither support nor influence. Such discredit had for a decade been the strategic goal of resolute Western leaders, principally President Reagan. The Moscow coup collapsed, and within months the Soviet empire dissolved.

Operations

First Deployments

After commissioning, the *Spruance*-class destroyers underwent testing, an overhaul to add their full complement of weapons, and training with the fleet. This took about two years. After that the ships deployed on operations for about six months every two years to the Sixth Fleet in Europe and the Seventh Fleet in the western Pacific and the Indian Ocean. The *Kidd-* and *Ticonderoga*-class ships deploy on the same pattern as the *Spruance* class. Ships based in Japan are considered to be deployed almost all the time. The LAMPS helicopters are permanently assigned to light helicopter ASW squadrons (HSLs) and embark aboard ship as detachments from the parent squadron. Most of the notable operations of the *Spruance* class have occurred during these deployments.

USS *Spruance* (DD 963), first of the class, was the first to deploy. She embarked a LAMPS Mark I Seasprite helicopter in August 1977, began working up, and sailed for Europe in October. First she exercised in the Bay of Biscay with Canadian, British, and other American ASW ships. The Royal Navy found *Spruance*'s large helicopter deck useful as a distant base for Wessex helicopters flying from HMS *Hermes,* extending their range and increasing the amount of flight time spent on ASW work.

Spruance continued east to the Baltic Sea for ASW operations with the German Navy. Soviet and East German missile boats and patrol craft trailed the new destroyer. The ship's advanced sonar and silencing worked very well in the cold depths of the Skagerrak. *Spruance* detected exercise submarines passively at "standoff" range, probably meaning the first convergence zone, or

30–40 nautical miles from the ship. The Seasprite localized and classified contacts with passive sonobuoys. After the Baltic exercise, *Spruance* visited Oslo and took Norwegian officials and journalists for a six-hour voyage down Oslo Fjord. This combination of tactical exercising, wary confrontation, and diplomacy is typical for a deployment.[1]

In 1978 the first three Pacific Fleet DD 963s, *Paul F. Foster* (DD 964), *Kinkaid* (DD 965), and *Hewitt* (DD 966), deployed to the Seventh Fleet for routine operations in the western Pacific. The third Atlantic DD 963, *Peterson* (DD 969), trailed a Soviet Kashin-class destroyer and two other warships that sailed from Cuba along the Florida coast in the Gulf of Mexico in April 1978.

Elliot's Surveillance Mission against the *Minsk* Task Group

USS *Elliot* (DD 967) performed both carrier escort duty and independent surveillance missions on her maiden deployment in 1979. She was the first DD 963 in the Middle East, soon to become the most active theater of operation for the entire class. *Elliot* was commissioned in 1977 under Commander D. L. Gurke, visited her namesake's home state of Maine, then sailed through the Panama Canal to her new home port of San Diego. She was outfitted with NATO Sea Sparrow, Harpoon, WLR-1 radar detection gear, chaff rockets, satellite communications, and LAMPS Mark I equipment at the Long Beach Naval Shipyard.

Full-fledged, *Elliot* reported to the commander of Destroyer Squadron 31, first Captain J. T. Howe and then Captain J. M. Poindexter; both were men of future prominence. Commodore Poindexter promptly drafted *Elliot* as squadron flagship. An empty space behind the bridge reserved for the Outboard system was jury-rigged for staff berthing and offices. She embarked an SH-2F helicopter detachment from the HSL-33 Seasnakes. In February 1979 Commander S. S. Clarey took command, and *Elliot* sailed for the western Pacific with an aircraft carrier battle group centered on USS *Ranger* (CV 61).[2]

The battle group stopped at Pearl Harbor, where at a remote berth the naval shipyard quietly outfitted *Elliot* with a communications intelligence intercept unit. *Elliot*'s first encounter with Soviet forces came when she detected a Soviet Bear-D bomber approaching the task group at 200 nautical miles. The battle group stopped at the huge Subic Bay naval base to prepare for the next operation. *Elliot* got the best berth there, where the base commander could look out from his headquarters at the handsome new destroyer.

Critics had derided the emphasis on habitability in the DD 963 design, but with it *Elliot* berthed the squadron staff, the helicopter detachment, and nearly 30 communications technicians and other intelligence personnel. One was a Marine whose duty, Captain Clarey wryly explained to an inquiring sailor, was to shoot the others if the ship were sunk, lest they be captured.

The battle group departed for the Indian Ocean. *Elliot* shot a shore-bombardment exercise for proficiency and tested her electronic warfare systems, including firing a chaff rocket to check its bloom time and duration. The Seventh Fleet commander hailed the force as "the cutting edge of the 7th Fleet," an accolade that became grimly humorous when *Ranger* rammed a Taiwanese oil tanker off Singapore.[3] *Elliot* and other escorts continued through the Straits of Malacca as a surface action group and anchored at Diego Garcia, where the submarine *Los Angeles* joined up. Diego Garcia was then just a radio station with a short runway, but it was important for logistics. The Air Force

consolidated cargo for the battle group at Travis Air Force Base, California, and flew it to Naval Air Station Cubi Point at Subic Bay. Mail, cargo, and personnel flew from Cubi Point to Diego Garcia. The surface group doubled back to the northeast to meet the aircraft carrier *Midway* and then headed for the Gulf of Aden to replace the carrier battle group there.

The fall of the shah of Iran had cost the United States its strongest regional military ally and left Southwest Asia vulnerable to Soviet intervention. Iranian militants menaced Americans still in the country, which stopped neither government from completing the sale of the *Kidd* class to the U.S. Navy. Many Arabs resented the American-sponsored Camp David accord. At sea the Navy battle groups' primary mission was reportedly to protect Americans in any troubled spot,[4] although the ships carried few helicopters or Marines for an evacuation. Meanwhile, the most powerful fleet the Soviet Navy had ever deployed to that time entered the Indian Ocean with the clear mission of exercising Soviet influence (see table 14.1). This task group was centered on the new aircraft carrier *Minsk*.[5] Soviet Bear and May patrol aircraft flying from Ethiopia and Yemen harassed *Midway*'s flight operations.[6]

The battle group commander (Rear Admiral Robert Kirksey) decided to detach an escort to keep watch on the *Minsk*. He required a surveillance ship with endurance, a helicopter, and an intelligence collection system. *Elliot* had all three. Among the other escorts, *England* and *Downes* each had two of these capabilities; *Robison*, of the DDG 2 class so extolled by critics of the DD 963 class, had none of them. Admiral Kirksey's choice of *Elliot* showed the value of her capabilities for modern military missions.

Elliot was further suited for the mission of offsetting Soviet naval diplomacy by a concentration of talent in her wardroom: commanding

Table 14.1   Soviet and U.S. task groups in the Indian Ocean, May 1979

Soviet Indian Ocean Squadron	U.S. Navy Task Group 77.4
Minsk (*Kiev* class)	*Midway* (CV 41)
Petropavlovsk (Kara class)	*Los Angeles* (SSN 688)
Tashkent (Kara class)	*England* (CG 22) (*Leahy* class)
Vladivostok (Kresta I class)	*Elliot* (DD 967) (*Spruance* class)
1 Krivak-class destroyer	*Robison* (DDG 12) (*Charles F. Adams* class)
Ivan Rogov (LPD)	*Downes* (FF 1070) (*Knox* class)
1 Foxtrot-class submarine	*Camden* (AOE 2) (*Sacramento* class)
1 (+ ?) oiler	1 MSC oiler
IL-38 Mays land-based aviation	

In 1979 *Elliot* became the first *Spruance*-class destroyer to operate in the Indian Ocean and the Arabian Sea. Here a Soviet IL-38 ASW patrol aircraft, operating from Aden or Ethiopia, inspects her near Socotra. (U.S. Navy photo)

officer, Commander Steve Clarey, an aristocratic Harvard Business School graduate and future admiral; executive officer, Lieutenant Commander Roger Miller; weapons officer, Lieutenant Commander Tom Reeves, who had risen from seaman and whose sister, Martha, was a pioneer Detroit soul singer; air officer and LAMPS detachment commander, Lieutenant Commander Mike Coumatos, a LAMPS test pilot; the well-organized chief engineer, Lieutenant Mark Edwards; operations officer, Lieutenant Dave Lind, an expert ship handler; and supply officer, myself, also a lieutenant. Captain Clarey called this group *Elliot's* board of directors.

Elliot made three sorties to the Soviet task group. *Midway's* aircraft had first located the Soviet force north of the Seychelles. *Elliot* closed on the group and maneuvered close astern of the hybrid aircraft carrier *Minsk* for close observation. *Elliot* launched her Seasprite. *Minsk* launched a Hormone ASW helicopter for a combination of competition and harassment. Coumatos in the Seasprite outmaneuvered the Hormone. After watching *Minsk* for a while *Elliot* signaled, "I intend to pass up your starboard side," using the incidents-at-sea code agreed upon between the navies. *Minsk* acknowledged, "Your signal understood. Don't pass too close." *Elliot* passed slowly

alongside, then went over to inspect the Kara-class cruiser *Petropavlovsk.*

Petropavlovsk's crew was standing in ranks on her helicopter deck, listening to an animated harangue by an older man, perhaps her political officer. Soviet discipline was impressive: few heads turned even to glance at *Elliot* as she passed by at 200 yards. The Soviet sailors wore fresh tropical uniforms that looked better, but offered less fire protection, than the dungarees the U.S. Navy provided its sailors.

On her second mission *Elliot* located *Minsk* off Socotra Island. *Minsk* launched two Forger vertical takeoff/landing fighters in an apparent diplomatic display. *Elliot* maneuvered almost alongside so that Soviet cameras could not avoid her. During her approaches toward the Soviet force, *Elliot* attempted to locate Soviet ships passively with her electronic support measures (ESM) system. An intercepted radar signal revealed the Soviet ship's identity and provided a line of bearing toward her but not range. *Elliot* launched her Seasprite to get a second line of bearing or active radar contact. When the LAMPS crew succeeded, *Elliot* simulated an over-the-horizon Harpoon attack at 60 miles.

Elliot and the Soviet warships exchanged signals by flag hoist and flashing light, all in Russian or incidents-at-sea codes. *Elliot* announced her

In the Arabian Sea on her 1979 maiden deployment, *Elliot* joined up with the largest Soviet naval task group ever to deploy up to that time, centered on the new aircraft carrier *Minsk*. Closing in dead astern of the *Minsk*, *Elliot* obtained the West's first detailed look at the underside of the YAK-38 Forger V/STOL fighter when *Minsk*'s aircraft overflew her to land aboard the carrier.

intended maneuvers. *Minsk* signaled bearings and ranges (in the old nautical measure of cables) where she wished *Elliot* to take station. *Elliot* had no obligation to conform to these, and she ignored them. Once she asked *Minsk* to explain the function of the unusual ribbed door in her transom, but the response by flashing light was unintelligible. Among *Elliot*'s intelligence assignments was to learn more about the then-new Forger. Cruising directly astern of *Minsk*, *Elliot* obtained the West's first photographs of the underside of a Forger when the aircraft came in low overhead to land aboard the Soviet carrier. The crew on *Elliot*'s signal bridge felt the exhaust from the Forgers' lift engines. At such proximity the Forgers felt LM-2500 exhaust shooting out of *Elliot*'s stacks. *Minsk* soon flashed a signal demanding that the destroyer move out from under her landing path. *Elliot* shifted to *Minsk*'s quarter, the only time she complied with a Soviet formation signal.

Elliot spent a total of six days on these surveillance missions. Forger vertical takeoffs and landings were interesting, but Soviet naval aviation

was tactically immature. *Minsk* slowed to 4 knots, bare steerageway, for flight operations. She never flew more than two Forgers at once, and neither Forgers nor helicopters flew beyond sight. Damage control and flight deck safety precautions were inadequate by U.S. standards.

Surface warships often assist diplomacy. After the first visit to the Soviet task force, *Elliot* and the frigate *Downes* anchored at Port Victoria in the Seychelles, near another Soviet Kara-class cruiser, *Tashkent*. Commodore Poindexter sent a boat with a letter in Russian inviting her captain to dinner aboard *Elliot*. A Soviet cutter returned with a note, in almost indecipherable Russian, apparently declining. "We fought a great war together" could be made out. *Tashkent* weighed anchor and left. American sailors donated blood and toys to the Seychelles and left a better impression than had *Tashkent*, whose crew had annoyed the citizens by brandishing submachine guns (probably disarmed ceremonial pieces) on deck. The Seychelles government owned no aircraft, so *Elliot*'s Seasprite took officials aloft to survey the islands.

Elliot, *Downes*, and the American consulate held a reception for the diplomatic community. Navy funds offset the puritanical Carter administration's cuts in the State Department budget for social activities. Guests included the Soviet chargé d'affaires. He said he had visited *Tashkent* but could not keep up with her officers' drinking during movies of Brezhnev's speeches. He read American news magazines and was surprised at American suspicions that the Soviets would cheat on SALT II. He thought Diego Garcia must be an intercontinental ballistic missile base and was curious as to why the Russian Backfire bomber was so called: "Does it fire back?" He fled upon mention of the Helsinki Accords.[7] This may have been the last informal colloquy between an American naval officer and a Russian official before President Carter forbade such contacts in response to the Soviets' invasion of Afghanistan on Christmas Day 1979. Back in the United States an *Elliot* officer's wife came to a type of national prominence as a *Playboy* centerfold, although her association with the Navy was not disclosed.[8]

In June *Elliot* and *Downes* made a sonar sweep of the Straits of Malacca, and the battle group departed. Back in the Pacific *Elliot* operated a large SH-3 Sea King helicopter for a week during a missile exercise in the China Sea. The fleet staff planned to exchange her Seasprite for two Sea

Kings. Captain Clarey preferred to keep the Seasprite with its LAMPS capability, so *Elliot* embarked just one Sea King. With rotors and tail folded, the Sea King fit snugly within *Elliot's* hangar. The Seasprite was parked on the flight deck. Seas were calm, and the crew maneuvered the big helicopter by hand. Exchanging a grounded Sea King in the hangar for another on the flight deck would have been a challenging aircraft-handling exercise in higher sea states.

During the final phase of her deployment *Elliot* operated in the southern Pacific. Local officials in Western Samoa and New Caledonia came aboard for formal wardroom dinners. As in the Seychelles, Mainland Chinese diplomats attended in clear approval of the American destroyer's patrol as a counter to Soviet power. A case in the wardroom displayed a collection of presentation silverware. *Elliot* held the set from the old light cruiser USS *Phoenix* (CL 46), a Pearl Harbor veteran then serving in the Argentine Navy as *General Belgrano*.

Elliot joined warships from the United States, Great Britain, Australia, and New Zealand at Auckland in the largest fleet to visit New Zealand since World War II. Antinuclear protesters showed up in kayaks upon the arrival of the American ships, but the attendance of 3,000 visitors to the new destroyer revealed the enthusiasm of New Zealand's people for their naval guests. Auckland was the last cargo stop, the key delivery being a new helicopter engine flown in from Hawaii. *Elliot* operated with the County-class destroyer HMS *Norfolk* (D 19), a cousin by profile, in the Tasman Sea before returning to San Diego in September.

Arthur W. Radford in the Mediterranean

USS *Arthur W. Radford* (DD 968; Commander W. W. King) entered the Mediterranean for the first time in 1979. She was immediately ordered to patrol the Straits of Gibraltar for a Soviet Type II (Charlie/Victor/Yankee) nuclear submarine suspected to be en route. The frigate *Pharris* (FF 1094) with an SQR-18 short towed-array sonar patrolled outside the straits. She lacked the Naval Tactical Data System (NTDS), so her contact reports were difficult to correlate aboard *Arthur W. Radford*.

The Sixth Fleet staff planners of this operation were overeager about early SQS-53A passive sonar capability. It could find loud passive bearings (broadband). But inside the straits, 500 ships and fishing craft blanketed the SQS-53A scopes. That

made it impossible to identify which bearing might include a submarine. Correlating radar and sonar contacts "was like playing goalie at a bowling alley," recalled one officer. *Arthur W. Radford* did not have the SQR-17 LAVA processor, so could not separate specific low-frequency sound patterns (narrow band), below 150 Hz, that might distinguish characteristic submarine motors and pumps. The destroyer resorted to active sonar search with the SQS-53A. She checked an area of the straits, went silent, sprinted to another area, then went active again. Eventually she detected a possible submarine contact astern of *Pharris*. The contact took off into the Mediterranean at a speed consistent with a Victor-class SSN.[9]

In July *Arthur W. Radford* sailed to the Gulf of Sidra to launch the first Harpoon in the radar conditions of the Mediterranean. Radio traffic carried the plans for the missile shoot. No doubt alerted via the Walker-Whitworth spy ring, a Soviet intelligence trawler (AGI) was waiting to record telemetry from the Harpoon. Short endurance of supporting aircraft required *Arthur W. Radford* to fire the Harpoon on schedule. A P-3 patrol aircraft passed the target location to the destroyer by Link-11; two F-14 fighters would chase the Harpoon; and an EA-6B electronic warfare aircraft would fly down the reciprocal bearing away from the destroyer toward the Soviet trawler to jam her telemetry receivers. For the jamming to work without disrupting American reception of missile telemetry, the trawler had to be kept at least 5 nautical miles away.

The accompanying cruiser *Richmond K. Turner* (CG 20) also carried an instrumented Harpoon, and she made actual missile launch preparations in an attempt to lure the trawler away or, failing that, to launch her Harpoon. The Russians did not take the bait. *Arthur W. Radford's* captain remained determined to launch his missile and, further, to set a distance record for this weapon. The combat information center crew formed a plan for getting rid of the trawler and promptly put it into effect. The key to the plan was their ship's fast-response gas turbine propulsion.

Arthur W. Radford turned north and steadily accelerated. The trawler eagerly followed, revealing in the process that her best speed was about 22 knots. That was what the destroyer needed to know. About 10 nautical miles from the true launch point *Arthur W. Radford* suddenly slowed to 2–3 knots and hoisted signal flags indicating that she was about to fire a missile. Diesel-powered ships, such as the trawler probably was, often have a minimum speed of about 8 knots and must turn their engines off to stay below that. *Arthur W. Radford* waited for the trawler to idle her slow-reacting plant and for the clock to tick down.

At roughly T minus 20 minutes the destroyer suddenly brought all gas turbines to full power for flank speed. She raced south at 32 knots toward the launch point with the trawler far behind. She turned to point the Harpoon directly toward the target and, still making 32 knots, steadied on that course long enough for the missile's gyroscopes to settle. She crossed the launch point and fired the Harpoon on time. The F-14s roared over her in pursuit, and the EA-6B aimed its jammers at the distant trawler. The target, formerly the old *Gleaves*-class destroyer *Lansdowne* (DD 486, a sister of USS *Murphy*, mentioned in chapter 12), returned from loan to Turkey, was over the horizon and therefore invisible to surface search radars. The P-3 fed target location data to *Arthur W. Radford* by NTDS data link (Link-11). Range to the target was reported as 60 nautical miles but was actually greater because of navigational deviation between the P-3 and *Arthur W. Radford.* Even without a warhead, the impact and residual fuel fires from the Harpoon sank the target.[10]

Caron Deploys Outboard

Among the first ships to mount the Outboard communications intelligence system, USS *Caron* (DD 970) made two cruises to the Black Sea in 1979. Soviet warplanes, including the new supersonic Tu-22M Backfire bombers, made over 30 mock missile attacks on *Caron* and another destroyer in the Black Sea.[11] In March 1980 she sortied from Norfolk for a two-month surveillance operation. During this mission she became the first DD 963 to operate north of the Arctic Circle. She developed over-the-horizon targeting tactics using Outboard against a Soviet battle group centered on the aircraft carrier *Kiev* in the Barents Sea.[12] Reportedly she patrolled within 30 miles of the Soviet naval base at Murmansk.[13]

In September *Caron* departed Norfolk for northern Europe again. She sortied from Rotterdam on September 29 for about 10 days of operations in the Baltic Sea. Poland's Solidarity movement, which had originated in a shipyard in Gdansk, was strengthening, and there was concern that the Soviet Union or the Polish Communist regime might suppress it by military force. Reportedly *Caron* patrolled 14 miles off Gdansk,[14] probably to detect short-range radio traffic from covert military activity. The Air Force claimed that surveillance flights deterred a Soviet move against Solidarity in 1980 by preventing secrecy.[15] *Caron* was apparently part of this operation. Why her patrol was so brief is unknown.

During the 1973 Yom Kippur War Libya declared the Gulf of Sidra (Sirte) to be its own internal waters and thus closed to transit by foreign ships and aircraft. It was as if Israel or Egypt had declared the entire Mediterranean Sea east of Cyprus to be a river. To be legitimate under international law, "such a claim must be long-standing, open, and [generally known]—with effective and continuous exercise of authority by the claimant—to which other states acquiesce."[16] The Nixon, Ford, and Carter administrations all rejected Libya's claim. In 1981 the new Reagan administration scheduled a fleet exercise in the Gulf of Sidra and described it specifically as a freedom-of-navigation operation to defy Libya.

Three principles of international law underlay the American strategy. First, any open peaceful sortie into the Gulf of Sidra showed that the United States did not acquiesce in Libya's claim, and thus that the claim was disputed. Second, since American warships and aircraft were within their rights to exercise in international waters, they had the right to defend themselves if Libya attacked them. Third, self-defense had to be just that, not a facade for bullying and aggression. It was important not to give Libyan forces an opportunity to fire in self-defense themselves. If Libya backed down from confrontation, it would prove the American contention that Libya's claim on the Gulf of Sidra was a fraud.

The United States anticipated trouble, and *Caron* was sent to detect Libyan reaction to the exercise. U.S. forces were to follow standard peacetime rules of engagement in responding if Libya attacked them. The intent of these rules is "to permit the commander responsible maximum flexibility to respond to threats with appropriate options, to limit the scope and intensity of any armed confrontation, and to discourage escalation, while terminating hostilities quickly and on terms favorable to the United States."[17]

Libyan aircraft often flew toward the aircraft carriers, whose F-14 fighters intercepted them and

Table 14.2 Commanding officers of USS *Caron*

Cdr. Earle G. Schweizer	1 Oct. 77–26 Jul. 79
Cdr. David P. Yonkers	26 Jul. 79–23 Jun. 81
Cdr. David G. Kaiser	23 Jun. 81–22 Aug. 83
Cdr. James S. Polk	22 Aug. 83–14 Dec. 85
Cdr. Louis F. Harlow	14 Dec. 85–22 Apr. 88
Cdr. John Z. Stepien	22 Apr. 88–18 Oct. 90
Cdr. Brent B. Gooding	18 Oct. 90–

escorted them docilely from the area. Libya apparently hoped to lull the U.S. Navy aircrews into carelessness. A few days into the exercise, two Libyan Su-22 fighters flew toward the force and fired a missile at close range directly at the leading F-14 of a pair sent to meet them. The F-14s dodged the missile and shot down both Su-22s with Sidewinders. Ten other Libyan fighters were in or near the exercise area at the time of the dogfight. *Caron*, escorted by a cruiser for long-range air defense, detected and tracked the Su-22s completely passively with Outboard.[18]

Libyan patrol boats watched the ships. A surfaced Libyan Foxtrot-class submarine moved off after being warned not to submerge. A Libyan Soviet-supplied Osa-class missile boat was evidently aware of U.S. rules of engagement, again probably via the Walker-Whitworth ring, and directed her missile-targeting radar toward *Caron*. *Caron* kept the missile boat off her bow, to present a small radar cross-section to Styx missile seekers while keeping her Sea Sparrow firing arc open. The missile boat opened a launcher canister door so that the destroyer would detect the Styx missile's radar seeker, then quickly closed it again.

Caron targeted weapons on the missile boat and reported these events to the fleet commander. The Osa boat's clear hostile intent gave *Caron* the right to open fire in self-defense. Sensing that Libya was seeking to provoke an incident that it would construe as aggression, the fleet commander ordered *Caron* and a cluster bomb–armed A-7 overhead to hold their fire unless they were actually fired upon. When the missile boat opened a Styx canister door again, *Caron* fired her chaff rockets, creating a bloom near the cruiser. Perhaps taking that as a warning shot, the missile boat retreated at high speed.[19]

Operation Jittery Prop: Surveillance of Nicaragua

The United States was at first friendly to the Nicaraguan Sandinista movement that overthrew the dictator Anastasio Somoza Debayle in 1979. Sandinista leaders toured *Moosbrugger* (DD 980)

Caron (DD 970) visits Toulon, France, after operations in the Black Sea in 1979. Soviet aircraft made at least 30 mock missile attacks on her in the Black Sea. She later made several short deployments to the Nicaraguan coast. (U.S. Navy photo)

in November 1979 with an eye to joint training to help them to build a new military organization independent of Somoza's National Guard.[20] During 1980 an extremist Communist faction took over the Sandinista movement and drove all democratic politicians from the regime. Nicaraguan opponents to the Sandinistas took up arms and set up bases in exile in Honduras and other bordering nations. Soviet and Cuban military activity in Nicaragua increased. The Sandinistas smuggled arms to an independent rebellion in El Salvador, whose people had reason enough to revolt already.

Early in January 1982 *Deyo* (DD 989) arrived outside the Gulf of Fonseca, an inlet of the Pacific bordered by Nicaragua, Honduras, and El Salvador, on Operation Jittery Prop. In mid-February *Caron* arrived to replace *Deyo* on this duty. *Deyo* transferred her LAMPS helicopter to *Caron* and departed for her home port, Charleston, via the Panama Canal and the Bahamas.[21] The Reagan administration revealed that the destroyers were on an intelligence collection mission using Outboard.[22] The stated purpose was to detect Sandinista arms smuggling into El Salvador. Along the way it would detect Sandinista military action against raids by Nicaraguan resistance forces from their bases in Honduras.

Each destroyer operated independently on Operation Jittery Prop, without material support except to refuel every two weeks at Rodman Naval Station in the Panama Canal Zone. *Deyo* returned in April, transited the Panama Canal at night, and replaced *Caron*. *Deyo* was on Operation Jittery Prop duty until late May 1982. In August 1982 the Nicaraguan regime complained that another *Spruance*-class destroyer was anchored 10 nautical miles off the coast.[23]

Caron (now under Commander J. S. Polk) began an independent "special operation" in the Pacific off Central America on November 11, 1984, again with occasional calls at Rodman. At first she was one of two Navy ships in the area. On November 13 the Sandinista regime charged that unidentified American warships had just trailed a Soviet freighter into Nicaraguan waters near the Pacific port of Corinto. A Pentagon official said that the freighter was suspected of carrying MiG-21 fighters, but that the threat of discovery and sharp U.S. reaction had kept them from being delivered.

Blind beyond their coasts, the Sandinistas mobilized to repel a feared amphibious assault. Pentagon sources denied that an invasion was planned but said that a "full-scale quarantine" was under study as an option to block Soviet offensive arms shipments. *Caron* left on December 18. She made another five-week sortie to Central America in April 1987. The Sandinista regime fell in 1990.[24]

Operation Wagon Wheel: Drug Interdiction in the Caribbean

Operation Wagon Wheel used Navy warships as *posse comitatus* assets in the Caribbean to block drug smuggling. In July 1983 *Kidd* (DDG 993), bound for Venezuela, encountered a decrepit 70-foot trawler far out to sea. The trawler identified herself as the Honduran motor vessel *Ranger*, sailing empty to a new owner in the Bahamas—close to Florida. If so, she was hundreds of miles off any reasonable course, and, far from being empty, she was visibly bow-heavy and mounted large new radio antennas. That afternoon the U.S. Coast Guard radioed that no such vessel was registered in Honduras. A State Department message declared M/V *Ranger* to be stateless and subject to U.S. law.

Ordered to heave to, *Ranger* tried to outrun *Kidd*, her master complaining that *Kidd* had no right to board. *Kidd* fired the first shots of the new drug war at sea, .50-caliber machine-gun volleys ahead of *Ranger* as warnings to obey. Sensing that *Kidd* was not authorized to use direct force, *Ranger* sailed on, but dared not dump contraband overboard. After a second day a Coast Guard flag was air-dropped to *Kidd*. Duly deputized, *Kidd* hoisted these colors and fired three more .50-caliber warning volleys. Accusing the United States of aggression and *Kidd* of piracy, the *Ranger*'s master still refused to halt. The Coast Guard authorized *Kidd* to disable her by force.

At dawn *Kidd* sounded reveille for *Ranger* by firing a 5-inch salvo overhead. She warned her crew to gather forward because the next shots would hit aft. *Ranger*'s master dared the destroyer to sink his ship. *Kidd* obligingly raked *Ranger* astern with .50-caliber gunfire. With that, *Ranger*'s crew mutinied and forced the master to surrender. A Coast Guard team from *Kidd* boarded the vessel and found 28 tons of marijuana in her hold. *Kidd* took the smugglers to Puerto Rico, where they were convicted and imprisoned.[25]

Stump (DD 978) was another successful hunter. In 1989 she detected and tracked a suspicious aircraft off Florida, which, when it landed, was seized and found to be carrying 650 kg of cocaine. A few weeks later she intercepted a fishing vessel carrying 6.5 tons of marijuana and arrested the crew of six.[26] These experiences proved useful as preparation for enforcing the blockade of Iraq since 1990.

Elliot and *Callaghan* Search for Flight KE-007

In 1983 *Elliot* (DD 967) was still largely in her 1979 configuration. The Iran-Iraq War made the Persian Gulf too risky an area for a ship without the Mark 23 TAS (target acquisition system) missile-warning radar and Phalanx guns for anti-ship missile defense. Fleet planners assigned her to operate in Japan and Korea, where hostile encounters were reckoned to be less likely. The estimate was wide of the mark.

Saturday, September 3, found *Elliot* in routine upkeep at the shipyard in Sasebo, Japan, and preparing for a change-of-command ceremony for the destroyer squadron commodore later that morning. A Boeing 747 airliner en route from New York to Seoul with 269 passengers and crew, Korean Air Lines flight KE-007, was in the news as missing since about 3:30 A.M. local time the previous day.[27] Electronic surveillance systems in Japan had recorded that Soviet Air Defense fighters had attempted to intercept it, but the systems had not triggered an alarm because the aircraft did not approach Japanese air space. Their revelation that a Soviet Su-15 fighter had shot down the airliner was discovered by Saturday morning. *Elliot* received immediate orders to sail in four hours for the crash site, 16 nautical miles west of Sakhalin Island in the Sea of Japan.

Shore patrol parties rounded up men on Saturday liberty. The change-of-command ceremony was hurriedly held on the flight deck while the ship prepared to get under way. Guests were sent ashore, a muster showed all but three of the crew to be aboard, and *Elliot* cast off on time.

She proceeded west through the scenic islands of Kyushu to open water, went to 32 knots, and turned northeast through Tsushima Strait. In four hours she had gone without notice from weekend liberty and cold iron in a shipyard to full speed and combat readiness. *Elliot* raced through the Sea of Japan at 32 knots for over 1,000 miles. The officers reviewed rules of engagement and Soviet territorial water boundaries. The ship set general-quarters conditions of watertight integrity (mate-

rial condition Zebra) below the main deck and manned all weapons, including machine guns on deck. She maintained those levels of combat readiness from then on.

Elliot arrived at the crash site early on September 5, the first American ship there. She found no survivors, bodies, or wreckage. She did find the Soviets there in force, as well as normal commercial shipping and many fishing boats. Analysts thought that the wreckage of KE-007 lay near Moneron Island, which was Soviet-held territory, so the American ships could search no closer than 12 nautical miles from it. *Elliot*, joined by other ships, searched by sonar for the flight data recorder and cockpit voice recorder. The recorders emitted beeps precisely so that a sonar-equipped ship could find them. The Soviets planted an intermittently beeping decoy and masked their own recovery effort with fishing craft.[28] Whether *Elliot* ever detected the actual recorders is unknown.

Soviet aircraft and missile boats circled and harassed *Elliot* closely and almost continuously. The Kara-class large ASW ship *Petropavlovsk*, which *Elliot* (under a different crew) had encountered in 1979, appeared. *Elliot* used this as an opportunity for surveillance of Soviet activities. As U.S. Navy forces on the scene increased, she became flagship for the salvage mission. On September 14 she turned over her flagship duties to a cruiser and left for an exercise in Korea. USS *Callaghan* (DDG 994) of the *Kidd* class joined the search. After searching 250 square miles of seafloor outside the 12-nautical-mile boundary around Moneron Island, the Navy found nothing. Soviet divers recovered the recorders around October 18.[29] The U.S. operation ended on November 8.[30]

Diplomatic pressure surrounded the salvage attempt. The United States accused the Soviets of mass murder. The Soviets first lied that they knew nothing about the shootdown, then claimed that the airliner had really been on an espionage mission. That assertion was preposterous, but American officials in Washington spread imaginative rumors that led reporters to wonder whether the off-course airliner really had been involved in surveillance.[31] At best those officials had only the information received in situation reports from *Elliot* and other ships in the search group. As with the Tonkin Gulf incident nearly 20 years earlier, excessive secrecy and inexpert analysis of naval message traffic had political costs. In 1993 the Russian government released the recorder tapes, which showed that KE-007 was flying north of its intended route simply because its crew had connected the autopilot to the magnetic compass instead of to the inertial navigation system.[32]

Operation Urgent Fury: *Caron* and *Moosbrugger* at Grenada

In 1979 Cuban-aligned Communists seized power on Grenada and began turning the Caribbean island into a military base. Trained and armed by Cuba, Grenada's People's Revolutionary Army (PRA) was the most powerful military force in the southern Caribbean. By 1983 Cuban troops were completing a large airfield there, at Point Salines, which featured bunkers identical in size to those where Soviet forces in Europe stored bomber-launched cruise missiles and ground-launched ballistic missiles. The Soviet Union's "ambassador" was an active-duty general. Grenada was clearly being prepared as an offensive air base to transport Cuban and other hostile troops to Africa and South America. *Caron* was engaged on "special operations—southern Caribbean" for eight days in April 1983, shortly after the Grenadian regime mobilized to repel a feared invasion. Grenada blamed the invasion scare on a U.S. naval exercise it called "Ocean Venture 1983," but there was no such exercise, nor any other significant U.S. naval operation in the Caribbean that spring.[33]

In mid-October 1983 a Moscow-aligned terrorist faction overthrew the Grenadian Marxist leader, Maurice Bishop. The U.S. State and Defense departments considered evacuating 600 students at an American medical school on the island. On October 18 an amphibious task force carrying the 22nd Marine Amphibious Unit (MAU), composed principally of the 2nd Battalion, 8th Marine Regiment, had departed from Norfolk on a scheduled rotation to Lebanon. Facing popular opposition, on October 19 the terrorists murdered Bishop and other political opponents and declared a 24-hour shoot-on-sight curfew. On October 20 *Caron* (Commander J. S. Polk), *Moosbrugger* (Commander D. A. Dyer), and *Ticonderoga* (Captain R. G. Guilbault) deployed with an aircraft carrier battle group centered on USS *Independence* (CV 62) for the Mediterranean to cover the amphibious force off Lebanon. In the White House almost simultaneously Rear Admiral John Poindexter, *Elliot's* destroyer squadron commander in 1979, was convening the first of several Washington staff meetings regarding Grenada. At midday General John Vessey, chairman of the Joint Chiefs of Staff, ordered the *Independence* battle group toward Grenada for a possible rescue.

Late that afternoon *Caron* and *Moosbrugger* received flash-precedence orders diverting them southward in radio silence. *Caron* detached from the battle group and began a 2,000-mile dash to Grenada. She carried additional van-mounted electronics in her hangar and did not have a LAMPS helicopter embarked. Her Outboard system would obviously be valuable in intercepting radio traffic around Grenada and in verifying that U.S. forces approached in radio silence. *Ticonderoga*, commissioned in January 1983 and on her first deployment, led several other ships to the Mediterranean as a decoy force meant to divert Soviet ocean-surveillance satellites.

Moosbrugger initally stayed with *Independence*. Before deployment she had received the prototype SQQ-89 modernized ASW system, including the SQR-19 tactical towed-array sonar, but not the LAMPS Mark III systems to support SH-60B helicopters. Instead she had embarked a LAMPS Mark I SH-2F Seasprite helicopter from HSL-34. On October 21 an intelligence planning package for the Marines was flown out to *Independence*, which transferred it by helicopter to *Moos-*

Moosbrugger's commemorative emblem shows her crew's view of Operation Urgent Fury. (Courtesy Rear Adm. D. A. Dyer, USN)

USS *Moosbrugger* (DD 980) refuels a hovering SH-60B Seahawk helicopter, very similar to the Army Delta Force UH-60As she supported at Grenada. (Courtesy Al Lee)

brugger. The destroyer raced to rendezvous with the distant amphibious task group, which was steering southward now, too. After 14 hours at 32 knots she delivered the intelligence package to the Marines at sea, her Seasprite making the delivery over the horizon on the 23rd. *Caron* arrived off Grenada on October 23 and stationed herself 12 miles off the coast to gather intelligence. That evening in Washington, President Ronald Reagan signed the directive authorizing Operation Urgent Fury. Its objectives were to protect Americans on Grenada, to establish a democratic government, and to eradicate Cuban and Soviet influence.

Still at high speed, *Moosbrugger* headed for the Antigua Passage and rejoined *Independence* before dawn on the 24th. Anticipating a call for combat search and rescue (SAR) services, Captain Dyer ordered ASW gear removed from the Seasprite to make room for stretchers and M-60 machine guns. *Moosbrugger* was tentatively reas-signed to the amphibious squadron, probably for gunfire support, but it had been noticed that she was the only surface combat ship in the task force carrying a helicopter. *Independence* sent a large SH-3H Sea King ASW helicopter from HS-15 with Marine gunners to *Moosbrugger* to join the Seasprite. With two helicopters now, *Moosbrugger* was designated the primary combat SAR platform for Urgent Fury, and she detached from the battle group to join *Caron* off Grenada.

The destroyers rendezvoused at midnight. This was good navigating, since rain had obscured star sights and both ships were in radio silence. The force ran to Point Salines, the southwest tip of the island, at 25 knots in wide line abreast to arrive at first light on the 25th, when Army Ranger paratroopers would land on the airfield. Special Forces commandos (Navy SEAL [sea-air-land] and Air Force combat controller [CCT] teams) had parachuted into the sea off the coast late on the 23rd.

Four heavily loaded SEALs were lost. A boat from the frigate *Clifton Sprague* (FFG 16) towed the other Special Forces raft closer to the beach. It turned away to avoid a patrol and was swamped in the surf, killing the engine. For over 24 hours the boat crew and commandos drifted back out to sea on an offshore current. About 0430 lookouts on both destroyers spotted a red signal flare. *Moosbrugger*'s SH-3, airborne for SAR duty for the Ranger jump, identified men in the water. *Caron* recovered them by boat.

Urgent Fury had been put together very quickly, and there was no radio communications doctrine between Army and Navy units. *Caron* may have identified Army helicopter radio frequencies with her Outboard installation. Her large helicopter deck made her an impromptu Special Forces base. Helicopters, reportedly Army UH-60A Blackhawks, landed aboard *Caron*, picked up the SEAL platoon, and lifted it to the politically important Government House. There the SEALs protected an important ally, Governor-General Sir Paul Scoon, whose response to the terrorists' call to mobilize against invasion had been to put a "Welcome U.S. Marines" sign on the lawn, for the next 25 hours.[34]

The destroyers were 2 miles south of Point Salines at 0600 when the Army paratroopers hit the silk from C-130s over the airfield. Offshore, *Moosbrugger* had both her helicopters airborne to rescue anyone who ditched at sea. She nearly ran aground at 25 knots while her young conning officer watched the paratroopers and enemy machine-gun fire ashore in fascination, missing the navigator's notice to slow down and turn. Captain Dyer gave the necessary order.

In the critical first eight hours enemy forces outnumbered U.S. troops two to one, and they had armored vehicles and defensive positions. Better leadership, aircraft, and the destroyers' guns offset them. Cuban troops and the Cuban-led PRA fought fiercely. At 0615 Army Delta Force counterterrorist commandos attempted to rescue political prisoners at the Richmond Hill prison. Their helicopters ran into intense antiaircraft fire, and two UH-60As quickly diverted to *Moosbrugger*. Their pilots signaled by hand that they needed to land on her flight deck. *Moosbrugger* immediately waved the first Blackhawk aboard. Badly hit by .50-caliber fire, it had seven wounded men aboard, including the helicopter pilot and two commandos with serious chest wounds. *Moosbrugger*'s corpsmen set them up in the hangar bay and worked on their wounds. The copilot explained

the radio frequencies the Army used. The UH-60A took off, an impressive demonstration of this aircraft's combat survivability. Army practice is to move forward on takeoff, and the Blackhawk barely missed the superstructure. As *Moosbrugger*'s radiomen tuned to the Army frequencies they heard the pilot telling the other Blackhawks to take off vertically. A second shot-up Blackhawk landed, offloaded its troops including two more wounded, and literally went backward as it took off. *Moosbrugger*'s Seasprite ("Moose's Grey Goose") brought the destroyer squadron medical officer over from *Caron*. A large Marine CH-46E Sea Knight helicopter landed with a medical team at 0659 and evacuated the wounded men to the amphibious assault ship USS *Guam* (LPH 9) at 0725; all survived. The 1967 decision to enlarge the DX specification to operate helicopters of this size was paying off.

With communications established now and the wounded offloaded, *Moosbrugger* became an advance base for the Army helicopters to refuel and to receive orders via her multiple radio systems. She brought the Army helicopters in for refueling and gave them, without hesitation, over 25 tons of JP-5. The unwounded commandos regrouped aboard her and boarded helicopters to go back into action. At 0730 two Marine AH-1T Sea Cobra gunship helicopters escorted *Moosbrugger*'s SH-3H to a hot landing zone ashore, where it picked up 11 more wounded commandos and took them to the hospital-equipped *Guam*. Upon the SH-3H's return to *Moosbrugger* for refueling, a crewman handed down a penciled note for Captain Dyer:

CO—We just pulled 11 wounded out of zone. Flew em to *Guam*—All alive when we got em there.

The reverse side of the note was coated with blood. The wounded Blackhawk pilot stayed aboard his aircraft and reportedly was killed approaching Richmond Hill again. *Moosbrugger* made 43 landings of Army helicopters by 1300. Her Seasprite flew from *Clifton Sprague*, since the larger Blackhawks and Sea King needed the *Spruance*-class flight deck.[35]

The Rangers' parachute drop to seize the Point Salines airfield went better. The Cubans blew up their radios, lest their codes be captured, so a Cuban freighter anchored in the harbor, *Vietnam Heroico*, relayed radio messages between Cuba and the terrorists. *Caron*'s Outboard crew no doubt was listening, and *Caron* promptly ordered the freighter to leave Grenadian waters. *Vietnam Heroico* did so that morning but stood 12 miles offshore in international waters for the rest of the U.S. operation.[36] In the afternoon enemy rocket fire from Fort Frederick, an ancient harbor defense work, shot down two Marine AH-1T Sea Cobras that were repelling a PRA counterattack on the SEAL platoon holding Government House. A *Guam* Sea Knight helicopter recovered the crew of one Sea Cobra, but *Caron*'s boat searched for the other's two crewmen in vain. The destroyer stood in so close to shore to provide cover that enemy infantry fired at her with mortar rounds, which missed.

Further north along the coast, a PRA platoon under terrorist command counterattacked the SEAL commando platoon that had captured the "Radio Free Grenada" propaganda transmitter station at Beausejour Bay. The SEALs had killed 4 and captured 10 Grenadian troops that morning, but this new PRA force was larger. A big Soviet BTR-60PB armored personnel carrier with them fired 14.5 mm cannon shells right through the walls of the transmitter building where the SEALs had taken cover. The SEALs had no weapon that could penetrate the APC's armor. With assault clearly imminent and two men (including their commander) badly wounded, the SEALs called for support. *Caron*, nearby, responded. In coordination with the SEALs, she set up and controlled a strike by two A-7E bombers from *Independence* and opened fire on the PRA force with her 5-inch guns. The enemy unit retreated into defilade under this barrage, giving the SEALs their chance. The entire SEAL platoon broke out from the station and escaped into concealment up the coast.

After dark *Caron* silently approached within barely a ship length of shoal water to pick up the SEAL platoon. Her fast-reacting gas turbines and controllable propellers made this risky maneuver possible. If discovered she would likely draw enemy attention away from her boat, and she could withstand rocket and gunfire hits better than any helicopter. Explosions ashore showed that fighting continued. Linking up took time and made it a long swim. Ten SEALs swam out to *Caron*'s boat, eight towing the two wounded men from the radio station action. Aboard ship, *Caron*'s medical team treated their wounds. Confident now that the coast was clear, *Caron* vectored Army UH-60As to pick up the last two SEALs, a rear guard, from the beach.[37] All the SEALs had been recovered safely. Despite stories, *Caron*'s

gunfire and the air strike were neither prolonged nor aimed at the radio station or mast. They were a coordinated, fully successful fire support mission to cover the SEAL platoon's escape in daylight.[38]

Both destroyers stood on ready-fire alert for gunfire support all night. *Moosbrugger* linked up by radio with a ground unit and fired several star shells to illuminate its patrol area for spotting. She fired one high-explosive round upon request and received a report of "enemy squad destroyed," then requests for four more illumination rounds. Her officers suspected they were firing at shadows in the dark, but if they bolstered the ground troops' fortitude, the shells were well spent.

The destroyers spent the second day at Grenada, October 26, on call for gunfire support, looking for small boats that might be carrying fleeing terrorists or infiltrating Cubans, and searching for Cuban submarines in case any were around.[39] *Caron* detected a chartered sports fishing boat approaching from the north to land a TV news crew. After Vietnam many military commanders did not trust the press (and vice versa), and on the fleet commander's direct order an A-7E bomber from *Independence* drove the boat off with warning shots.[40] *Moosbrugger*'s crew called the sea off Grenada's northwest coast the "CBS Patrol Zone" after that. That night *Caron* fired three star shells (two were duds) to illuminate Fort Frederick for a Marine assault. Although no one could know it then, this assault was the last opposed action on Grenada. Later *Caron* picked up 11 Rangers who paddled out to her in a raft salvaged from a downed CH-46. They had stayed behind to guard the evacuation of 224 students, and then escaped despite enemy patrols.[41] Outboard probably contributed to the success of this rescue.

Moosbrugger rounded Point Salines so often that her crew began calling it "Moose Point." She often operated on one engine to save fuel; fast light-off of the gas turbines for full power if needed was a big advantage. Down to 39 percent of fuel and less than that of JP-5 for helicopters, she refueled on the 27th for the first time since leaving Charleston a week earlier. That afternoon she supported an Army 82nd Airborne helicopter assault on the Calvigny barracks on the southeast coast. Naval gunfire doctrine with the Marines is to keep shooting right until the helicopters land, so that shell bursts keep enemy heads down. *Moosbrugger* fired rounds to seaward for ranging

Caron suppressed a terrorist-led Grenadian army counterattack on the Radio Free Grenada transmitter, which a Navy SEAL commando platoon had captured. She coordinated air strikes as well. This let the SEAL platoon escape, and *Caron* picked them up from the coast that night. Although not shown in this watercolor, *Caron* mounted two Phalanx guns and the Mk 23 TAS radar during this operation. (U.S. Naval Institute collection)

and to warm up the 5-inch barrels (PACFIRE: pre-action calibration fire). She located precise navigational reference points ashore with her Mark 86 gunfire control system and calibrated range to the PACFIRE shell falls. These procedures would give very good shooting accuracy.

Authentication (validating that only the correct units are on the radio net) and deconfliction took three hours. An A-7E air strike scheduled to follow the naval gunfire mission happened beforehand. Finally *Moosbrugger* fired one high-explosive round at a barracks called out by an 82nd Airborne spotter in an Army helicopter. The shell went right through the roof of the target building. Such single-shot accuracy greatly impressed the 82nd Airborne spotter. Concerned that Army-Navy communications and deconfliction were weak, however, off-scene task commanders ordered naval gunfire support at Calvigny to stop 10 minutes before the landing and pulled the spotter helicopter to deconflict from another air strike. When the eight UH-60A assault helicopters came in to land

10 minutes later, one pilot thought he saw muzzle flashes. He swerved to disrupt enemy gunners' aim but collided with another helicopter, killing three Rangers, seriously wounding six others, and destroying both helicopters. This forced a third Blackhawk to veer off so sharply that it crashed and was destroyed, too. The base was deserted. A Naval War College analysis in 1986 hinted that this disaster showed that the Navy and the Marines, who were well practiced at coordinating supporting fire, were more skilled at airborne assaults than the Army.[42] A later version of the Naval War College study deleted mention of the Calvigny assault.

Both destroyers patrolled off Grenada for several more days to cover Marine advances north along the coast. *Moosbrugger* made an active sonar search for a Cuban Foxtrot submarine one night until an assault ship in the harbor, where many were probably getting their first sleep in days, called her and demanded, "Knock off the noise!" Her helicopters dropped psychological warfare pamphlets, and she supported a SEAL mission. On November 1 she covered the Marine landing on Carriacou Island, which turned out to be unopposed, so no firing was required. Meanwhile all the terrorists were caught and imprisoned at Richmond Hill.

Two days later *Caron* and *Moosbrugger* departed for Lebanon. Their 5-inch guns were the heaviest (and only naval) artillery employed during Operation Urgent Fury. *Caron* fired 17 5-inch rounds; *Moosbrugger* fired two high-explosive 5-inch shells plus illumination and PACFIRE rounds.[43] This may seem light, but the operation order required the invaders to avoid unnecessary damage to Grenada. Despite the initial lack of coordination, Army helicopters and Navy destroyers had found each other very useful. Wider knowledge of this might have cooled subsequent disputes over joint operations. Months later, when *Moosbrugger* was on a routine gunnery exercise near Puerto Rico, 82nd Airborne spotters attended and had her fire 200 rounds for them to practice spotting and coordination.[44] The liberation of Grenada has been criticized for inefficiency, but it stands as the most strategically successful American combat operation since Vietnam, Operation Desert Storm not excepted.

Support of the Marines at Beirut, 1983–1984

Israeli and Syrian combat in Lebanon during 1981–82 threatened another Mideast war that might widen and involve the Soviet Union and the United States. To contain the war after Israel invaded southern Lebanon, the U.S. Marines went ashore in August 1982 to supervise the evacuation of Palestine Liberation Organization (PLO) guerrillas from Beirut. This was a successful operation, and the Marines departed in September.

Afterward, the Israeli-backed president of Lebanon was assassinated. The Israelis did nothing to protect Palestinian camps behind their lines from revenge attacks by their Phalangist allies, who killed over 300 men and 35 women and children. President Reagan ordered a Marine battalion with 1,200 Marines back to Beirut. Along with some European forces, they were expected to protect the Lebanese civilians amid a civil war aggravated by 80,000 foreign troops. The United States began training the Lebanese Armed Forces, which the Syrians and the Lebanese Muslims regarded as Israeli-aligned. Muslim terrorists began attacking the Marines and other Americans in Lebanon.

Navy ships cruised off the coast to deliver gunfire support if the Marines called for it. The *Spruance*-class destroyers *Arthur W. Radford* and *John Rodgers* (DD 983) had deployed to the eastern Mediterranean in April 1983. (The third Soviet aircraft carrier, *Novorossiysk,* exited the Black Sea, and *John Rodgers* had observed her for a week.) *Arthur W. Radford* and *John Rodgers* patrolled off Lebanon for two weeks on call for gunfire support but did no shooting, then left when other ships relieved them.

John Rodgers returned to the Lebanon coast in late July. The commander of the Sixth Fleet's amphibious forces noticed that she was well equipped for command, control, and communications (C^3) and made her the flagship for the squadron off Lebanon. This was the first use of a destroyer as an amphibious force command post since World War II.[45] An assault force commander should be in the area of amphibious operations, but a warship needs freedom to maneuver and to maintain a seaward barrier against counterattacks. (During Operation Torch in 1942 General George Patton was aboard the heavy cruiser *Augusta* [CA 31], Admiral Kent Hewitt's flagship. He was preparing to go ashore when the cruiser suddenly took off to engage Vichy French destroyers, delaying him from joining his forces for five hours. To add insult to injury, muzzle blast from *Augusta*'s 8-inch gunfire destroyed the boat with his baggage, which fell into the sea. An almost identical event had disrupted the failed 1940 Anglo-French landing at Dakar.)

Ashore, matters were getting worse for the Marines in the 24th MAU. In July 1983 a Muslim Druze militia unit shelled the Marine compound at the Beirut airport. The Marines moved into a large building for protection from snipers. Another Druze bombardment on August 29 killed two Marines, the first Marine combat fatalities. Spotting with Army artillery-locating radars, the Marines fired back in anger for the first time, using 155 mm guns. The White House took a strong interest in the Marines at Beirut. The Israeli Army withdrew from Beirut to southern Lebanon, leaving the Marine battalion to protect the weak Lebanese government. Frequent visits from Washington-based officials obscured a dangerous underlying situation. No higher field commander held responsibility for the battalion ashore, so no staff planned what the Marines were to accomplish or what assets and support they needed. The impossible goal of pacifying Lebanon forgotten, the Marines concentrated on survival.

Rocket attacks on September 6 killed two more Marines in the airport compound. Off the coast, *John Rodgers* was at general quarters ready for gunfire support. Support was not requested, and later that day she detached from the squadron off Beirut. President Reagan personally authorized the force off the coast to provide naval gunfire support. Two days later Marine 155 mm guns and the *Knox*-class frigate *Bowen* (FF 1079) silenced a Druze rocket battery, the first use of naval gunfire support.[46] On September 15 *John Rodgers* was ordered back to the gun line and arrived the next day. That night, artillery (it was not clear whose) behind Syrian lines bombarded the Lebanese Ministry of Defense and the American ambassador's residence. *John Rodgers* responded by firing 36 rounds of high-explosive shell at two targets simultaneously.

Lebanese Armed Forces units with Marine advisers were holding a nearby ridge against an assault by two Palestinian battalions. On September 19 the Lebanese called for naval gunfire support against a column of tanks approaching the ridge. *John Rodgers* fired 154 rounds, and the cruiser *Virginia* (CGN 38) fired 184. The attackers broke and ran under the barrage, turning the course of the battle. Naval gunfire became the weapon of choice for the Marines. *John Rodgers*, *Arthur W. Radford*, and *Virginia* fired two missions against rocket launchers attacking the Ministry of Defense on September 21. The ships fired 90 rounds; *John Rodgers* fired 30 of these. The next day she fired 45 rounds in response to shelling of the airport and the ambassador's residence.

On September 24 the modernized battleship *New Jersey* (BB 62) arrived. Before she fired a shot, the Syrians, Druze, and Lebanese government forces agreed to a cease-fire.[47] The situation was quieter for four weeks. Then on Sunday, October 23, an Iranian terrorist working for Syria drove a truck with a 12,000-pound bomb into the Marines' building and killed 241 troops. The United States did not retaliate for this action. In December a poorly planned retaliatory air strike on Syrian positions in Lebanon cost two aircraft and a pilot. A bombardier was captured, raising the old diplomatic question from Vietnam about whether he was a prisoner of war if there was no war. Syria released him a month later. An advantage of Tomahawk cruise missiles is to eliminate risks to aircrews.

In mid-November Amphibious Squadron 4 arrived from Grenada, and the 22nd MAU replaced the battered 24th. Off the coast the new Aegis cruiser *Ticonderoga* began to earn renown, as discussed in chapter 13. *Ticonderoga* bombarded Syrian antiaircraft artillery batteries from about 1,000 yards off the coast on December 13, her first shots fired in action (15 rounds), and again on the 14th (30 rounds) and the 18th (about 30 rounds).[48] The Aegis displays even showed projectile impact.

Caron and *Moosbrugger* joined the Sixth Fleet after Operation Urgent Fury and proceeded to Beirut in January 1984. *New Jersey* bombarded Israeli-designated targets ashore with her 16-inch guns. Her old radars could not locate markers on the coast, so a Mark 86–equipped *Spruance*-class destroyer did this. The destroyer pinpointed her position with high-precision radar (SPQ-9A) and relayed the target's range and bearing from her. *New Jersey*'s gunnery system tracked the destroyer and triangulated to the target.[49] Without forward spotters, accuracy and effect were unknown. Secretary Lehman claimed eight Syrian artillery batteries were destroyed; rebels showed empty houses that they said were damaged.[50] *Moosbrugger* spent six weeks on call for instant gunfire support 2,000–3,000 yards off the beach, just outside infantry weapons range, as frequent tracer rounds and projectile splashes showed. Her ammunition rings were preloaded so that she could fire 50 5-inch rounds per gun before reloading from the magazines. Later the carrier group commander ordered her to stand near the aircraft carrier to defend against possible small-boat raids. The amphibious group commander, however, wanted her near the shore on call for gunfire support. *Moosbrugger* settled this by showing that the rapid-start feature of gas turbine propulsion let her achieve 32 knots very quickly, so that from the carrier screen she could reach her gunfire support station within the Marines' specified time to respond. She fired several support missions of about 10 rounds each.

Naval capability off the coast could not offset the impracticality of the Marines' mission ashore. On February 7 President Reagan announced that the Marines would be withdrawn from Beirut but that naval gunfire support for the Lebanese Armed Forces would increase. On February 8 *New Jersey* and *Caron* bombarded antigovernment artillery positions for seven hours. *Caron* fired 450 5-inch rounds, sometimes from only 1,000 yards off the coast. *Moosbrugger* fired 150 5-inch rounds the

During February 8–9, 1984, *Caron* and the battleship *New Jersey* bombarded Lebanese rebel positions called out by Israeli forces. *Caron* fired 600 rounds of 5-inch ammunition, leaving her gun barrels scorched from the heat. (U.S. Navy photo courtesy USS *Caron*)

following day. The United States withdrew the Marines from Beirut during February 21–26. During the final two days *Caron* fired another 141 5-inch rounds. This was the last naval action of the Lebanon campaign.

The Capture of the *Achille Lauro* Hijackers

In October 1985 four PLO terrorists hijacked the Italian cruise ship *Achille Lauro* off Egypt. The *Kidd*-class destroyer *Scott* (DDG 995) sortied from Haifa, Israel, to locate the liner but did not find her, not surprising given the large area of sea she had to search and the short range of her surface search radar. *Caron* with her Outboard system was already in the area on a NATO exercise in the central Mediterranean. *Caron* had an unusual propensity to be where the action was. Nothing has been revealed about whether this incident involved her, but what follows is consistent with Outboard-based tracking. The Aegis cruiser *Yorktown* (CG 48) sortied from Taranto, Italy, toward Egypt. SEAL commandos were available to board the liner at sea. Under fear of terrorist retribution, no nation within range would provide the Navy a base for use in a rescue.

The terrorists murdered an American passenger before the liner put in at Alexandria, Egypt. Egypt turned them over to the PLO's chief terrorist, Abu Abbas, and provided an EgyptAir Boeing 737 to take them to Tunis, PLO headquarters after its eviction from Beirut in 1982. An Israeli intelligence agent kept the U.S. military informed of this and gave the 737's identity and takeoff time. The aircraft carrier *Saratoga* (CV 60), beyond radar range in the Adriatic Sea, launched seven Tomcat fighters, two Hawkeye warning-and-control aircraft, and an EA-3B ELINT aircraft. High-precision radio direction intercepts, possibly from *Caron,* would explain how *Saratoga*'s aircraft knew approximately where they would find the airliner.

Yorktown was in the vicinity of the intercept. She and the E-2Cs maintained tight tracks of all aircraft that might be the airliner against the clutter of all other aircraft in the night sky. They guided fighters to each suspect aircraft for visual identification. The fifth aircraft checked was the 737. *Yorktown* and an E-2C vectored four F-14s to it. One of the E-2Cs raised the airliner on radio and upon getting the pilot's attention had the F-14s switch on their lights to show him he was surrounded. The 737 requested permission to land at Tunis, which refused, and then at Athens, which refused, too.[51] The airliner landed as ordered at the U.S. Naval Air Facility at Sigonella, Sicily. Italian authorities let Abu Abbas go but imprisoned the other terrorists.[52]

Operations Prairie Fire and El Dorado Canyon: Counterattack against Libya

A month after the *Achille Lauro* incident, terrorists hijacked an Egyptian airliner on a flight from Athens and murdered a U.S. Air Force civilian employee. The airliner, the same 737 that had carried the *Achille Lauro* hijackers, was destroyed and almost 60 passengers killed during an Egyptian commando raid at Malta. That Christmas, terrorists attacked airports in Rome and Vienna, killing five more Americans. Lot numbers found on residue of grenades recovered from the Malta, Rome, and Vienna attacks matched those on Libyan Army weapons captured in Chad, proving that Libya was behind these crimes.[53] Either by coincidence or to deter retaliation against their client, the Soviets stationed their large new missile cruiser *Slava* off Libya after the terrorist attacks.

Two Sixth Fleet exercises conducted near Tripoli early in 1986, Attain Document and Attain Document II, were surveillance operations against Libya. *Caron* with her Outboard communications intercept system participated in both exercises. She embarked an H-3 Sea King helicopter for search and rescue. *Yorktown* and *Scott* joined the battle group off Libya.

For retaliation the United States planned Operation Prairie Fire, a naval patrol by DD 963–family ships inside the Gulf of Sidra to lure out Libyan aircraft and warships for destruction. Despite the 1981 incident, Libya still claimed that the gulf was its own internal waters. The covering exercise was Attain Document III, an announced freedom-of-navigation exercise in international waters, including the gulf. Under modified peacetime rules of engagement, the supporting carrier battle group outside the gulf could use proportionate force to maintain freedom of navigation in international waters if Libyan units attacked ships or aircraft inside the gulf. As in 1981, it was important for the task force not to give Libyan units any pretext to claim that they were using weapons in justifiable self-defense themselves.

As in 1981, Libya could challenge the strong American legal position that the gulf was an international waterway by evicting the U.S. aircraft and surface warships to show that it really did control the gulf. Muammar al-Qadhafi shrieked that his claimed northern boundary for the gulf, 32°30′N latitude, was the "Line of Death." Agreeing with him on that, the Soviet Union moved its submarines well away from the area lest the U.S. Navy mistake them for Libyan submarines.[54] President Reagan approved Operation Prairie Fire on March 14, 1986. Vice Admiral Frank Kelso, commander of the Sixth Fleet, was authorized to act independently of his usual superior, the land-locked Army commander in Europe (CinCEur). CinCEur acquiesced but demanded use of three aircraft carriers for cover. Only two carriers were planned, and bringing in the third delayed the operation.

Ticonderoga led *Scott* and *Caron* as a surface action group across the Line of Death on March 24. Libyan antiship capability consisted of missile-armed fast attack craft covered by shore-based aircraft and antiaircraft missile batteries. Two U.S. Navy combat air patrols and an antiship surface combat air patrol were aloft. In the first and last activity by the Libyan Air Force, two large MiG-25 interceptors approached the task group but turned back upon discovering Navy fighters behind them, where they had been for several minutes, holding them in their gunsights. A Libyan battery at Surt fired two SA-5 long-range antiaircraft missiles at an air patrol but missed. *Ticonderoga*'s Aegis tracked the missiles, useful both for combat intelligence and as legal evidence that the Libyan attack had been unprovoked. After the Libyans continued firing SA-5 and SA-2 missile salvos at the air patrols, Admiral Kelso ordered his force to attack any Libyan ships or aircraft departing the 12-nautical-mile limit. Aircraft destroyed several Libyan fire control radars with antiradiation missiles.

That night two Libyan fast attack craft entered the gulf to attack the surface action group from opposite directions. South of the 32°30′N line, *Ticonderoga* detected the 311-ton *La Combattante II*–class missile boat *Waheed* near Misurata at the western end of the gulf. Surface-combat air patrol bombers from *Saratoga* sank her with cluster bombs and attacked the second missile boat at the eastern end of the gulf, again with cluster bombs. In standard terrorist fashion, the damaged boat took a merchant ship hostage by tying up next to her to ward off a follow-up attack while returning to base in Benghazi.

At 0115 local time *Yorktown* fired two Harpoons toward a surface contact detected by her SPY-1A radar, but they hit nothing. It is possible that the contact was a flight of birds, perhaps in line with a distant radar whose signals might have been detected by the cruiser's SLQ-32 system.[55] The 675-ton Nanuchka II–class fast attack craft *Tariq Ibn Zayid* sortied from Benghazi in daylight and was sunk by a Harpoon from a *Saratoga* A-6 bomber.

After that, the sole Libyan activity was search and rescue, which the Navy did not oppose. *Ticonderoga, Scott,* and *Caron* operated inside the gulf further south than Benghazi. The ships controlled air operations over the gulf. Forty-eight hours after the last Libyan opposition, the United States claimed that Libya's sustained failure to defend its claim to the Gulf of Sidra constituted acknowledgment that the claim was illegal. The surface action group departed 75 hours after the start of the operation.

Qadhafi responded with more terrorism. Intercepted radio messages proved that Libya instigated an arson attack in Berlin that killed an American soldier. On April 18 U.S. Navy A-6 bombers and Air Force F-111 bombers raided military bases in Libya (Operation El Dorado Canyon). North of Tripoli, *Ticonderoga* broadcast combat information to F-111s inbound to their targets. No aircraft was lost to enemy fire, although one F-111 was lost with its crew in an accident on its approach to Libya. The strikes barely missed killing Qadhafi himself along the way and may have wounded him, since he disappeared for weeks and a harridan described as his wife appeared on crutches. Libya continued to be a menace, but terrorism did decline after the air strike.

Freedom-of-Navigation Exercises in the Black Sea

The Soviet Union was another regime that made illegal claims on international waters. The Soviets said they had a law that let foreign warships proceed on innocent passage through the Soviet territorial sea within the 12-nautical-mile limit "only in specially assigned offshore areas. . . . There are no such areas in the Black Sea

off the Soviet coast."[56] This violated the 1982 United Nations Convention on the Law of the Sea (UNCLOS) treaty. The United States declined until 1994 to sign the UNCLOS treaty over a seabed-mining issue but agreed to abide by its other provisions.

If other coastal nations followed the Soviet precedent, international waterways such as the straits of Malacca, Hormuz, and Gibraltar could be closed to neutral warships. U.S. warships periodically sailed in the Black Sea to demonstrate freedom of navigation. *Caron* drew a sharp reaction there in 1979, as mentioned above. *David R. Ray* (DD 971) cruised near Novorossiysk in February 1984. A Soviet aircraft fired cannon rounds into her wake, and a helicopter buzzed her 30 feet over her deck.[57] However, available records do not substantiate this press account.

Caron and *Yorktown* cruised in the Black Sea in December 1985 without Soviet response. Early in 1986 the planners of Operation Prairie Fire received a legal brief about the right of innocent passage through the territorial sea off Libya. It apparently suggested an opportunity to assert the same international rights in the Black Sea and possibly to embarrass the Soviets, who could be counted on to act criminally in a legal confrontation. Secretary of the Navy John Lehman and Under Secretary of Defense for Policy Fred Iklé strongly backed the long-term strategy to drive the Soviet Union out of business. Warships would now enter the Soviet territorial sea itself.

In March 1986, while waiting for a third aircraft carrier to join the force off Libya, *Caron* and *Yorktown* were ordered to the Black Sea again. They departed from Sicilian ports and passed through the Bosporus two days later, on March 12. On March 13 they cruised off the south coast of the Crimea, very close to the Soviet naval base at Sevastopol. The Soviet Krivak-class patrol ship *Ladnyy* warned them to leave. A Soviet source said that the warships were within 6 nautical miles of the coast for two hours.

In Moscow the Soviet Foreign Ministry objected that the cruise was "of a demonstrative, defiant nature and pursued clearly for provocative aims." The commander in chief of the Soviet Navy, Fleet Admiral V. N. Chernavin, suggested that the U.S. ships might have been attacked had they remained longer in Soviet waters:

Taking into account the openly provocative nature of the actions of the U.S. ships, the command issued an order that the combat readiness of the strike forces of the fleet be enhanced. Ships and planes were promptly prepared for performing a combat mission.[58]

Another Soviet official called the exercise an attempt to test the Soviet Union and warned that Soviet forces would shoot the next time it happened.[59] The world press reported the Soviet tantrum in stories of how it matched terrorist Libya's claim on the Gulf of Sidra.[60] The Soviets were humiliated. *Caron* and *Yorktown* were in the Black Sea until March 17, then returned to Sicily before sailing for the Libyan operation on March 22.

In February 1988 the Navy again asserted the right to freedom of navigation in the Black Sea. *Yorktown* and *Caron* went again. One suspects that their return engagement was no coincidence and that the planners intended to goad the Soviets. The ships passed from the Mediterranean through the Bosporus on February 10. On February 12 they crossed the 12-nautical-mile limit into the Soviet territorial sea off the Crimea. Within minutes a Soviet Mirka-class corvette and the Krivak-class frigate *Bezzavetnyy* approached them. *Bezzavetnyy* announced by bridge-to-bridge VHF radio, "Soviet warships have orders to prevent violation of territorial waters. Extreme measure is to strike your ships with one of ours." *Caron* replied, "I am engaged in innocent passage consistent with international law."

About 15 minutes later the corvette approached *Caron,* and *Bezzavetnyy* approached *Yorktown,* both from the American ships' port quarters. Both Soviet ships swung their bows to starboard and struck the American ships' port quarters, apparently attempting to damage their propellers. *Bezzavetnyy* hit *Yorktown* twice, wrecking two Harpoon missiles in their canisters on her fantail. *Caron* was undamaged. *Yorktown* continued on an eastward course 9.8 nautical miles south of the Crimean coast; *Caron* was about 7 nautical miles off the coast. The American ships exited the territorial sea two hours later. They departed the Black Sea on February 15.[61]

The Soviet Union charged that the American warships had maneuvered dangerously and caused the collisions themselves. It failed to mention that the Soviet ships had announced they were acting under orders to collide. The Soviets protested that the exercise had undermined attempts to improve relations between the United States and the Soviet

Soviet frigates deliberately collided with *Caron* and the Aegis cruiser *Yorktown* in the Black Sea in 1988. Bubbles from *Caron*'s Masker and Prairie underwater acoustic silencers leave a very visible wake behind her. The wake shows she was steering a straight course, belying the Soviet propaganda claim that the U.S. ships caused the collision. (U.S. Naval Institute collection)

Union and that the ships had collected intelligence. The first count blamed the United States for inducing the Soviets to behave badly. On the second count, the Pentagon denied that *Yorktown* and *Caron* collected intelligence or did anything else inconsistent with the governing law, the UNCLOS treaty. Whether passive threat-warning systems such as SLQ-32 were used is unknown. Not to use them might be irresponsible, but they would unavoidably collect information that the Soviets no doubt would call military intelligence.[62]

These cruises established that the coastal nation has the burden of proving that passing warships are violating the convention; warships are presumed to be in compliance unless they give evidence of proscribed activity.[63] In 1989 the Bush administration agreed not to send warships back into the Soviet territorial sea, provided that the

Soviets amended their "law" to accord with the UNCLOS treaty. Since the UNCLOS treaty already required them to do that, the American concession looked like appeasement. President Bush may have hoped to strengthen Soviet chairman Gorbachev against opponents. Russia and Ukraine have disputed ownership of the Crimea after the collapse of the Soviet Union in 1991.

Operation Earnest Will: The Tanker War

In 1948 Admiral Richard Conolly (namesake of *Spruance*-class destroyer *Conolly* [DD 979]) established a Mideast Force in the Persian Gulf for naval presence. Before the Iran-Iraq War it consisted of an auxiliary ship, based in Bahrain, as flagship and two destroyers on deployment from the Atlantic Fleet. *Peterson* was the first *Spruance*-class destroyer on this duty. She transited the Suez Canal in July 1979. She watched two Soviet submarines and their tender off Socotra Island for several days, then passed through the Straits of Hormuz (which Lloyds of London had designated as a war zone because of terrorist threats) and entered the Persian Gulf on August 5. Operating from Bahrain, *Peterson* was flagship for the Mideast Force for a month until the regular flagship returned.[64] *John Young* (DD 973), *Briscoe* (DD 977), *Stump*, and *Conolly* were other *Spruance*-class destroyers to patrol the gulf early in their careers.

On duty in the gulf, *Spruance*-class destroyers maintained a Link-11 data link with U.S. Air Force E-3A AWACS aircraft. A detachment of four E-3As from the 552nd Airborne Warning and Control Wing from Tinker Air Force Base, Oklahoma, was based in Oman. They were known to the Navy as ELF One, probably their radio call sign. *Spruance*-class destroyers provided a surface radar picture and could identify contacts with their SLQ-32 radar detection systems, and perhaps with Outboard. These were vital capabilities. Link-11 capability was important, and the original Mideast Force destroyer patrol became a cruiser-*Spruance* patrol, since these ships all had the Naval Tactical Data System.

After the outbreak of the Iran-Iraq War in 1980, the destroyers watched dogfights over the northern gulf on the Link-11 plot. The ships were usually in the gulf for two months at a time. Crews did not like this duty with its risk of random attack and lack of liberty ports. Gas turbine plants have cooler engine rooms than steam, suiting *Spruance*-class destroyers for this duty. By 1983 the risk of

attack required that *Spruance*-class destroyers without the Mark 23 TAS missile-warning radar or without Phalanx missile-defense guns were not normally to be assigned inside the gulf.

Iran attempted to use its navy, inherited from the shah, to block trade to other Persian Gulf nations, which all supported Iraq. In January 1986 an Iranian warship stopped and boarded the American merchant ship M/V *President Taylor*. After that the Navy and the Maritime Administration instructed U.S.-flag merchantmen to let U.S. Navy warships act as intermediaries for Iranian challenges.

During the night of May 12 an Iranian *Saam*-class frigate challenged SS *President McKinley*, which replied with her destination. USS *David R. Ray* was listening nearby and closed, probably at high speed, to "within a mile of the *President McKinley* when the Iranian ship came into view,"[65] which suggests that the Iranian ship was close by and planning to board but not expecting company. Weapons ready, the destroyer came up on the radio circuit and told the Iranian ship to stand clear and not to interfere with freedom of navigation; this warning was repeated several times. The surprised Iranian warship backed off from this coiled and rattling snake and left.[66]

Iran had never attacked American-flag ships in the Persian Gulf, perhaps because of the clandestine arms traffic revealed in the Iran-contra affair. In May 1987 Kuwait registered 11 tankers under the U.S. flag so that the Navy would escort them past Iranian interference. The United States accepted this so as to safeguard freedom of navigation by neutrals and not to yield responsibility for guarding the West's oil routes to the Soviet Navy, which Kuwait had also asked to register the tankers.[67] This was the strategic turning point of the Iran-Iraq War: the United States would not allow Iran to win. "It was in those few early weeks in 1987 that Iran lost all hope of victory," wrote two well-informed British journalists.[68] The key was naval surface escort. Several joint-service bureaucracies and a few of the Navy's own competed to run Operation Earnest Will, the convoys sailing between the Gulf of Oman and Kuwait. Under a recent law known as the Goldwater-Nichols Act, the land-oriented Central Command staff planned the operation.

The plan was tactically weak. It assumed that presence of American escorts would deter Iran from attacking convoys. Apparently influenced by the Iraqi Exocet missile attack on the frigate USS

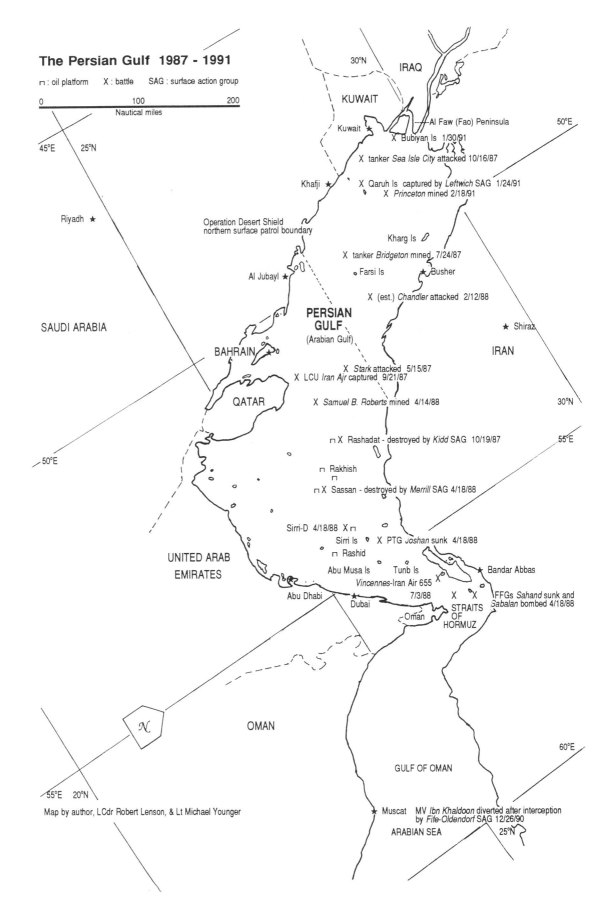

The Persian Gulf 1987 - 1991

⊓ : oil platform   X : battle   SAG : surface action group

0   100   200
Nautical miles

45°E   25°N

IRAQ

KUWAIT

30°N

Kuwait ★

Al Faw (Fao) Peninsula

X Bubiyan Is 1/30/91

X tanker *Sea Isle City* attacked 10/16/87

50°E

Khafji ★

X Qaruh Is captured by *Leftwich* SAG 1/24/91
X *Princeton* mined 2/18/91

Riyadh ★

Operation Desert Shield
northern surface patrol boundary

Kharg Is

X tanker *Bridgeton* mined 7/24/87

Al Jubayl ★

Farsi Is   Busher

X (est.) *Chandler* attacked 2/12/88

**PERSIAN
GULF**

(Arabian Gulf)

Shiraz ★

SAUDI ARABIA

BAHRAIN

IRAN

X *Stark* attacked 5/15/87
X LCU *Iran Ajr* captured 9/21/87

30°N

QATAR

X *Samuel B. Roberts* mined 4/14/88

55°E

50°E

⊓ X Rashadat - destroyed by *Kidd* SAG 10/19/87

⊓ Rakhish

⊓ X Sassan - destroyed by *Merrill* SAG 4/18/88

Sirri-D 4/18/88 X ⊓

UNITED ARAB
EMIRATES

Sirri Is   X PTG *Joshan* sunk 4/18/88

⊓ Rashid

Abu Musa Is   Tunb Is   Bandar Abbas

Vincennes-Iran Air 655 X

Abu Dhabi

Dubai   7/3/88   X   X   FFGs *Sahand* sunk and
Sabalan bombed 4/18/88

Oman   STRAITS
OF
HORMUZ

N

OMAN

60°E

GULF OF OMAN

55°E   20°N

Map by author, LCdr Robert Lenson, & Lt Michael Younger

★ Muscat   MV *Ibn Khaldoon* diverted after interception
by *Fife-Oldendorf* SAG 12/26/90

ARABIAN SEA   25°N

Stark (FFG 31) on May 15, the planners requested three destroyers to reinforce the half-dozen ships already in the gulf. Three Phalanx-armed guided missile destroyers could defend a convoy against attack by about 10 aircraft or antiship missiles. The plan disregarded threats of mines and boat raids, the risk of mistaken-identity attacks by Iran or Iraq, and tactical needs for secrecy, deception, and route surveillance.

The chairman of the Joint Chiefs of Staff acted to correct some of its shortcomings. Unfortunately the Navy had a weak leader as Chief of Naval Operations, Admiral Carlisle Trost, appointed without Secretary Lehman's approval. His aloofness disappointed even his supporters. Admiral Trost was oblivious to the Persian Gulf operation. Inside the Navy only one command apparently prepared for mine warfare: For a month helicopter minesweeping squadron HM-14 in Norfolk stood ready to deploy to the Persian Gulf on 24 hours' notice. The planners told it to relax.[69]

Kidd (Commander Daniel Murphy) deployed from Norfolk for convoy escort duty. Additional armament for the mission included machine guns, hand-held Stinger antiaircraft missiles, a LAMPS SH-2F helicopter from HSL-34, and SSQ-74 and SLQ-33 deception systems to generate false radar and acoustic signals. Kidd made the 15-hour transit of the Suez Canal on June 24, observing burned-out tanks and other relics of the Arab-Israeli wars along the Sinai bank, and entered the Persian Gulf on July 6. Designed for gulf duty, her air-conditioning plant easily handled the tropical heat.

Kidd commanded the first Earnest Will convoy (number 87001). Kidd and two other ships escorted two reflagged tankers, which were returning empty from the Gulf of Oman to Kuwait. The tankers were Gas Prince and Bridgeton, the latter the largest tanker in the Arab world. The convoy sailed on July 21. Three laden foreign tankers had passed down the narrow deepwater channel the previous day. As home for the television and newspaper reporters covering the convoy, Kidd became the best-known destroyer in the Navy. This was the first use of the press pool, which had been set up to cover military operations after reporters complained about being excluded from Grenada.

About 120 nautical miles southeast of Kuwait on July 23, Bridgeton was steaming at 16 knots in the channel when she struck a moored contact mine 20 feet beneath the surface. "It felt like a 500-ton hammer hit us up forward," said the master. "The shock wave move[d] back and up the ship and into the bridge and superstructure. . . . We all had to hold on so that this shock undulation didn't knock us off our feet."[70] Bridgeton was 2,200 yards off Kidd's starboard bow, and the crew and reporters aboard Kidd heard the explosion clearly. Damage was minor, and the convoy continued at 10 knots. Unabashedly using her as a mine bumper, the warships sailed in line behind Bridgeton, with Kidd in the rear. Lookouts and riflemen watched for floating mines, and Kidd alerted a passing Soviet warship to the mine danger. The convoy arrived in Kuwaiti waters that evening. Kidd anchored overnight and at sunrise returned south at high speed to Bahrain, keeping outside the shipping channel.[71]

The mining was a major embarrassment and revealed the need to keep the shipping channel under constant surveillance. Kidd and her SH-2F Seasprite helicopter kept an eye on choke points, narrows where mines would be most effective. The Navy quickly outfitted two oil field workboats with paravanes to sweep moored contact mines. Minesweeping helicopters arrived in mid-August. The Aegis cruiser Valley Forge (CG 50) joined one convoy to provide missile cover over the Straits of Hormuz. On August 24 Kidd fired warning shots at dhows approaching outbound Earnest Will convoy 87005.[72]

Kidd led six convoys during the period from August 1 to September 13. Attacks on tankers had been occurring at the rate of about three per week, but except for the Bridgeton mining, attacks dropped to zero after convoying began through mid-August. On one convoy, Soviet merchant ships tagged along for protection. On another, a Soviet warship flashed to Kidd, "Good luck on your convoy, from your friends."[73] Under Admiral Trost the Navy accumulated no list of "lessons learned" from Operation Earnest Will, and antimine precautions were forgotten before Operation Desert Storm.

The Capture and Sinking of *Iran Ajr*

Kidd sortied from anchorage at Bahrain on September 20 for a surveillance mission south of Abu Musa, an Iranian-held island in the southern gulf 300 miles east of Bahrain and 50 miles south of the Iranian mainland. She kept a watch on the straits against minelaying.[74] Iran denied laying mines in international waters, but American intelligence became aware that Iran was loading mines on the *Iran Ajr*, a large landing craft.[75]

The *Iran Ajr* sneaked west to lay a minefield 50 miles northeast of Qatar in an anchorage in international waters used both by the Navy and by tankers waiting to move on to oil-loading terminals.[76] Unseen company tracked her. While she prepared to sow her mines in the predawn darkness of September 22, two silenced Army AH-6 night-attack helicopters from the frigate *Jarrett* (FFG 33) approached and raked her with machine-gun fire. The Iranians were stunned, having no idea what had shot at them. Thinking that their assailants had left, they resumed working on the mines. The AH-6s strafed the ship again, killing three. Terrified, the Iranians abandoned not just their mission but their ship, which a SEAL squad boarded by helicopter and captured.

Kidd came up north to protect Navy ships towing the *Iran Ajr* to Bahrain. A large Iranian naval hovercraft approached but was driven off by 3-inch gunfire from *Jarrett*. The *Iran Ajr*, 26 Iranian prisoners, and 10 North Korean M-08 mines (the type that damaged *Bridgeton*) still aboard the minelayer were shown to the world press and to naval intelligence. On September 25 *Kidd* and the frigate *Hawes* (FFG 53) towed the *Iran Ajr* out into the gulf, rigged her with explosives, and blew her up. *Kidd* then resumed her patrol south of Abu Musa. Iran abandoned minelaying in the gulf for months after this debacle.[77]

After the *Iran Ajr* incident all Mideast Force ships maintained a heightened state of air defense readiness against possible Iranian retaliation. For the first three weeks of this period *Kidd* was the air defense command post. The United States issued a Notice to Airmen in the Persian Gulf region. This notice told all pilots to be prepared to identify themselves on specific radio circuits and to state their intentions. It was still in effect in July 1988.[78]

Countering Cruise Missiles: Nimble Archer and Tomahawk

Frigate-based Army OH-58 and AH-6 attack helicopters fought Iranian speedboats on October 8. On October 15 Iran fired Silkworm antiship cruise missiles from the Fao Peninsula into the tanker anchorage near Kuwait. The Silkworms, Chinese-made copies of the Russian SS-N-2 Styx missiles that sank the Israeli *Elath* in 1967, damaged two tankers, including a U.S.-flag Kuwaiti tanker with 18 casualties.

In retaliation the Mideast Force sent four destroyers to attack the Rashadat oil platforms on October 19 (Operation Nimble Archer). Iran used these platforms, 80 nautical miles east-northeast of Qatar, for surveillance in the central gulf and as a command post and base for small boats. The destroyers were *Kidd*, the *Spruance*-class destroyers *John Young* and *Leftwich* (DD 984), and the old destroyer *Hoel* (DDG 13). After giving the Iranians 20 minutes to abandon the platforms, which they did, the destroyers opened fire at 6,000 yards. The

In October 1987 *Kidd*, *Leftwich*, *John Young*, and the old destroyer *Hoel* destroyed the Rashadat oil platform, which Iran had used to direct attacks on Persian Gulf shipping, by gunfire. *John Young* is seen here firing at the platforms. This is one of the few combat photographs in military history to show units of both sides in the same frame. (U.S. Navy photo)

first platform exploded in seconds. Under orders to bombard it for 90 minutes, the destroyers fired 1,065 5-inch shells at the platform, aiming at its legs to knock it into the water. It was still burning a week later. SEAL commandos landed on the second platform, captured intelligence material, and blew it up.[79]

Kidd left the gulf after Operation Nimble Archer, but other DD 963 types had arrived. The new Aegis cruiser *Bunker Hill* (CG 52) and the battleship *Missouri* (BB 63) covered six Earnest Will convoys, releasing the aircraft carrier *Ranger* for other missions.[80] The Navy became provocative toward Iran. The Soviet *Sovremennyy*-class destroyer *Boyevoi* reported that *Missouri* frequently steamed slowly along the deepwater channel at night with lights burning brightly, pretending to be a tanker. "The intention," according to the *Boyevoi*, "was to provoke an attack by Iranian speedboats and then open fire."[81] *Bunker Hill* was right behind *Missouri* for air cover. Iranian forces did not take the bait, if such it was.

USS *Chandler* (DDG 996) arrived in the Persian Gulf in November 1987. On February 12, 1988, *Chandler* was escorting Earnest Will convoy 88013 inbound to Kuwait. That night she detected an aircraft flying toward the convoy. Probably using her SLQ-32 radar detection system, *Chandler* identified the aircraft as an Iraqi Badger bomber and, more alarmingly, as one that was on the attack. The Iraqis often shot at anything they detected on their radar, as they had done in the *Stark* incident. Their usual announcement of strikes on "large naval targets" revealed that they rarely knew anything specific.

Chandler fired two parachute flares from her 5-inch guns, at once warning the Iraqi aircraft that she was a warship, not a tanker, and showing that she was not hiding, thus was not Iranian. This logic was lost on the Iraqi aircrew, who launched two C-601 Silkworm missiles at *Chandler* and banked away.

"This," reported *Chandler*, "was the high point of the deployment." She tracked the Silkworms on radar and prepared to shoot them down. The first splashed into the water on its own. It appears she fired chaff decoy rockets and turned away to open the range. The second Silkworm did not find *Chandler*, streaked past her several miles to starboard, probably into a chaff cloud, and exploded. *Chandler* departed, or fled, from the gulf in March, her crew glad to be rid of the place.[82]

The Navy targeted Iranian land-based Silk-

The destroyers had been ordered to shoot for 90 minutes, but their first rounds set the Rashadat platforms afire. They aimed at the platform legs, attempting to weaken the structure or to knock it into the water. Several of the legs have been hit in this view. (U.S. Navy photo)

worm bases for destruction by Tomahawk land-attack missiles (TLAM-C). Iran removed its batteries from the Fao Peninsula before a strike with the early, slowly targeted TERCOM Tomahawks. Nimble Archer was substituted. Battleships *Iowa* and *Missouri* were the Tomahawk platforms until February 1988. The departure of *Iowa* without replacement by another battleship hinted that the United States might be reducing its military commitment to the gulf, a charge that the Office of the Secretary of Defense denied without giving details.

The *Spruance*-class destroyer *Merrill* (DD 976; Commander Roger Miller, *Elliot*'s executive officer in 1979) was the new Tomahawk shooter. *Merrill* mounted two four-barrel armored box launchers for eight Tomahawk cruise missiles, probably all the TLAM-C conventional-warhead land-attack version. She arrived in the North Arabian Sea in February 1988.[83] When Earnest Will convoys ran the gulf, *Merrill* stood outside the Straits of Hormuz within Tomahawk range of the Silkworm sites. If a strike were ordered, her mission was to aim the Tomahawks for designated points ashore where the missiles would begin terrain contour matching for their flight to the Silkworm bases.

On March 1 a clutch brake on *Merrill*'s port main reduction gear broke down. Her speed was

Chandler stands by the tanker *Ariadne* in December 1987 to rescue crewmen wounded in an Iranian speedboat attack. She also rescued injured crewmen from another damaged tanker, *Pivot*. Two months later she herself was the target of an Iraqi cruise missile attack. (U.S. Navy photo)

reduced to 12 knots, suggesting that the propeller and shaft had to be locked for repair at sea. She remained on station and kept the Silkworm sites covered and could make full speed again on March 5.[84] *Merrill* rendezvoused with the destroyer tender *Cape Cod* at Masirah in Oman on March 10 for a week. She waited at anchor for the convoys to start running again and on March 19 headed for patrol station in the North Arabian Sea.

Merrill covered the Silkworm sites during Earnest Will voyages until late March. She departed for the Seychelles for a change of command (Commander C. C. Covington relieved Captain Miller), then returned to station.

Merrill entered the Persian Gulf on April 11. She promptly encountered the Iranian old *Allen M. Sumner*–class destroyer *Babr*. *Merrill* watched the *Babr* for three hours, until the *Babr* changed course and headed away. Shortly afterward the lookouts heard a loud explosion and saw smoke. *Merrill* headed toward the explosion and found the Saudi Arabian cargo ship *Saghera* on fire at the bow. The *Babr* had hit her with a Standard missile. Rules of engagement authorized Navy protection against nautical mugging only for American-flag ships. *Saghera* radioed for tugs but ignored offers of help from *Merrill*.

Merrill also encountered the Iranian *Saam*-class frigate *Sabalan*, a ship with a particularly nasty reputation for attacking merchant ships. The *Sabalan* would signal "Good morning" to victims and come close alongside, then fire into the crews' quarters where an attack would cause the most casualties. When she met *Merrill* her career had just one more week to run.

Operation Praying Mantis

Iran, strategically desperate, resorted again to mining the shipping channels to put pressure on Iraq's supporters. On April 14, 1988, the new frigate *Samuel B. Roberts* (FFG 57) discovered a minefield northeast of Bahrain. She detonated a mine while backing down her wake in an attempt to get clear. The explosion beneath her single main engine room nearly broke her in half, but her well-trained crew kept her afloat. In this and in the Iraqi Exocet attack on her sister ship *Stark* the year before, modern American warships and their crews showed their toughness. In contrast, during the Falkland Islands War no ship survived a large internal explosion or fire.

Marine helicopters from the amphibious transport *Trenton* (LPD 14) discovered more moored mines that evening. They were North Korean M-08 contact mines from the same production series that the *Iran Ajr* had carried, proving Iranian responsibility for the damage to *Samuel B. Roberts*. Washington ordered retaliation. Three surface action groups were to sink the recently seen *Sabalan*, if she could be found, and to destroy two large gas/oil separation platforms (GOSPs) that Iran was using for surveillance in the southern gulf. These platforms were so large that maps of the

Mideast show them with their names. The objective of the retaliatory strikes, Operation Praying Mantis, was to destroy Iranian assets of equal value to the damaged *Samuel B. Roberts.* The strike was executed on April 18.

Merrill and *O'Brien* (DD 975), which had entered the gulf on April 13, were assigned to Praying Mantis. *Merrill* was the flagship for Surface Action Group Bravo (SAG B), which included herself, the old destroyer *Lynde McCormick,* and a Marine Corps air-ground task force embarked on the transport *Trenton* (see table 14.3). SAG B's mission was to destroy the Sassan oil platform while minimizing environmental damage and Iranian casualties.

The Marine force commander found *Merrill's* extensive communications facility highly useful and set up his command post for the raid in *Merrill's* combat information center. Command post functions included the landing force operations center, the helicopter coordinator section, and the supporting-arms coordination center.[85] The supporting-arms coordination center assigns priorities to enemy units and designates specific weapons, such as naval gunfire or aircraft, to attack them.

Merrill's SAG B approached the seven-section Sassan oil platform before dawn on April 18. The destroyers took station 4,000 yards east of the platform. *Merrill* warned the Iranians in Arabic, Farsi, and English over several radio circuits that they had five minutes to get off. About 30 men fled in a tug. At 0803 *Merrill* opened fire on the southern platform with 5-inch air-burst (VT frag) shells to destroy radio antennas. Large hydrogen sulfide tanks had to be avoided during bombardment so that the Marine assault team could land.

The Iranians remaining on the platform manned a 23 mm gun mount and fired at *Merrill* and at Marine AH-1T Cobra gunship helicopters from *Trenton.* With shells falling nearby, *Merrill* located the gun mount and destroyed it with counterbattery fire. The destroyers fired 50 5-inch shells into the southern platform, setting it afire and setting off secondary explosions of 23 mm and 12.7 mm ammunition. The Iranians ("converted martyrs," as one account put it) jumped into the water 50 feet below or huddled near the undamaged chemical tanks. *Merrill* radioed for the Iranian tugs to rescue them, which they did.

The destroyers resumed fire to suppress any remaining defenses while Marine helicopters ap-

proached. In a closely coordinated action the ships fired until 0926, when the Marine Cobras took over to cover troop landings at 0931 by rope from three assault helicopters. The raiders found no live defenders. They collected weapons and intelligence material, set 1,300 pounds of plastic explosive, and evacuated at 1303. At 1310 the charges detonated. Four sections collapsed in a terrific explosion, leaving only the sulfide tanks sticking above the surface. *Merrill* fired 103 5-inch shells during the raid.[86]

Saudi Air Force F-15 fighters, a civilian news helicopter, and a United Arab Emirates patrol boat came to the area to watch. A LAMPS Mark III SH-60B Seahawk helicopter relayed its longer-range airborne radar picture to *Merrill* over Link-11. (This was *Samuel B. Roberts's* Seahawk, now operating from *Trenton. Merrill* was not equipped to operate the SH-60B until 1992.) SAG B headed north to attack a second platform, Rahkish. A 25-knot surface contact was detected, obviously a warship. A Marine Cobra reported that it was

Table 14.3   Operation Praying Mantis

Surface Action Group B
Merrill (DD 976)—flagship, ComDesRon 9, Capt. J. B. Perkins
Landing Force command post—Col. W. M. Rakow
1 SH-2F
Lynde McCormick (DDG 8)
Trenton (LPD 14)
1 SH-60B (from *Samuel B. Roberts*)
Marine Light Attack Helicopter Squadron 167
—4 AH-1T
—2 CH-46E
—2 UH-1N
2 Marine Corps assault teams

Surface Action Group C
Wainwright (CG 28)—flagship, SAG C
Bagley (FF 1069)
1 SH-2F
Simpson (FFG 56)
1 SH-60B
1 UH-60A (Army)
SEAL platoon

Surface Action Group D
O'Brien (DD 975) with SEAL platoon
1 UH-60A (Army)
2 SH-2F
Jack Williams (FFG 24)—flagship, Capt. D. A. Dyer
2 SH-2F
Joseph Strauss (DDG 16)
Air Wing 11 from *Enterprise* (CVN-65)

possibly an Iranian *Saam*-class frigate. *Merrill* targeted the contact for attack by Harpoon but checked fire while attempting to identify it. It was a Soviet *Sovremennyy*-class destroyer coming to photograph the action.

No DD 963s were in SAG C, but its operations were interesting. This group attacked the Sirri-D oil platform, which blew up under bombardment before SEAL commandos boarded it. An Iranian 275-ton *La Combattante II*–class fast attack craft, the *Joshan*, approached SAG C. The *Joshan* ignored warnings to stand clear and at 26,000 yards fired a Harpoon. A LAMPS Mark I SH-2F Seasprite helicopter detected it and launched chaff. The Harpoon missed, decoyed by chaff.[87] The three SAG C ships disabled the *Joshan* with five high-speed Standard SM-1 missiles fired in line-of-sight mode. A Harpoon missed, possibly unable to detect the sinking target. The *Joshan* was finished off with gunfire.

Five Iranian Boghammer gunboats attacked gulf-state ships and oil platforms. *Merrill*'s SAG B was redirected to guard an oil platform. Two A-6s from *Enterprise* sank one Boghammer, and the others fled to Abu Musa and ran aground there.

The mission of the third surface action group (SAG D), consisting of *O'Brien* and two other ships, was to locate and sink the British-built Iranian frigate *Sabalan*. With her large flight deck, *O'Brien* was a SEAL base and embarked an air wing of two SH-2F Seasprites and the SEALs' UH-60A Blackhawk in case an opportunity arose to board and take down an Iranian ship. The force patrolled the Straits of Hormuz, using electronic intelligence to watch for Iranian warships.[88] In midafternoon SAG D obtained intelligence that warships were getting under way from their base at Bandar Abbas.[89] The *Sabalan* and her sister *Sahand* emerged from the coast. SAG D and *Enterprise*'s A-6s attacked them. A laser-guided bomb disabled the *Sabalan*. Three Harpoons and six bombs set the *Sahand* afire from stem to stern and sent her to the bottom. Monitoring the action by radio in Washington, Secretary of Defense Frank Carlucci decided that proportionate retaliation was complete. He canceled SAG D's follow-up strikes on the *Sabalan*. SAG B had already been diverted from attacking Rahkish.

Late that afternoon, uncertain about Iranian retaliation, the commander of the Joint Task Force Mideast (CJTFME) ordered two Marine Cobra attack helicopters from *Trenton* to join SAG C, 50 miles east, to provide local air cover. The Marine commander feared that this demanded too much from the pilots after a long day of combat flight operations. As the helicopters arrived at 2015, *Wainwright* vectored one to investigate a new surface contact. The Cobra reported radar contact at 2101 but then disappeared. *Merrill* joined in an unsuccessful search for the missing helicopter, an apparent victim of aircrew exhaustion. Its two Marine aviators were the only American casualties during Praying Mantis, the biggest American sea battle since World War II.

The *Vincennes* Incident

Concerned about Iranian retaliation after Praying Mantis, the commander of the Joint Task Force Mideast (Rear Admiral A. J. Less) suspended Earnest Will convoys and requested an Aegis cruiser to protect against possible Silkworm attacks. Iran had set up a Silkworm battery on Abu Musa inside the Straits of Hormuz.[90] On April 20 USS *Vincennes* (CG 49; Captain W. C. Rogers) was ordered from San Diego to the Persian Gulf. Earnest Will convoys resumed on April 27 with *Merrill* as convoy commander. She escorted two convoys and then left the gulf. *Vincennes* joined the task force on May 31 and entered the Persian Gulf on June 1.

On June 2, her first full day in the gulf, she watched the Iranian frigate *Alborz* send a boarding party to a merchantman to inspect it (legal, according to observers), or perhaps to seize it (not). *Vincennes* ordered the frigate USS *Sides* (FFG 14; Captain R. Hattan) to take station 1,500 yards astern of the *Alborz*. Captain Hattan on *Sides* thought that this close maneuver, barely six weeks after Praying Mantis, might provoke the *Alborz* to open fire. He further disliked the tactical position, which evidently used *Sides* as bait so *Vincennes* could retaliate. CJTFME agreed, detached *Sides* from *Vincennes*'s operational control, and ordered the cruiser to back off.[91]

This seems to have started a feud between the commanding officers of *Vincennes* and the frigates. Frigate crews objected that *Vincennes* was a cruiser doing a frigate's job. They called *Vincennes* "trigger-happy" and suspected that Captain Rogers was seeking to provoke an incident in which his ship could fire on Iranian forces.[92] Combat experience as in Praying Mantis might help officers in Navywide competition for promotion. Cap-

Gunfire drives Iranian defenders from the Sassan oil platform during a raid led by *Merrill*, part of Operation Praying Mantis in 1988 to retaliate for Iranian minelaying in international waters. Sassan, like Rashadat, had been an armed base for Iranian surveillance units. Part of the platform has been burned out by destroyer gunfire. (U.S. Naval Institute collection)

tain Rogers claimed he had every confidence in his crew, but events suggest that they were poorly trained and badly led, two sides of the same coin. Videotapes show *Vincennes* sailors at general quarters wearing T-shirts instead of protective clothing and appearing to be distracted from their jobs.

During 1984–86 Iranian fighter aircraft had often attacked shipping in the gulf. Attacks declined in 1987 and stopped in February 1988. That might have reflected a policy change in Iran, or consumption of pilots and weapons. During June 1988 Iran transferred American-made F-4 Phantoms and F-14 Tomcats, bought by the shah, to Bandar Abbas. *Vincennes* frequently sighted these aircraft. On June 18 naval intelligence warned, "All units are cautioned to be on the alert for more aggressive behavior."[93]

On July 3 *Vincennes* was patrolling near Abu Musa. Armed Iranian Boghammer speedboats assembled about 20 nautical miles to her north. The day before, speedboats had attacked a Danish freighter with rockets and gunfire near Abu Musa but had made no hits.[94] The frigate *Elmer Montgom-*

ery (FF 1082) reported that the speedboats were preparing to attack a Pakistani freighter, and that explosions were audible but no merchant ship had requested assistance.[95] *Vincennes* went to general quarters, launched her SH-60B helicopter to investigate, and turned north at full speed. Coastal authorities in Oman reprimanded *Vincennes* by radio that 30-knot speed was inconsistent with innocent passage and ordered her to leave Omani waters.[96]

At 0840 the commander of the Joint Task Force Mideast learned that *Vincennes* had left her surveillance station and angrily ordered her back to Abu Musa. *Vincennes* secured from general quarters and turned back but left her Seahawk airborne. Violating CJTFME orders, the SH-60B approached within 2 or 3 miles of the Iranian boats but kept out of small-arms range. The Iranian boats fired tracer shots that passed about 100 yards ahead of the helicopter.[97] It is speculation whether they were fired as bravado or to warn the Seahawk not to approach closer. The helicopter left undamaged. In April, shots at an A-6 had triggered the unscheduled attack that sank the *Sahand*.

Upon hearing about these shots at 0910, *Vincennes* went to general quarters again and turned north at flank speed. Captain Rogers pressed CJTFME for permission to fire on the speedboats. He had authority to shoot in self-defense if the Boghammers threatened *Vincennes*. Since they were over 15 nautical miles distant, they did not. Without a Link-11 tactical data link to *Vincennes*, the CJTFME staff in Bahrain could not see this tactical picture. At the time the *Spruance*-class destroyer *John Hancock* (DD 981) with Link-11 was anchored at Bahrain, but her link was inactive. At about 0918 CJTFME approved *Vincennes*'s request.[98]

The Boghammers retreated north to Iranian territorial waters at high speed. *Vincennes* pursued them for 20 minutes over the nautical battlefield of Operation Praying Mantis. At 0940 she entered Iranian waters, violating international law, since it was hardly innocent passage.[99] Reporting incorrectly that the Boghammers were closing on her, *Vincennes* repeated her request for permission to fire, which CJTFME approved again. At 0943 she opened fire with her 5-inch guns inside Iranian waters on targets that had not fired at her. The Boghammers returned fire toward the cruiser, but none approached within 2.5 nautical miles, and they made no hits on *Vincennes* with any

Vincennes in the Persian Gulf on the day she shot down the Iranian Airbus. The boat alongside may be bringing investigators. (U.S. Navy photo)

weapon. They were out of range of the cruiser's 25 mm Bushmaster guns.

A misfire downed Mount 51, the forward 5-inch gun, during the first minute of firing, so the cruiser changed course frequently and sharply to keep Mount 52, dead aft, bearing on the targets. High-speed turns cause any ship to heel, and Aegis cruisers heel steeply. Putting the rudder over 30 degrees at 30 knots caused *Vincennes* to heel 32 degrees. The CIC crew was tossed around and pelted with falling manuals and equipment. This situation was not conducive to cool thinking aboard *Vincennes*. "Stress, task fixation [shooting at the Boghammers], and unconscious distortion of data may have played a major role in this incident," Navy investigators wrote.[100]

At 0947 an Iran Air Airbus 310 airliner took off on a scheduled flight from Bandar Abbas across the gulf to Dubai. *Vincennes*'s Aegis system promptly detected it and designated it as track number (TN) 4474. A chain of errors began. A sailor misused an identification-friend-or-foe (IFF) interrogator, one not connected with Aegis, and received a response from a military aircraft. His error was in not having adjusted the IFF unit to the decreasing range to the contact. In the surface radar duct, the IFF unit still reached the airfield at Bandar Abbas, and it interrogated an Iranian military aircraft still on the ground there.[101] Aegis recognized that the nearby frigate *Sides* (now under Commander David Carlson) had designated the approaching aircraft as TN 4131 and adopted this designation. *Vin-*

cennes's crew assumed it was an F-14 she had seen before and labeled it as hostile.

The Aegis IFF system, however, was already correctly identifying the contact as a commercial aircraft. The Airbus would pass in the commercial air corridor directly over *Vincennes* and was invisible above the clouds. *Vincennes*'s SLQ-32 system detected no radar signal, either military or commercial, from it. The aircraft was flying "cold nose," so had no radar image that *Vincennes* was even present. An F-14 had never attacked a ship in the gulf and had no known weapon to do so beyond a 20 mm gun. Aegis displayed that TN 4131 was climbing. *Sides* accepted *Vincennes*'s assumption that it was an F-14 but still evaluated it as a nonthreat. Minutes later *Sides* identified it as an airliner.[102]

But *Vincennes* was getting every answer wrong. A hysterical sailor in the combat information center screamed repeatedly that the aircraft was descending, as if to make a kamikaze dive. Either he was ignoring accurate data on the Aegis display about TN 4131, or he was looking at the wrong aircraft track number, probably TN 4474, originally assigned to the airliner but now reassigned to a Navy aircraft out in the gulf.[103] Fearing that his officers were untrained and inattentive to their duties, he was inattentive to his own.[104] Another sailor announced that TN 4131 had identified itself as a military aircraft. In fact all Aegis IFF responses showed civilian codes.[105] Stuck on a notion of loyalty to the *Vincennes* "team," Captain

Rogers and other officers swallowed the panic-stricken sailors' false reports without challenge.

Vincennes attempted to contact the aircraft by voice radio. The Airbus may not have understood that the messages were directed to it, since three of *Vincennes*'s seven warnings addressed the Airbus as a fighter. Perhaps it was not tuned to the correct frequency (announced in the September 1987 Notice to Airmen). Perhaps transmission quality was poor; only one of three foreign allied warships in the area heard any of the warnings.

At 0951 a CIC watch stander told Captain Rogers that the aircraft was possibly a commercial airliner. Captain Rogers still assumed that it was an F-14 and that a distant Iranian P-3 patrol aircraft was guiding it. *Vincennes* fired two Standard SM-2 missiles from her forward launcher at 0954 at a range of about 10 nautical miles. Both hit the Airbus at 8 nautical miles, altitude 13,500 feet.[106] There were no survivors from the 290 passengers and crew, many of them Gulf Arabs returning home.[107] If the Aegis system had run the engagement in fully automatic mode, it would not have fired the missiles.

Vincennes had fired 72 5-inch rounds at the Boghammers. She claimed five sunk, based on SPQ-9 track termination, which could also result from boats getting out of range. After the shoot-down *Vincennes* anchored at Bahrain for two weeks for investigation of the disaster, then left the Persian Gulf, never to return.[108] USS *Spruance*, equipped with Outboard and LAMPS Mark III, arrived to patrol the Straits of Hormuz as gate guard ship. Her SH-60B and the digital data link (ARQ-44 to SRQ-4) proved highly useful in tracking all the merchant and naval ships in the area. *Spruance*'s SH-60B patrolled at dawn and dusk. Iranians looked the destroyer over daily from a P-3 or C-130. *Spruance* warned the aircraft if it approached closely but felt no need to open fire.[109]

The Reagan administration blamed Iran for sending the airliner over the gunfight. Iran, imagining that the incident signaled that the United States now was a full belligerent, quickly settled its eight-year war against Iraq.[110] Critics later accused President Reagan and Vice President Bush of misleading the world, because Iran had not started the action. However, Iran's fundamental culpability should not be ignored. Iranian authorities knew their gunboats had long been up to no good beneath the air corridors. They operated military aircraft from the airfield at Bandar Abbas.

Their support for terrorists made a kamikaze attack or treacherous use of IFF codes plausible.

Vincennes's severe heeling during high-speed maneuvers raised again the misunderstanding about the stability of the *Ticonderoga* class. The Aegis system came in for scrutiny, and improvements have been made; it is an advantage that Aegis can be changed quickly by routine updates to the computer software. USS *Leyte Gulf* (CG 55) received an experimental full-color map-display system using commercial computers that worked very well in trials and in actual service during Operation Desert Storm.

Postmortem inquiries into the *Vincennes* incident are not over. Journalists incorrectly assumed that the Aegis displays showed unambiguously that the aircraft was an airliner.[111] Chief of Naval Operations Admiral Trost did nothing to help with the political effects. He blocked any office in the Department of the Navy from elucidating the error-laden and incomplete investigation report. Instead it went through the Army-oriented Central Command to the chairman of the Joint Chiefs of Staff, Admiral William Crowe. Admiral Crowe, another political submariner appointed against Secretary Lehman's judgment, put an incomplete and misleading endorsement on it. Admiral Trost made sure that the Navy took no disciplinary action against Captain Rogers.[112] These actions conveyed the view that *Vincennes*'s actions were officially consistent with the rules of engagement, standards of accountability, and all other regulations. Together Admirals Crowe and Trost hung the albatross of Captain Rogers's alibi around the Navy's neck. Admiral Crowe was called to testify before Congress in 1992 after *Newsweek* exposed more about the incident, in particular about his statements officially construing *Vincennes*'s attack on the gunboats as self-defense. Probably fearing recall for court martial under an administration not indebted to him like President Bush's, Admiral Crowe finessed the situation by endorsing the likely new President, Bill Clinton, during the 1992 campaign. He now holds high office in the Clinton administration as ambassador to Great Britain.

Operation Sharp Edge: *Peterson* off Liberia

In May 1990 *Peterson* was visiting Tunisia when she received urgent orders to join Task Force 62, the Sixth Fleet amphibious squadron.

This consisted of USS *Saipan* (LHA 1), *Ponce* (LPD 15), *Sumter* (LST 1181), and the 22nd Marine Expeditionary Unit (MEU) embarked aboard them. The amphibious force was sailing for West Africa for a possible evacuation of Americans from Liberia, where civil war had broken out. *Peterson* went along to provide communications intelligence from her Outboard system and gunfire support if needed.

While the force was en route, fighting increased around Monrovia, Liberia's capital, making it urgent to get a helicopter-borne security force to the American embassy. At 20 knots the amphibious ships would arrive in three days. *Peterson*, however, could get there in two. At 30 knots she would use about 18 percent of her fuel capacity to cover the distance. Her flight deck had the room and strength to operate a large CH-46 assault helicopter. After quick discussion the Marines decided to outfit *Peterson* as a fast destroyer transport and to send her ahead with a landing force.

Marine helicopters transferred troops and weapons to her from *Saipan*. *Peterson* embarked a 71-man reinforced rifle platoon with infantry packs and weapons, an 8-man SEAL commando team, a 4-man gunfire support control team, and a CH-46 helicopter with crew. This took four hours: one hour for the well-prepared Marines, three for the SEALs. An MEU staff officer was in charge of the landing force. For two days *Peterson* sprinted south at 30 knots with the helicopter chocked and chained to the flight deck. Landing force equipment was stored in the hangar and the ship's armory. *Peterson*'s crew hot-bunked with the Marines and SEALs.

She arrived off Monrovia on June 2 and stood by, ready to land her Marines and SEALs to reinforce the embassy or to evacuate noncombatants, and to cover either operation with gunfire. *Peterson* stayed in radio contact with the embassy, the amphibious force commander, and the 22nd MEU. If a landing were required, the 22nd MEU commander would control it from *Saipan* via *Peterson*'s radios and the NTDS data link. The situation ashore held. The landing force and the CH-46 returned to *Saipan* when the amphibious task force arrived 24 hours later.[113]

The force remained off Liberia ("Mamba Station") for two months. Fighting ashore worsened. On August 5 the assault ships closed to 6 nautical miles off Monrovia and sent helicopters with ground security forces ashore to evacuate Americans. *Peterson* closed to 3 nautical miles off Monrovia for gunfire support if called for. During the next two weeks the Marines evacuated 1,174 people from Monrovia to the ships offshore, which then flew them to Freetown, Sierra Leone. On August 13 *Peterson* and *Saipan* sailed south to Buchanan to support the evacuation of 99 diplomats and family members who had been unable to reach the U.S. embassy in Monrovia. By August 22 the total number of evacuees was over 1,600, most of them not American.[114]

Operations Desert Shield and Desert Storm

THE BLOCKADE OF IRAQ

On August 2, 1990, Iraqi forces invaded Kuwait and within two days began massing along the frontier with Saudi Arabia. *David R. Ray* was on patrol in the gulf, probably maintaining the surface picture for the AWACS detachment. The first U.S. reinforcement sent to the area was an aircraft carrier battle group, which needed no diplomatic permission to go into the area and begin operations. *Antietam* (CG 54) entered the Persian Gulf on August 6 to provide electronic surveillance and Aegis cover for other ships and forces there. New target assignments were sent to her to replace the original Iranian targets for Tomahawk land-attack missiles in her vertical-launch system (VLS) cells. A second carrier battle group quickly followed via the Suez Canal.

This began the buildup under Operation Desert Shield of U.S. forces to defend Saudi Arabia, an important if unofficial ally. In the 1980s Saudi Arabia had driven down oil prices to dry up Soviet

Spruance-class destroyers can operate helicopters as large as the Marine Corps CH-46E Sea Knight, a capability that proved valuable at Grenada and Liberia. (Lt. [jg] Michael Mitchell, USS *Fife*)

export earnings, contributing to the Soviet Union's incipient collapse. Journalists called the crisis the War in the Gulf, an interesting popular identification with the Navy, considering that almost all the activity was on land or in the air over Iraq. Navy strategic planners had anticipated during the 1980s that the most likely theater for a major war was the Mideast. The Army and the Air Force agreed only that war might break out there; their planning assumed that a conflict would quickly shift to the European front.[115] The Reagan administration did not choose between these strategies. Events proved the Navy right and the land services wrong. Marine brigades and Army airborne units almost certainly could have seized a defensible beachhead in Kuwait within a few weeks, to be covered by aircraft carriers and Aegis cruisers. Such a move could have destroyed Saddam Hussein's illusion that he could annex Kuwait and would have prevented much of the environmental catastrophe that Iraq caused later. This was not tried, probably for the same reason that other forward-defense tactics were forbidden: fear that it would provoke Iraq to invade Saudi Arabia.[116] Transporting Army and Air Force units to the Mideast took six months.

Eventually 24 ships from the *Spruance*, *Kidd*, and *Ticonderoga* classes participated in Operation Desert Shield, the defense of Saudi Arabia, and Operation Desert Storm, the campaign to liberate Kuwait. *O'Brien* served as flagship for antiship warfare in the northern Persian Gulf until early December 1990, when *Leftwich* relieved her. In addition to their normal armament and LAMPS Mark I helicopter, these destroyers were outfitted with extra night vision scopes; a mast-mounted sight with a low-light-level high-magnification TV camera (see chapter 9); two Army OH-58 night-attack helicopters; and ceramic armor bulwarks to protect deck-mounted machine guns. Ostensibly to avoid provoking Iraqi action during his ponderous buildup ashore, Army General Norman Schwarzkopf forbade naval operations above 27°30′ north, 55 miles south of the Kuwait–Saudi Arabia border, until January. Left free and unobserved in the northern 200 miles of the Persian Gulf, the Iraqis laid over 1,000 mines off the Kuwaiti coast.[117]

Iraq attempted to break the United Nations' trade sanctions against it, which coalition navies implemented as a close naval blockade. *O'Brien* trailed an outbound Iraqi cargo ship during October and fired warning shots from her .50-caliber,

Fife (DD 991), *Oldendorf* (DD 972), and other units take down the Iraqi blockade runners *Ibn Khaldoon* and *Ain Zalah* during Operation Desert Shield. (Lt. [jg] Michael Mitchell, USS *Fife*)

25 mm, and 5-inch guns to induce her to stop. When the ship did not comply, *O'Brien* turned her over to other coalition ships, which surrounded and boarded her. In December the Iraqi merchant ship *Ibn Khaldoon* attempted to run the blockade with 60 "peace activist" passengers and a cargo of contraband. Iraq's blockade runners' voyages coincided with Iraqi approaches to the Soviet Union for assistance. The "peace activists" apparently hoped to disarm the boarding party and take it hostage. The Iraqi regime hoped the activists would suffer casualties, which it could use to weaken coalition fortitude. The countermeasure was to identify which of the hundreds of ships being stopped was the blockade runner and to surprise her with enough force to seize her over the violent opposition of the "peace activists," without inflicting casualties for Iraq to parade on TV.

The *Ibn Khaldoon* departed the Red Sea bound

for the Persian Gulf late in December. The *Spruance*-class destroyer *Fife* (DD 991) located her with Outboard when the blockade runner exited the Red Sea through the Bab al-Mandab strait, which was over 1,000 miles away, assuming that *Fife* was in the Arabian Sea with other blockade forces. *Fife* tracked the blockade runner's voyage accurately with Outboard and sprung the trap on her in the Gulf of Oman on December 26. During the night *Fife, Oldendorf, Trenton,* and HMAS *Sydney* converged where *Fife* predicted they would find the blockade runner. At dawn *Ibn Khaldoon* found she had company. Marine helicopters hovering off her bridge aimed machine guns at her decks. *Sydney*'s Seahawk landed SEALs and Marines on her. One woman suffered a heart attack (she was treated successfully aboard *Oldendorf*). The activists tried to steal the boarders' guns and to form a human chain to induce violence. Instead warning shots, smoke grenades, and stun grenades corralled them. The ship was diverted to a coalition-controlled port. *Fife* detected another Iraqi blockade runner, *Ain Zalah,* by Outboard and she too was boarded by surprise. After that Iraq abandoned its "peace ship" gambit.[118]

TOMAHAWK STRIKES

Paul F. Foster and *Princeton* (CG 59) deployed in December 1990 for the gulf, both carrying Tomahawk conventional land-attack missiles in their VLS cells. *Paul F. Foster* loaded additional Tomahawks at Subic Bay. The ships sailed ahead of their battle group and entered the gulf on January 14. On January 17, the first day of hostilities, *Spruance*-class destroyers in the gulf were *Paul F. Foster, Leftwich,* and *Fife;* Aegis cruisers present were *Valley Forge, Bunker Hill, Mobile Bay* (CG 53), and *Princeton.* Their primary duties were Tomahawk attacks and air defense over the gulf (see table 14.4). (Without VLS, *Valley Forge* carried no Tomahawks.)

San Jacinto (CG 56) in the Red Sea fired the first shots of Desert Storm, a salvo of Tomahawks. Distant ships fired first so that all missiles would arrive on target simultaneously. *Paul F. Foster* fired the first Tomahawks from the Persian Gulf, followed by *Fife* and *Leftwich,* both stationed farther up and thus closer to Iraq. Within the first 24 hours, seven Navy ships fired 116 Tomahawks at targets in or near Baghdad (*Fife* alone fired 22). These missiles hit 16 heavily defended power stations and command-and-control facilities and were coordinated with Air Force F-117 stealth

Table 14.4   Tomahawks fired during Operation Desert Storm, January–February 1991

Ship	Missiles Fired
Caron	2
Fife	60
Leftwich	8
Paul F. Foster	40
Spruance	2
Spruance-class subtotal	112
Bunker Hill	28
Mobile Bay	22
Leyte Gulf	2
San Jacinto	14
Philippine Sea	10
Princeton	3
Normandy	26
Ticonderoga-class subtotal	105
Rest of navy	71
Coalition total	288

fighter raids.[119] They left Baghdad blacked out and silent for the rest of the campaign.

Firing a Tomahawk from a VLS cell on a *Spruance*-class destroyer created a huge wall of fire and a very loud explosion as felt by the bridge watch. Lookouts watched for the booster to drop and the fins to deploy, indicating a successful launch.[120]

Presence of *Spruance*-class destroyers in the gulf, the Red Sea, and the Mediterranean suggested use of Outboard to cross-fix Iraqi radio transmitters. The attack plan targeted Iraq's military and civilian telephone systems, to force Iraqi commanders to use radio. Radio traffic intercepted by Outboard was perhaps deciphered for intelligence. Transmitters were jammed and then destroyed. Tomahawk salvos were coordinated with reconnaissance satellite passes to assess damage before the Iraqis could conceal or fake it. New Tomahawk launch orders were radioed to the ships for additional strikes. Almost all the Tomahawks were fired during the first two weeks of the campaign. *Fife* fired 27 more Tomahawks on days 2–4 of the campaign (January 18–20) and 11 during January 24–31, for a total of 60. *Spruance* fired two Tomahawks from the Mediterranean on January 22, in company with HMS *Ark Royal* and HMS *Sheffield.*

OPERATIONS INSIDE THE GULF

After the initial Tomahawk strikes, *Paul F. Foster* took over from *Leftwich* as antisurface command flagship. *Leftwich* was assigned to combat search and rescue duty in the northern gulf. She embarked two SH-3 Sea Kings and transferred her OH-58Ds to a frigate. In addition she carried a 10-man SEAL platoon, extra medical personnel, and a Coast Guard SAR expert. *Leftwich* was involved in 16 combat SAR recoveries. On January 24 carrier air strikes sank an Iraqi minelayer and a patrol boat near Qaruh Island, 30 miles off the Kuwaiti coast. When Iraqi forces on the island fired at helicopters assisting the Iraqi survivors, *Leftwich* launched her helicopters to land her SEAL platoon and a smaller SEAL unit from a frigate. In a six-hour operation covered by helicopter gunships and two frigates, the SEALs captured Qaruh, the first Kuwaiti territory to be recovered in the war, along with 29 Iraqi defenders and a large cache of arms that was sent to *Leftwich*.

On January 29 Iraqi land and naval forces attacked the evacuated Saudi port of Ras al-Khafji but were mauled by coalition defenders. The next day most of the surviving Iraqi fleet attempted to flee to nearby Iran but were ambushed at the start off Bubiyin Island. Only two ships reached Iranian waters. *Leftwich* captured 22 Iraqis from a sinking *Polnochny*-class landing ship used as a minelayer. The following day she recovered 15 more survivors from another sunken ship who were on an oil platform, and she controlled a Royal Navy helicopter strike on an Iraqi TNC-45 patrol boat that attempted to interfere. By February 2 the Iraqi Navy had no combat-effective ships left, and coalition naval attention shifted to minesweeping to prepare for an amphibious assault.

Kidd, now with the New Threat Upgrade to her antiaircraft missile system, arrived in the gulf on February 14. After embarking a cryptological team and a Stinger missile team, she began duty as the air defense guard for the aircraft carrier *Ranger.* When the ground campaign began, *Kidd* and *Bunker Hill* tracked and identified aircraft over Kuwait, Iraq, and the northern Persian Gulf to catch intruders and to prevent blue-on-blue (friendly-fire) engagements. This was successful, and the allied coalition had no friendly-fire losses.

To jump ahead to postwar patrol, many ships stayed in the Persian Gulf after Desert Storm to enforce the cease-fire and the continuing sanctions against Iraq. *Kidd* embarked two Army OH-58D helicopters, two SH-2F Seasprites, and a Coast Guard detachment. She had 415 men aboard, about 50 more than she could normally accommodate. One Seasprite was outfitted with minehunt-

Army OH-58D attack helicopters operated from destroyers both during and after Operation Desert Storm. The device above her rotor is a stabilized mast-mounted sight for night surveillance. (Bell Helicopter/Textron, in U.S. Naval Institute collection)

ing equipment. It was not uncommon for *Kidd* to perform the following missions simultaneously:

- Provide two helicopters for mine warfare missions
- Lead merchant ships through the swept channels into Kuwait
- Control the northern gulf air picture as force antiaircraft warfare coordinator and force electronic warfare coordinator
- Control the Persian Gulf surface picture as force over-the-horizon track coordinator and force antisurface warfare coordinator

MINE DAMAGE TO *PRINCETON*

Following Soviet practice, the Iraqis placed Silkworm antiship cruise missile batteries ashore to attack minesweepers. On February 18 *Princeton* (Captain E. B. Hontz) was the antiair warfare commander for the northern Persian Gulf. Earlier in the campaign she had coordinated 11 Royal Navy helicopter attacks that sank or disabled six Iraqi ships and gunboats. The minesweeping force advanced cautiously, with major ships staying 5 miles outside the estimated locations of minefields, some of which were shown on charts that *Leftwich*'s SEALs had captured on Qaruh Island. Intelligence estimates proved incorrect. At 0430 on February 18 USS *Tripoli* (LPH 10), carrying minesweeping helicopters, hit a moored contact mine. *Princeton* steered to position herself between the damaged *Tripoli* and an Iraqi Silkworm battery ashore.[121]

Princeton proceeded warily at 3 knots with both shafts turning at 55 rpm, and with bow lookouts searching for buoyant mines, but she crossed a line of Italian-made Manta influence-fuzed ground mines in 11 fathoms of water.[122] At 0716 a mine on the seabed detonated directly under the cruiser's fantail, just forward of Mount 52. Her fantail flexed 4 feet upward from the blast, hinging at the break in the hull at the end of the 01 level (frame 472). A gunner at the 25 mm gun amidships glimpsed the mooring chocks on the fantail as it kicked upward. It immediately sprang back almost into its normal position, sending severe harmonic mechanical shocks whipping through the entire ship. These were felt on the bridge as 4- to 6-foot vertical jolts. After 6–7 seconds the whipping translated into violent side-to-side shuddering. The mine explosion detonated a second ground mine 300 yards off the starboard bow, but all damage to the ship was from the first mine.

Structural damage was widespread. The fantail was left bent upward slightly and twisted sideways. Steel longitudinal hull beams had fractured where the fantail kicked up. Hull strength at the stern was suspected to be 80 percent destroyed. The hull upper strength girder, an 8- by 10-inch beam, was distorted with a 7-inch-deep crease. The port rudder was jammed at right 30 degrees. All weapons aft of number 2 VLS were disabled.

The first stringer is welded to *Princeton*'s hull to stiffen it after mine damage broke internal structural beams. She was repaired in dry dock at Dubai. (U.S. Navy photo)

Princeton (CG 59) was towed to Bahrain for emergency repairs to her mine-damaged fantail. (U.S. Navy photo)

Three Harpoon missiles poked through their canister covers. Spring bearings on both propeller shafts were knocked out of alignment, but the shafts and propellers kept turning, though noisily. Water leaked in through the port shaft, so it was stopped and secured by an inflatable boot seal. Hull plating at frame 472 was buckled from the main deck down to the inner bottom. Fuel tank covers popped loose and flooded ammunition in the Mount 52 magazine with oil, a formidable fire hazard. Flooding from fractured fire mains and chilled-water lines shut down the aft generator switchboard. Along the 01 level about 10 percent of the deckhouse seam with the hull separated, where explosively bonded steel-aluminum coupling unzipped. Further forward the shock transients split the deckhouse laterally forward of the mainmast right down to steel deck (01 level) and caused additional hull plate buckling at frame 380. Despite her damage, however, *Princeton* was still mobile and had no primary seawater flooding or significant hull ruptures.

The crew quickly reconfigured the Aegis SPY-1B radar to operate with a reduced cooling water supply. Less than two hours after the mining, *Princeton* could fire her Standard and Tomahawk missiles from the forward (number 1) VLS, and Mount 51 was ready for action. *Princeton* resumed duty as antiaircraft warfare commander. All four main propulsion gas turbines and both reduction gears remained serviceable although the number 2 plant was unusable because of the locked port shaft. The crew had been at breakfast on the mess decks amidships, with few men at the extremities. Crew casualties were only three wounded. The shock vibrations kicked two bow lookouts, the men farthest from the actual blast, 10 feet upward and threw a gunner's mate in Mount 52, closest to the blast, against the overhead. Captain Hontz had ordered the crew to keep all gear well secured, and this good if time-consuming practice prevented any injuries from missile hazards.

Meanwhile, the fact became apparent that *Princeton* had unknowingly passed through an outer minefield before reaching the line of Manta mines. Her most immediate problem was that she now had to go back through it to escape. *Princeton* straightened her port rudder and could make headway on her starboard shaft, but she could not maneuver well enough to steer clear of mines. Further, noise and pressure from the propeller might trigger other influence mines. Late that day the salvage ship USS *Beaufort* (ATS 2) took the cruiser in tow. They proceeded behind the Naval Reserve minesweeper *Adroit* (MSO 509), which was busy marking mines with flares and bunches of chemical lights. Throughout all this *Princeton* still provided area surveillance and air defense, until *Valley Forge* relieved her over 30 hours after the mining.

Princeton arrived under tow at Bahrain the next day. Helicopters flew men and equipment from the repair ship USS *Jason* (AR 8), a 48-year-old veteran of World War II. Welders attached 20-foot-long girders to the main deck and hull plating where the hull had flexed, then steel tubes when they ran out of girders, then padeyes for wire rope and turnbuckles when they ran out of tubes. Navy engineering officers debated whether to make minor immediate repairs, or even to send the cruiser home on a heavy-lift transport. They decided to

repair her permanently in theater to let her return seaworthy and on her own power. A side benefit was to prevent antimilitary congressional staff lawyers from showing up, second-guessing the repairs, and declaring the ship to be beyond salvage. The cruiser was dry-docked in Dubai. The British- and German-run shipyard there often repaired grounding damage to tankers and considered the cruiser's mine damage to be very similar.

Asked for advice, Ingalls Shipbuilding recommended leaving the stern askew and realigning the shaft bearings and A-frame struts. The Navy sent Ingalls a list of specialized material, such as new steel-aluminum couplings to reattach the deckhouse where it had sprung loose. Less than 96 hours later a C-5 arrived with 70 tons of material, diverted from Aegis cruisers under construction in Pascagoula. *Princeton* was in dry dock for five weeks. She passed sea trials on April 23 and returned home proudly. Benefiting from combat experience and a succession of effective commanding officers, during the next two years *Princeton* won every significant battle efficiency award in the Pacific Fleet. Within a year both main reduction gears failed, probably from shaft distortion. Additional repairs seem to have fixed that problem.

A few days after *Princeton* was mined, *Paul F. Foster* was cruising in the gulf when men at the stern felt a slight but distinct concussion, as if she had hit something. A noise came from the propellers. The damage control surveillance network reported no fires or flooding, and repair teams sent to inspect the stern compartments found nothing amiss. Later that day divers in the area found a floating contact mine, knocked loose from its mooring cable. A fresh, deep gash from a propeller blade passed within inches of three detonator horns.[123]

Operation Provide Comfort

Late in 1992 the Bush administration ordered U.S. military forces to assist famine-relief operations in Somalia. *Valley Forge* was sent to cover the operations, providing air surveillance for the U.S. ships offshore and standing by to provide gunfire support for Marines on land. Relief agencies flew in supplies, despite gunfire from one Somali faction or another. Between civil war and severe equipment limitations, the Mogadishu airport could not handle the load. Even the runway lights did not work. *Valley Forge* took over as the Mogadishu International Air Traffic Control Facility. She used Aegis to sort out the flights, identifying which were military aircraft supporting the ground troops and which were relief flights.[124]

Retaliatory Strikes against Iraq, 1993

Iraq violated the terms of the cease-fire that had been imposed by Operation Desert Storm. *Spruance*-class destroyers and *Ticonderoga*-class cruisers fired retaliatory strikes. On January 17, 1993, *Cowpens* (CG 63), *Hewitt*, and *Stump* launched Tomahawk salvos from the Persian Gulf, and *Caron* fired a salvo from the Red Sea. All 45 missiles were BGM-109C conventional unitary-warhead land-attack Tomahawks. Four missiles failed at launch, and six missed. The missiles arrived in two waves. Iraqi air defenses shot down one Tomahawk from the second wave, which crashed into the Al Rashid Hotel in Baghdad, stunning war correspondents staying there. The remaining 35 missiles, 78 percent of those fired, demolished a uranium-enrichment laboratory near Baghdad. All seven targeted buildings were hit.[125]

On June 18, 1993, in the first military action ordered by the Clinton administration, *Chancellorsville* (CG 62) fired 9 Tomahawks from the Persian Gulf, and *Peterson* fired 14 from the Red Sea. A 24th missile failed and was not fired. All missiles were again BGM-109C unitary-warhead types. The target was the Iraqi terror-police headquarters, which was hit by 16 missiles. Attack at night reduced Iraqi antiaircraft defense effectiveness. However, the target was probably unoccupied then. The early TERCOM Tomahawks were less vulnerable to air defenses at night.[126] The good news was that the target was hit hard, and no American or allied aircrews were ever at risk. This was the first U.S. naval strike since World War II to be executed without involvement of an aircraft carrier group commander.

In October 1994 Iraqi forces again advanced toward the Kuwait border. *Hewitt* and *Leyte Gulf* were already on station in the Persian Gulf as the first-line offensive strike-back force. Ground troops began flying in, but Iraq backed down even before they all arrived. This successful deterrence showed that the surface fleet is in its highest political stature since the Cuban missile crisis.

Haiti

In 1993–94 Outboard-equipped *Spruance*-class destroyers patrolled off Haiti as part of the Clinton administration's pressure to reinstall a deposed politician. *Caron* was active off Haiti in 1993. American diplomacy at times encouraged Haitians to attempt to immigrate illegally into the United States by sea. Often when American ships approached, the Haitians scuttled their craft because survivors were promised a landfall outside Haiti. In July 1994 *Spruance* rescued 900 refugees in one day, nearly three times above her accommodation capacity. Standard procedure was to take them to Guantánamo Bay while the United States government tried to figure out what to do next.

Arthur W. Radford and *Conolly* also were active off Haiti. In September 1994 *Comte de Grasse* arrived off Haiti in advance of the occupation force, which landed without incident. A new DD 963 mission was to support the new *Cyclone* (PC 1)-class coastal patrol ships as tenders and to provide fuel and fresh water. The destroyers' flat vertical sides made them the easiest ships present for the patrol ships to tie up alongside.

Conclusion

A Weapon That Works

Strategy is a plan to force a decisive series of tactical defeats on an enemy so that he sues for peace terms. *Spruance*-class destroyers influenced modern strategy and tactics. Their development brought to fleet service propulsion gas turbines, LAMPS helicopters, and all-digital combat systems. In service they introduced submarine-style passive surveillance tactics to the surface Navy. Before the *Spruance* class, surface warships were defensive screening ships for aircraft carriers and convoys. Tomahawk cruise missiles entered service on the *Spruance* class. Their sisters in the *Ticonderoga* class introduced Aegis and vertical-launch systems for surface missiles. With these strike weapons, in particular Outboard and Tomahawk cruise missiles, surface warships are first-line offensive forces.

Following the collapse of Soviet control over Eastern Europe and elsewhere during 1989–90, President George Bush announced a new military strategy for the post–Cold War world. He listed four strategic demands for the military: nuclear deterrence, forward presence, crisis response, and ability to reconstitute large forces. The Navy publicly stated its post–Cold War missions in a pamphlet titled ". . . From the Sea." Implicitly rejecting the Army–Air Force belief that the Navy should limit itself to ASW patrols, it emphasizes naval strike operations and amphibious raids against land-based forces along the ocean littoral, the regions within 200 miles of the coast where 80 percent of the world's population lives.[1] It proved that a long line of naval planners had been right to emphasize naval strike capabilities despite political pressure to focus on protecting shipping.

It appears in 1995 that most or all *Spruance*-class destroyers will continue in service under the new national strategy until after 2010. When ordering the *Spruance*-class destroyers, the Navy expected them to have a service life of 20 years. Now 20 years after the USS *Spruance* was commissioned, the class will likely serve for another 20 years. In the force reductions following the collapse of the Soviet Union, almost all earlier ships have been phased out. That much is unsurprising. However, nuclear cruisers built simultaneously with the *Spruance* class are being phased out too, as will be most of the newer FFG 7 class. Without exception, all the ships being retired were either smaller or more heavily armed than the *Spruance* class as built. They were the warships extolled by critics of the *Spruance* class. What went right for the *Spruance* class?

In his study of naval tactics Captain Wayne Hughes described historical trends and constants in naval warfare. These can be virtually summed up with the observation that in battle, scouting and communications pay off better than raw firepower.[2] The increasing range of modern weapons necessitates scouting over wider and more diverse areas. In the larger modern battle space, more contacts will exist and must be simultaneously evaluated as friendly, hostile, suspicious, neutral, unknown, or false. The *Spruance*-class destroyers brought highly advanced scouting capabilities to naval operations, in particular passive surveillance through long-range sonar and communications intelligence.

Captain Hughes listed the reasons why new weapons have often been disappointing.[3] The converse of his list defines the characteristics of a good weapon:

- The weapon must be available, implying affordable, in quantity

Hayler and the Coast Guard training bark *Eagle* visit
New York for a naval review. (U.S. Navy photo)

- It must be reliable
- It must be easy to integrate into tactical doctrine
- It must be difficult for an enemy to copy
- It must provide a military advantage

The *Spruance* class has passed this five-part test.

Success and Opposition

Some weapons have individual champions who
bring a paper project to life. Admiral Rickover for
the nuclear submarine is the most famous case.
Admiral Meyer for Aegis is certainly another. The
Spruance project had many champions. Admirals
Weschler and Sonenshein were strong proponents
of the evolving DD 963 project, but both left
Washington long before the ships entered service.
Admiral Zumwalt kept it going.

No weapon has a life of its own. What the
Spruance-class project did have was an attraction
for disparate decision makers that motivated them
to keep the project going from its start. The terrific
advantages that these ships would bring to the
fleet were their attraction. Other projects have

been similar in continuously motivating individuals toward success of the project: NASA's manned
space program and Data General's Eagle computer
are examples.[4] Ease of modernization was another
attraction. The ships remain, today, easier to
modernize than newer, tighter designs.

The *Spruance* class was designed with empty
spaces, strength and stability for future weapons,
and large reserves of power, cooling, berthing
space, and other utilities. These margins increased
their construction cost by about 5 percent. But by
extending the destroyers' life by 20 years, that 5
percent expenditure is paying back at a compound
interest rate of over 25 percent, the saving from not
building replacements today. The ships have
proven themselves to be a good investment. Ships
such as the FFG 7 class that will be retired after
fewer than ten years of service are no bargain. Yet
subsequent warship designs, including the *Ticonderoga* class, reverted to traditional small modernization margins. The history of the *Spruance* class
suggests that the Navy reverted to tight margins to
avoid repeating the *Spruance* controversy within
the service.

Navy partisans often warn that domination by

the Office of the Secretary of Defense since 1958 and legislated "jointism" since 1986 have led to bad decisions for the sea services. The Navy's autonomy and independence are legendary but, like most legends, partially mythological. The historical record suggests that national strategy has defined the missions and configuration of the modern Navy more strongly and for a longer time than the Navy as an institution realizes. At issue is whether the Navy would really have made better decisions independently, and if so, how. In the case of the *Spruance* class, several significant decisions made outside standard channels were distinctly better than those reached inside. Two examples are the push toward gas turbines in the late 1960s and the adoption of the Tomahawk-capable vertical-launch system in the early 1980s.

Naval analyst Norman Friedman noticed that within the Department of Defense, weapons emerge from a struggle among technology-oriented designers, quantity-oriented force planners, and the weapons-oriented fleet. Despite the later controversy over their quality, the *Spruance*-class destroyers were very much the product of fleet-oriented officers. The Navy's technologists of the era did not have much influence in decision making about the DX/DD 963 project, notably in propulsion and missile armament. The ships' promised quality was so high that force planners went along with them despite their cost. Later, the last of the World War II fleet was sacrificed to pay for them. The judgment of history is that the project managers (principally Admiral Weschler, Admiral Sonenshein, Dr. Reuven Leopold, and later Admiral Meyer) chose well.

Entry of the *Spruance* class into service was not trouble-free. Sudden introduction of so many new systems entailed problems: the sonar dome coating, low-pressure steam piping, scarcity of replacement parts for new systems, and others. These were costs of the disruption of shipbuilding from budget cuts and from disputes over nuclear propulsion and the Total Package contract. If new technology had been introduced gradually under a sustained shipbuilding program, problems could have been remedied before they hit the fleet all at once. If in future a similar attempt is made to hold up development while attempting to achieve integrated perfection, similar problems should be anticipated. In the case of the *Spruance* class, the advantages of systems integration and contractor procurement of subsystems were worth the difficulties. That may not always be true.

Total Package Procurement was the most ambitious attempt ever made to reform defense contracting. It attempted to eliminate contractor claims, to infuse commercial creativity into warship design, to modernize industry, and to reduce costs for construction and operation of ships. For shipbuilding, it succeeded in four significant goals. First, it produced a creative and well-integrated ship design. The *Spruance*-class destroyers were as good technically as anything the Navy could have designed itself. They were far better than anything the Navy, split by dissension over nuclear propulsion and hamstrung by conservatism, actually would have designed. Advances included weapons modernization margins, gas turbines, acoustic silencing, and appearance. Second, Total Package Procurement strengthened project management offices. That facilitated the accomplishments of Admiral Meyer in creating Aegis. Third, it produced a standardized ship class that was easy to support. Fourth, it produced a modern shipyard, although it did not revive the American shipbuilding industry.

Since 1989 the two final competitors for the DD 963 Total Package contract, Bath Iron Works and Litton-Ingalls, are the only U.S. shipbuilders still constructing large surface warships. In 1995 politicians are asking whether the nation can afford to build enough surface warships to sustain even those two yards. This ties into warship longevity. If good surface warships can serve 40 years, then building three per year yields a fleet of 120 warships, which would support about 25 forward-deployed ships worldwide.

In other areas, Total Package Procurement failed. First, the contract did not discipline the contractor to stay within price. Second, no contractor had the Navy's expertise or insight about warship requirements. Without that, Litton guessed wrong on corrosion, durability of internal fittings, and reliability of some minor equipment. Third, design time was much shorter than under traditional methods, but that did not reduce time to delivery. One cost of the short time for design was that it prevented Litton from discovering that it had been too cautious about the narrow hull.

The lessons of Total Package Procurement seem to have been forgotten. In the late 1980s the Navy issued a contract for the A-12 carrier-based bomber that was a Total Package contract in all but name. The contractors could not make the trade-offs to keep technology and performance within project budget. When they began preparing to sue to cover their cost overrun, Defense Secretary Dick Cheney killed the A-12. President Clinton's health care planners in 1994 made many assumptions that Total Package Procurement showed to be invalid, such as the hope that coercive contracting will uncover huge savings in industry, and the belief that gigantic contracts are manageable and enforceable.

Congress has sometimes suggested centralizing arms procurement in a new agency, separated from the military departments. The *Spruance* procurement shows that such an undertaking would fail. An agency without operational responsibility could not possibly make the trade-off decisions necessary for modern weapons, such as whether a new feature warrants raising the project budget to buy it, or at the other extreme, that no proposed design is worth the cost.

Testing and Evaluation

During the 1970s and 1980s Congress became very confused over military acceptance testing of Aegis and other weapons. Legislators understood neither the objectives nor the results of the tests.[5] They would have preferred to see weapons perform successfully in combat simulations. The Mark 71 8-inch gun lost approval after tests appeared to show that it was inaccurate. The Aegis shipbuilding program advanced after *Ticonderoga* proved in a costly demonstration that Aegis could shoot down multiple targets simultaneously.

It is facile to criticize expensive and technologically unnecessary testing as a cost of Congress's declining talent and lack of real-world experience. Or was Congress right in criticizing military weapons testing as inadequate for practical purposes? At Grenada, and initially off Lebanon, the fleet made no use of *Ticonderoga* and Aegis. In Operation Desert Storm, strike planners doubted that Tomahawk flight-testing had represented combat conditions, so they used scarce ordnance unnecessarily. In all cases confidence grew when real results were seen. Conversely, destroyer crews put confidence in the ability of their well-tested 5-inch guns to hit antiship missiles. Navy planners have no confidence in this tactic and have not funded antiaircraft or antimissile enhancements for 3- and 5-inch guns.

Congress might approve funds for really intense testing to measure, for example, how a ship might withstand surprise or terrorist attack. Per-haps remotely controlled, recently decommissioned warships could be attacked by strafing aircraft and infantry weapons fired from small boats. Phalanx and other short-range weapons should be tested in detecting and hitting bombs and small rockets launched at close range. The purpose should be to test both weapons and training, and to evaluate the effectiveness and limitations of both. Another payoff would be greater public confidence about the Navy. The Air Force has profited from publicity demonstrations of weapons since the days of Billy Mitchell. A question is how to sustain crew motivation and perseverance in sudden, clandestine, or weakly delimited operations ("low-threshold operations" such as peacekeeping).[6] As this book went to press, a test ship was placed in service.

The *Vincennes* Incident

Political interest in the *Vincennes* incident continues. Captain Carlson of the frigate *Sides* believes that he, Captain Rogers, and Admiral Less escaped a court-martial because the proceedings would have exposed embarrassing errors in the official report.[7] The Navy should revisit the investigation of the *Vincennes* incident, correct the errors in the report of the Joint Chiefs of Staff, and release an accurate account. The United States wants to bring to justice the accused Libyan bombers of Pan Am flight 103, which in December 1988 exploded over Lockerbie, Scotland, killing 270. We should anticipate that at a trial of the Lockerbie bombers, the defense would attempt to connect this event to the *Vincennes* incident, to sow doubt about the defendants' culpability. Reasonable doubt in the minds of jurors means acquittal. A true account could, and should, highlight the contribution of Iranian terrorism to the 655 shootdown.

Conclusions from Operations

Recent destroyer operations have provided actual tests of ships and doctrines. The incidents are representative of:

- The naval presence mission. This is more important to warship engineering now than in the Cold War since with a smaller military, individual deployed ships have a larger responsibility for surveillance, self-defense, and flexibility.

- Transition from a crisis situation to warfare, and from warfare back to a postwar environment under an enforced armistice.
- Low-intensity conflicts, meaning those with limited political commitment: low intensity for politicians.
- Joint-service missions.

Before drawing conclusions from these operations, some caveats:

- Most incidents were under peacetime rules of engagement, which restricted weapons use.
- U.S. ships never faced deliberate or sustained attacks as in World War II or as the Royal Navy faced in the Falkland Islands War. This may have permitted heavier use of soft assets such as transport helicopters.

The significance of operations lies in the number of common factors:

- The ships were committed to action suddenly, usually without opportunity to rehearse or to augment on-board systems before action. The Navy's long-standing emphasis on continuous operational readiness was proved well-founded.
- In almost all operations, at least one destroyer present had the Outboard communications intelligence collection system. Furthermore, Outboard seems usually to have been the reason that the particular destroyer was sent on the mission. An exception was the Persian Gulf, which apparently was adequately served by equivalent shore-based systems.
- Helicopter usage was often intense. The ability to operate multiple helicopters was valuable. Special forces often used the destroyers as helicopter platforms and sometimes as command posts.
- Speed and endurance were important in enabling one destroyer to cover two assignments. Sometimes, notably at Grenada, maneuverability was useful.
- Naval gunfire support was very important. At close range it would have been the weapon of choice even if the ship had carried longer-range bombardment rockets too.
- Regardless of individual armament, ships were sent into potential opposed action in company, not singly. Pocket battleships and strike cruisers still seem unrealistic.

The Navy has not recognized the potential to offer destroyers as secure, armed, mobile helicopter landing zones to support Marine, Army, and Special Forces ground operations. With the decline in numbers of amphibious transports, availability of destroyers for some amphibious missions increases tactical flexibility. This would also support standard naval tasks such as blockade, boardings, raids, and combat search and rescue. It would be interesting to ask combat-experienced senior officers from the SEALs, Marine Corps, and Army Special Forces to list services they would find useful from a warship close offshore. They would probably want a fast, shallow-draft ship, mine-resistant, with a very large helicopter deck, at least as much gunfire support capability as the DD 963 class, medical facilities, and an operations planning space with TV and radio channels. Intelligence collection capability is vital for amphibious operations, including both communications intelligence and microwave ESM. Blue-water missions such as ASW would be further down on their list.

The Destroyer of the Future

The *Spruance* class ended the concern of the 1960s about whether cruisers and destroyers had a place in the future fleet. Today the U.S. Navy is the most mobile force of strategic significance in any military service in the world. *Spruance*-class destroyers mount guns and passive surveillance systems that will keep them in front-line service for many years. Naval gunfire, cruise missiles, and Aegis are very powerful and almost surgically precise weapons. The Navy needs to make itself part of ground force operational planning. Helicopter decks, gunfire support, electronic surveillance, medical evacuation, coastal flank attacks, and forward air control can sustain and amplify the power of special forces and paratroops, not just the Marine Corps.

Wide coverage of the 50th anniversary of the June 1944 Normandy assault was a reminder that littoral nations often are blind beyond the ocean horizon. Naval surprise, blockade, and assaults are feasible. Further, naval forces are essential to enforce the terms to keep a war terminated, using techniques such as coastal surveillance and blockade, and providing platforms to reintroduce ground troops quickly if necessary. Considerations of diplomacy, economics, and security make it impossible to leave garrisons based near every hot spot.

The Navy will again face a block-obsolescence problem around 2010–20 when the *Spruance*-class destroyers will need to be replaced. Their smaller cousins, the frigates, seem destined to become extinct in the U.S. Navy. Over time *Spruance*-class destroyers will take on frigate duties such as escorting amphibious task groups. Before designing the destroyer of the future, we must list its duties:

- Participate as a unit in multiservice, multination combat operations, providing coordinated use of weapons and surveillance systems
- Detect, target, and destroy enemy units moving by sea, land, or air
- Support amphibious assaults, raids, rescues, and evacuations
- Discover and stop enemy minelaying, raids, espionage, or other clandestine moves
- Detect and suppress enemy deployment of ships, submarines, aircraft, coastal ground forces, missile launchers, radar and radio equipment, and other weapons
- Conduct blockades, including detection and inspection of suspect shipping, to diminish enemy war efforts and to enforce armistice terms
- Block criminal encroachments such as piracy, poaching, dumping of pollutants, illegal immigration, smuggling of drugs and weapons, and the expulsion of populations from racist or other repressive regimes

Weapons and performance requirements follow:

- Sensors and weapons to control an assigned coastal or open-ocean sector against attacks by ground forces, aircraft, mines, cruise and ballistic missiles, submarines, other ships, and chemical or radiological agents
- Long-range weapons such as cruise missiles to deploy from remote positions, and medium-range weapons to support amphibious operations
- Self-defense weapons, decoys, countermeasures, and survivability features for use against attack while on independent missions, including mine countermeasures
- Stealth characteristics and decoys to exploit stealth
- Ability to deploy, monitor, and maintain remote automated sensors such as drones and sonobuoy fields, and defend them against sabotage
- A large landing platform for vertical-takeoff aircraft, including one CH-53E helicopter or MV-22A tilt-rotor transport, plus a hangar for two helicopters the size of the SH-60B Seahawk, plus magazines
- A docking well to deploy small submarines and stealth boats, both either robotic or with

crews, and to transfer equipment between the destroyer and air-cushion landing craft (LCACs)
- Seakeeping for military tasks to be performed at necessary levels of efficiency in specified sea conditions

Ship specifications should include the following:

- Features for survivability, modernization, and rapid outfitting for emergent operations
- Accommodations for the crew and embarked units such as commandos, intelligence analysts, command staffs, and crews for helicopters, small boats, and submersibles
- Hull, propulsion, auxiliary, and logistical systems for seakeeping, survivability, endurance, reliability, sustained high speed, and other functions
- Ship's work procedures and security spaces to allow naval reservists and in-port contract laborers to augment the crew while maintaining accountability of stores and equipment
- Good appearance and admirable names to support peacetime naval presence
- Distributed independent, networked computer display systems
- Standardized multinational software so that allies can afford to build warships and weapons locally, using the common software for combat systems integration

Armament, performance, and characteristics of this ship are left to the reader. No single ship needs to perform all missions. With almost 60 Aegis-equipped *Ticonderoga*-class cruisers and *Arleigh Burke*–class destroyers in service or under contract and more planned, a replacement design for the DD 963 class might not need an area-defense guided missile system. It might be enough for a replacement class to complement Aegis-equipped ships. A low-cost, mine-resistant destroyer armed for gunfire support, surveillance, and helicopter operations would be useful on coastal missions because damage or even her loss would not destabilize an operation in the way that loss of an Aegis cruiser might. By 2010, Aegis-type systems may be relatively inexpensive and perhaps could arm affordable new destroyers anyway.

Surface warships are capable and economical vehicles for performing these diverse offensive missions, most of them simultaneously. That was not an accepted fact before the *Spruance* class. Revival of the surface warship is the legacy of the *Spruance*-class destroyers.

Appendix A

Documents on the Origin of DX/DXG

Record of Decision

December 16, 1966

DRAFT

MEMORANDUM FOR THE PRESIDENT

SUBJECT: Major Fleet Escort Forces

I have reviewed our Major Fleet Escort Ship programs for FY68–72. As a basis for the FY68 budget, I recommend:

1. $30 million for contract definition of a new type of guided missile escort ship, tentatively called the DXG. This funding will also provide for contract definition of a new type of ASW escort ship, tentatively called DX. This joint funding will permit investigation of the desirability of various degrees of commonality between the DX and the DXG.
2. Denial of the Navy's request to build two conven-tionally-powered guided missile destroyers (DDG) at a cost of $167 million.
3. Denial of the Navy's request to build a nuclear-powered guided missile frigate (DLGN) at a cost of $151 million.
4. Continuation of the guided missile escort ship modernization program with one ship at a cost of $22 million.

Force, procurement, and financial summaries follow. Subsequent sections deal with the main consider-ations underlying my recommendations and with specific decisions.

OSD Control CCS X-6909
ASD(SA) Cont Nr. 5-4581

I. FORCE LEVEL

The guided missile escort ship force which I recommended last year comprised 12 cruisers, 30 frigates [DLGs], 31 destroyers, and 6 destroyer es-corts, for a total of 79 ships. A force of this size would provide, simultaneously, 4 ships for each of our 15 attack carrier groups, 2 ships for each of 4 anti-submarine carrier groups which might operate inde-pendently in areas subject to enemy air attack, and 11 additional ships for other missions such as the protection of amphibious groups, underway replen-ishment groups, and convoys, or independent opera-tions in high threat areas. If some of the 15 attack carrier groups operate together, they would benefit from mutual defense, and could release some of their escorts for other missions. In addition, it should often be possible to reallocate escorts between the various tasks noted above, since it is unlikely that peak requirements would occur in all simultaneously.

In view of these general capabilities of the 79-ship force, and in light of the fact that current analyses have not demonstrated that a force consisting of this many ships would be inadequate, I do not believe at this time that a larger force is needed. On the other hand, the Chief of Naval Operations believes that we should have a force of about 100 guided missile escorts, and that further analysis will justify that position. I am encouraging continued analysis, and will re-examine the issue in the light of any new information that may be developed.

II. FORCE COMPOSITION

While I do not currently see a need for more than the approved level of 79 ships, and while the majority of the approved ships are relatively new and capable, there are others which ought to be replaced in the early 1970's in approximately the following order:

1. Two guided missile heavy cruisers (CAG [1–2]) equipped with the earliest surface-to-air missile (SAM) systems in the Navy. We had planned to modernize these at a total cost of $63 million, but the estimate has now risen to $99 million. In addition the annual operating cost of these ships is over $10 million each, they have no ASW capability, and their hulls are already 23 years old. However, they each have six 8″ guns which are valuable for fire support purposes.
2. Six guided missile light cruisers (CLG [3–8]) equipped with early SAM systems (3 with TERRIER and 3 with TALOS), modernization of which is impractical. They also have a high annual operating cost ($9 million for the TERRIER ships and $12 million for the TALOS ships), no significant ASW capability, and their hulls are already 20–22 years old. However, 4 of the 6 ships have fleet flagship facilities.
3. Six guided missile destroyer escorts (DEG [1]) which, though new and inexpensive to operate, have a limited SAM capability by reason of their small size. For example, they carry only 15 TARTAR missiles, as opposed to the 40 carried by a guided missile destroyer (DDG). In addition, their maximum speed of 27 knots, though satisfactory for other missions, is inadequate for operations with attack carrier task groups.
4. Three guided missile cruisers (CG [10]) which, though having the very capable SAM complement of 2 TALOS and 2 TARTAR systems each, as well as a full ASW capability, have the very high annual operating cost of $13.5 million. Though extensively modified in the early 1960s, they were originally built in World War II.

I believe that we should start now on a program to replace these ships with a new class which would not only be far more effective, [but] would also be relatively inexpensive to procure and to operate. This program would follow the general pattern pioneered by the FDL program and followed by the LHA program. Its primary features would be as follows:

1. To minimize procurement cost, the design and construction of these ships would be centralized in a single prime contractor. Such an arrange-

ment, together with appropriate production contract cost incentives, will encourage design of both ship and shipyard for producibility. . . .
2. To minimize the operating cost, we would establish contractual incentives that will encourage careful consideration of manning during the design process. Through use of automated equipment, design for easy maintenance, and reconsideration of current manning criteria, I am hopeful that the complement could be reduced to less than half that of comparable current ships. . . .
3. Possible integration of this new class of guided missile ships, tentatively called DXGs, with a new class of ASW escort ships, tentatively called DXs. As one alternative, the two classes could share identical hulls, and differ only in weapon systems. This would take maximum advantage of standardization, and of the economies of large scale series production. However, if the requirements of the DXG were to result in such an oversized DX as to more than offset the advantages of commonality, it might be preferable to build two separate classes of ship. Between these two extremes there are possibilities of having various degrees of commonality between the two classes, such as common bow and stern sections with midbodies specialized for the DX or DXG. However, determination of the best alternative is one of the main purposes of the Contract Definition phase.
4. Possible use of a modular design so that major components, such as the surface-to-air missile system, could be installed *en bloc*. This would be particularly desirable if the DX and DXG share a common hull. However, even with separate classes, the modular concept should result in more rapid construction since the systems would be more nearly ready to operate prior to installation. In future years, the ships could be modernized more rapidly and inexpensively, since the replacement systems could be designed to fit the standardized module and mate with the existing interface connections.

. . . For planning purposes I am assuming a construction program of 18 DXGs and 75 DXs, starting with FY69 and completing with FY74 funding. This will allow us to retire all the World War II cruisers[1] from the major fleet escort force and to transfer the 6

1. The two CAGs would be phased out of the guided missile escort ship force in FY73. Depending on the status of the LFS fire support ship program, it might be desirable to revert these ships temporarily to heavy cruiser (CA) status and man them only with enough personnel to provide 8″ gunfire support, or to hold them in the reserve fleet.

small DEGs to the ASW escort force. Should we subsequently decide not to retire the capable (though expensive) CGs, the last 3 DXGs would be deleted unless further analysis indicates a need for more than 79 major fleet escort ships. This program should also allow us to retire all the World War II destroyers from the ASW escort force.

III. MODERNIZATION

In 1964 I approved a program to modernize 27 major fleet escort ships at a total cost of $614 million. Though the improvements varied among the individual ships, depending [on] what equipment each already had aboard, the program comprised installation of the Naval Tactical Data System and associated weapon direction system, better search radar, replacement of early missile fire control systems with improved models, etc.

Since that time, the program has undergone some changes. More attention is being paid to standardized configurations, plans have been worked out to minimize the time spent in the shipyard while undergoing modernization, and there have been some slippages in the original schedule. . . .

IV. NUCLEAR FRIGATES

The primary use of frigates [DLGs] is to escort attack carriers. A potential advantage of nuclear power in a frigate is a reduction in the amount of ship's fuel which must be delivered to the force. This could increase the percentage of time spent "on the line", depending on how far and how frequently the force must retire for replenishment.

In situations involving little or no enemy air or submarine opposition, replenishment can be conducted on the line, and adding nuclear propulsion to a frigate would have no substantial effect on the percentage of time spent on the line. This is the case in our current Southeast Asian operations.

In situations involving heavy air attack against the enemy, the frequency of replenishment is determined by the consumption of ordnance rather than ships' fuel. In this case (which is the primary determinant of our attack carrier force levels), the potential advantage of nuclear propulsion in a frigate [DLG] is realized only if the force must retire extreme distances (e.g., more than 1000n.mi.) for replenishment.

In situations involving a war only at sea, aviation ordnance expenditures might be small, and replenishment frequency could be determined by ships' fuel requirements. Furthermore, in the absence of attacks on enemy bases, the air and submarine threat might force long transits to the replenishment group. The potential advantage of nuclear propulsion in this event is illustrated in the following table:

FRACTION OF TIME SPENT ON THE LINE

Replenishment Distance (n. mi.):	500	1000
Conventional Escorts (1 DLG, 3 DDG)	.80	.65
3 Conventional DDG and 1 DLGN	.82	.70
2 Conventional DDG and 2 DLGN	.84	.73
Nuclear Advantage 1 DLGN	2.5%	7.7%
[2] DLGN	5.0%	12.3%

The importance of this potential advantage depends on whether we would want to maintain attack carrier groups in high threat areas during a war only at sea. However, nuclear propulsion's additional advantage on conducting intense air operations after a long, high-speed transit would be useful also if (a) we wanted to set up such an operation quickly in order to reduce early losses of shipping, or (b) we wanted to move distant carriers rapidly in response to an out-break of war on land.

In addition to escorting aircraft carriers, frigates can also be used on independent operations. Examples are quarantines, shows of force, rescues, protection of minesweeping operations, prevention of aerial minelaying, and submarine trailing and hold-down operations. A future possibility is the trailing of missile submarines in preparation for boost-phase missile interception. . . . Nuclear propulsion might offer some (perhaps 10%) cost savings for independent operations in areas remote from normal logistic support facilities.

. . . For the same cost as a 4-ship escort group consisting of two nuclear frigates (DLGNs) and two conventional destroyers (DDGs), we could support a 5-ship group consisting of one conventional frigate (DLG) and four destroyers. A comparison of the number of sensors and weapons systems would favor the all-conventional group, as would consideration of the "decoy effect" of the additional ship. On the other hand, the numerical disadvantage of the group with the two nuclear escorts might be less important than its superiority in high-speed tactics and long hold-downs in anti-submarine warfare.

V. SPECIFIC DECISIONS

DX/DXG: I am recommending $30 million in the FY68 budget to cover the cost of Contract Definition for the DX and the DXG. I am also including funding in FY69–74 for construction of 18 DXGs and 75 DXs as shown below. . . . All of these figures must be considered highly tentative pending completion of the Contract Definition phase. I will carefully review the program at that time to ensure that the effective-ness of the ships, as well as the refined estimates of costs and schedules, merit final commitment to the DX/DXG concept.

DX/DXG CONSTRUCTION PROGRAM

(No. Ships/$Million)[a]

FY	69	70	71	72	73	74	Total
DXG	3/180	3/158	3/150	3/147	3/143	3/140	18/918
DX	12/279	12/245	12/234	13/246	13/240	13/235	75/1479
Total	15/459	15/403	15/384	16/393	16/383	16/375	93/2397

a. Contract cancellation allowance not included in this table. Costs are based on 94% cumulative average learning curve, FY67 DDG as basis for DXG, and DE-1052 Class as basis for DX.

DDG. Approval of the Navy's request to construct two conventionally powered DDGs in FY1968 would create a unique class of two ships and compound the fleet's logistics and training problems. Since the DXG program promises ships of more modern design at lower cost I am recommending denial of the Navy's request.

DLGN. . . . If nuclear propulsion is ever to become competitive, we must not allow our technical expertise to atrophy. I have authorized the construction of a nuclear carrier in FY67 and recommend construction of additional CVANs in FY69 and FY71. Thus, with the ENTERPRISE, we shall have four nuclear carriers. Since we now have three nuclear escorts, I am tentatively recommending construction of a fourth in FY72. Delivery of this ship will coincide with the delivery of the fourth CVAN. Final approval of this ship awaits receipt of a satisfactory justification for the Navy's nuclear frigate program. . . .

SECRET

Record of Decision   **January 5, 1968**

DRAFT

MEMORANDUM FOR THE PRESIDENT

SUBJECT: Escort Ship Forces
I have reviewed our escort ship programs for FY69–73. For the FY69 budget, I recommend:

1. Authorizing procurement of two nuclear-powered guided missile destroyers (DXGNs) as part of a five-ship DXGN program with a total cost of $605 million. . . . These ships, together with the four other nuclear escorts already funded or built, will give us two all-nuclear attack carrier groups. . . . The Navy also recommends an additional eleven new nuclear ships. . . . I do not believe existing analyses support the Navy recommendation, but I will reconsider the issue before the five-ship DXGN program is completed.

2. Continuing [with] Contract Definition for the DX, a conventionally-powered destroyer with anti-submarine warfare and fire support capability, and upon completion of Contract Definition for the DX, starting with Contract Definition for the DXG, a conventionally-powered destroyer with a combined ASW and anti-air warfare (AAW) capability. (I am including $20 million for DXG Contract Definition.)

3. Procuring five DXs as recommended by the Navy at a cost of $246 million (lead ship cost plus the actual cost of the first five ships on the learning curve); and planning to build forty DXs during FY 69–73 at a cost of $1,364 million, rather than 49 as recommended by the Navy. . . .

4. Planning to build 24 DXGs during FY 70–73 at a cost of $1,154 million (the first six in FY 70 as recommended by the Navy), rather than 33 as recommended by the Navy. . . .

OSD Control CCS X-0352
ASD(SA) Cont Nr. [illegible]

SECRET
DECLASSIFIED by OASD(PA&E), 29 Mar 1991

A Tour of a *Spruance*-Class Destroyer

Crew and visitors board a *Spruance*-class destroyer from one of three accesses: the amidships quarterdeck forward of the boats; the flight deck for helicopters; or the fantail (the Royal Navy's quarterdeck) from small boats and the low piers of many foreign ports.

The DD 963 design is similar to that of a cruiser in that there is a continuous deck almost entirely inside the ship. This is the DD 963's main deck. The only weather section of the main deck is the fantail. The hull, in U.S. Navy parlance, is the body of the ship below the main deck. Each DD 963 hull module is accessible only from watertight doors and hatches along the main deck. The long weather deck from the bow to the aft missile launcher is the 01 level (one level above the main deck) and is technically part of the ship's superstructure rather than the hull. The 01 level of the DD 963 is the uppermost strength member of ship girder and helps to hold the ship together in severe weather or after action damage. (The lowest strength member is the keel.) The structures above the 01 level are deckhouses.

Visitors and crew usually board the ship in port at the amidships quarterdeck. On either side of this area are underway-replenishment (UNREP) stanchions, often a 25 mm Bushmaster gun, and platforms for the ship's boats and their davits. Refueling stations are further forward and aft along the 01 level. Further forward, a tunnel leads to the fo'c's'le and bow.

Here are the Mark 45 5-inch gun Mount 51 and the forward missile module. The large module was sized for a twin-arm missile launcher and 24-round magazine (Mark 26 Mod 0). The ship is broad here so as to provide adequate hull strength around the large missile magazine and passageways around it on the main deck below. This makes for a large ship, larger still for buoyancy and stability for the added weight of hull steel. As built, the DD 963 class mounted an ASROC launcher here with its magazine beneath it.

Armored box launchers for Tomahawk cruise missiles are adjacent to the now-vacant ASROC launcher spot on seven destroyers. On the other 24 DD 963s a vertical-launch system (VLS) has replaced the ASROC launcher or is planned to replace it. Other systems on the bow are the anchor chains and windlasses (ground tackle), the forward UNREP station, and deck-edge direction-finding antennas for Outboard. The large SQS-53 sonar dome is beneath the fo'c's'le.

Returning to the quarterdeck, internal ladders lead to the forward deckhouse. The large combat information center (CIC) is on the 02 level above the computer center. The bridge (pilothouse) is on the 03 level above the CIC. The signal bridge is the 04 level. The 04 level is the site of the SRBOC chaff rockets, the SLQ-32 electronic countermeasures antennas, a WSC-3 ("whiskey 3") satellite communication antenna, and the forward Phalanx close-in weapons system gun (CIWS, pronounced "sea whiz"). Aft of it one deck down is the Harpoon antishipping missile battery. SRBOC, CIWS, and Harpoon are modular weapons that implement the original DD 963 "podularity" concept.

The masts position communications and radar antennas for coverage of sky and surface. Masts and topside equipment can withstand 130 mph gusts and distant nuclear air bursts. The foremast supports the SPQ-9 ("spook 9") and SPG-60 gunfire control radars, the SQR-4 LAMPS Mark III data link antenna, SPS-55 and SPS-64 surface search radars, several radio antennas, and on most DD 963s the direction-finding array for the Outboard intelligence system. The mainmast supports the SPS-40 long-range air search radar, the Mark 23 target acquisition system (TAS) radar for Sea Sparrow, more radio antennas, a TV camera for gunfire control, and navigation-related antennas. The unique USS *Hayler* mounts an SPS-49 radar instead of the SPS-40. The

DDG 993 and CG 47 classes have very different mast configurations.

The large deckhouse aft of the mainmast contains the helicopter hangar. Atop it are the aft CIWS mount, the port SLQ-32 countermeasures box on some of the class (*John Young* is one), the aft WSC-3 SATCOM (satellite communications) antenna, chaff rockets, and the remotely controlled Sea Sparrow director. There is no starboard-side hangar bay. Aft of the flight deck is the Sea Sparrow launcher, the number 3 SSGTG gas turbine intake and uptake, and a clear deck area to receive cargo during underway replenishment. The Sea Sparrow director's location clear of the aft stack and the launcher's location well aft of the helicopter deck give the Sea Sparrow system clear firing arcs forward of the beam, although obviously it cannot hit a target approaching from dead ahead.

The fantail is the location of Mount 52 (the aft Mark 45 5-inch gun), the aft vertical-replenishment station, several more communications antennas, and the Harpoon launchers on Aegis cruisers. Depending on the ship's configuration, the broad transom stern has two torpedo decoy (Nixie) towing ports, a circular port for the towed-array sonar (SQR-19 TACTAS), and two radio signal sensors (part of Outboard). The original DD 963 design included wells in the corners of the transom for chaff rockets, and USS *Spruance* as built had the starboard well. The Navy rejected chaff rockets in favor of the more effective RBOC system, and *Spruance*'s well was plated in. The Nixie torpedo decoy winch is now in this space.

Inside the ship, two parallel internal main deck passageways extend forward from the fantail to the bow adjacent to the shell plating along the ship's sides. Entering the ship from the fantail, one passes through a personnel CBR (chemical-biological-radiological) decontamination station to an athwartships passageway. A monorail can be installed on this deck to trundle torpedoes between the torpedo rooms on either side of the ship here, and to a loader that transfers torpedoes to the 01-level UNREP station and to the flight deck. In the DD 963 class the magazine space is often used for berthing, a gym, workshops, and provisions.

Proceeding forward, the next hull module is for crew berthing, and above it is the flight deck. The next three hull modules forward are engineering and supply spaces. The aft main engine room (MER) contains the gas turbines for the starboard shaft, and the forward MER contains the gas turbines for the port shaft. Aboard U.S. Navy ships, even numbers denote port-side equipment and spaces, and odd numbers denote those on the starboard side. Spaces and systems are further numbered sequentially from bow to stern. Combining these rules, it happens that MER number 1 has gas turbine modules 2A and 2B, while MER number 2 has 1A and 1B. Auxiliary equipment rooms are at the lowest level between the MERs. Above them are supply storerooms, the engineering central control station, and the galley. The next hull module forward of MER number 1 includes two berthing compartments and ship's offices. The forwardmost third of the ship includes the forward missile module, the Mount 51 5-inch gun, and the SQS-53 sonar.

Table B.1   DD 963 hull modules from stem to stern

Forepeak
SQS-53 sonar
Mount 51
ASROC magazine or VLS magazine
Forward damage control station (Repair 2)
Administrative offices and forward berthing
Main engine room #1
Galley and mess decks
Supply
Engineering central control station
Auxiliary rooms #1 and #2
Main engine room #2
Aft berthing
Sea Sparrow
Reserved space for missile magazine
Torpedo rooms (port and starboard)
SSGTG #3
Aft damage control station (Repair 3)
Mount 52
Towed-array sonar and torpedo decoy

Name Histories

Hull	Name	Ships so Named	Historical Era	Source of Ship's Name
DD 963	*Spruance*	1	World War II	Adm. Raymond Spruance, aircraft carrier battle group commander at Battle of Midway, 1942; 5th Fleet commander, 1943–45, at Battles of Tarawa, Saipan, the Philippine Sea, Iwo Jima, and Okinawa.
DD 964	*Paul F. Foster*	1	World War I era	Vice Adm. Paul Foster, won Medal of Honor at Vera Cruz, Mexico, 1914; father of a Supply Corps admiral.
DD 965	*Kinkaid*	1	World War II	Adm. Thomas C. Kinkaid, commander in chief, 7th Fleet at Battle of Leyte Gulf, 1944.
DD 966	*Hewitt*	1	World War II	Adm. H. Kent Hewitt, commander in chief, U.S. naval forces in Mediterranean, 1943–44.
DD 967	*Elliot*	1*	Vietnam	Lt. Cdr. Arthur (Jack) Elliot, riverine boat squadron commander; killed in action, 1968.
DD 968	*Arthur W. Radford*	1	Korean War, Cold War	Adm. Arthur Radford, commander in chief, Pacific, 1949–53; chairman, Joint Chiefs of Staff, 1953–57.
DD 969	*Peterson*	1*	Vietnam	Lt. Cdr. Carl Peterson, riverine boat squadron commander; killed in action, 1969.
DD 970	*Caron*	1	Vietnam	Hospital Corpsman Wayne Caron, killed in action serving with USMC, 1968; posthumous Medal of Honor.
DD 971	*David R. Ray*	1	Vietnam	Hospital Corpsman David Ray, killed in action serving with USMC, 1969; posthumous Medal of Honor.
DD 972	*Oldendorf*	1	World War II	Adm. Jesse Oldendorf, battleship division commander at Surigao Strait, 1944, and Lingayen Gulf, 1945.
DD 973	*John Young*	1	Revolution	Capt. John Young, commanded first USS *Saratoga;* ship lost at sea with all hands, 1781.
DD 974	*Comte de Grasse*	1	Revolution	Lt. Gen. Count François Joseph Paull de Grasse, marquis of Grasse-Tilly; commander in chief of French fleet ("naval army") at Battle of Yorktown, 1781.

* Denotes that a previous warship bore the name but commemorated a person or locale different from that commemorated by the current ship.

Hull	Name	Ships so Named	Historical Era	Source of Ship's Name
DD 975	*O'Brien*	5	Revolution	Capt. Jeremiah O'Brien, commanded sloop *Unity* and warded off British raid on Machias, Maine, 1775.
DD 976	*Merrill*	1*	World War II	Vice Adm. Aaron Merrill, surface force commander at Battle of Empress Augusta Bay, 1943.
DD 977	*Briscoe*	1*	World War II, Korean War	Adm. Robert Briscoe, surface task force commander at Guadalcanal, 1942–43, and commander in chief, 7th Fleet, off Korea, 1952.
DD 978	*Stump*	1	World War II, Cold War	Adm. Felix Stump, escort aircraft carrier division commander at Battle of Samar, 1944; commander in chief, Pacific, 1953–58.
DD 979	*Conolly*	1	World War II, Cold War	Adm. Richard L. Conolly, task group commander for assaults on North Africa, Sicily, the Marshalls, and the Marianas; commander in chief, Naval Forces Europe and Mideast, 1946–50.
DD 980	*Moosbrugger*	1	World War II	Cdr. Frederick Moosbrugger, destroyer squadron commander at Battle of Vella Gulf, 1943.
DD 981	*John Hancock*	6	Revolution	John Hancock, first signer of Declaration of Independence and effectively first Secretary of the Navy. Most predecessor ships were named *Hancock*.
DD 982	*Nicholson*	4	Revolution– 19th c.	Named for five members of Nicholson family. Two commanded original USS *Hornet* and *Constitution*; another commanded U.S. Navy force evacuating Egypt, 1882.
DD 983	*John Rodgers*	6	Revolution, Civil War, World War I era	Cdr. John Rodgers, commanded famous monitor *Weehawken*; captured Confederate ram *Atlanta*, 1863. Name also honors an ancestor and a descendant, both naval officers. Three ships were named *Rodgers*.
DD 984	*Leftwich*	1	Vietnam	Lt. Col. William Leftwich, USMC; reconnaissance battalion commander; killed in action, 1970.
DD 985	*Cushing*	5	Civil War	Cdr. William Cushing, heroic commander of spar torpedo boat that sank armored ram *Albemarle* in raid far behind Confederate lines, then escaped, 1864.
DD 986	*Harry W. Hill*	1	World War II, Cold War	Vice Adm. Harry Hill, amphibious task force commander under Spruance for assaults on Tarawa, Tinian, and Iwo Jima, 1943–45, and early Cold War strategist.
DD 987	*O'Bannon*	3	Barbary Wars	1st Lt. Presley O'Bannon, USMC, fought with distinction at Tripoli, 1803.
DD 988	*Thorn*	2	Barbary Wars	Lt. Jonathan Thorn, fought with distinction at Tripoli, 1803; killed in battle with Oregon Indians, 1811.
DD 989	*Deyo*	1	World War II	Vice Adm. Morton Deyo, naval gunfire support commander for assaults on Normandy, Cherbourg, and Okinawa, 1944–45.

* Denotes that a previous warship bore the name but commemorated a person or locale different from that commemorated by the current ship.

Hull	Name	Ships so Named	Historical Era	Source of Ship's Name
DD 990	*Ingersoll*	1*	World War II	Adm. Royal Ingersoll, commander in chief, Atlantic, during campaign against U-boats, 1942–44.
DD 991	*Fife*	1	World War II, Cold War	Adm. James Fife, commanded 7th Fleet submarines, 1942–45; deputy CNO, 1952–53.
DD 992	*Fletcher*	1*	World War II	Vice Adm. Frank Jack Fletcher, task force commander at Coral Sea, Midway, and Eastern Solomons, 1942. Ship also honors *Fletcher* (DD 445)–class destroyers, backbone of fleet destroyer operations in World War II and Korean War. DD 445 commemorated the admiral's uncle.
DDG 993	*Kidd*	2	World War II	Rear Adm. Isaac Kidd, Sr., battleship division commander; killed in action aboard USS *Arizona* at Pearl Harbor (bombing), 1941; posthumous Medal of Honor.
DDG 994	*Callaghan*	2	World War II	Rear Adm. Daniel Callaghan; killed in action aboard USS *San Francisco* at Guadalcanal (shellfire), 1942; posthumous Medal of Honor.
DDG 995	*Scott*	2	World War II	Rear Adm. Norman Scott, commanded U.S. forces at Battle of Cape Esperance; killed in action aboard USS *Atlanta* at Guadalcanal (shellfire, some friendly), 1942; posthumous Medal of Honor.
DDG 996	*Chandler*	1*	World War II	Rear Adm. Theodore Chandler; cruiser division commander; killed in action aboard USS *Louisville* at Lingayen Gulf, 1945 (kamikaze; ship survived).
DD 997	*Hayler*	1	World War II	Vice Adm. Robert Hayler, cruiser commander in eight battles from Solomons through Leyte Gulf, 1942–44.
CG 47	*Ticonderoga*	5	French & Indian War, Revolution	Fort Ticonderoga on Lake Champlain, the water invasion route from Quebec to New York and key to the North American continent. Captured five times by British, French, and American forces during mid-18th c.
CG 48	*Yorktown*	5	Revolution	Gen. Washington's victory over British Army, which Count de Grasse's fleet had blockaded by sea, 1781; ended British colonial rule of United States.
CG 49	*Vincennes*	4	Revolution	Lt. Col. George Rogers Clark defeated British unit at Vincennes, Indiana, gaining effective control of Old Northwest, 1779.
CG 50	*Valley Forge*	2	Revolution	Washington's army's winter quarters near Philadelphia, 1777.
CG 51	*Thomas S. Gates*	1	Cold War	Thomas S. Gates, Secretary of the Navy, 1957–59, and Secretary of Defense, 1959–61. His feisty widow demanded a warship namesake.

* Denotes that a previous warship bore the name but commemorated a person or locale different from that commemorated by the current ship.

Hull	Name	Ships so Named	Historical Era	Source of Ship's Name
CG 52	*Bunker Hill*	2	Revolution	British troops under Gen. Howe drove rebels from fortifications on hills overlooking Boston, 1775, but suffered heavy casualties and were too weakened and demoralized to exploit their tactical victory.
CG 53	*Mobile Bay*	1	Civil War	Union Navy force under Adm. Farragut captured Mobile, Alabama, 1864. Farragut's famous "Damn the torpedoes . . . full speed!" order at this battle reflected his knowledge that his naval raiding parties had neutralized the "torpedoes" (moored mines) before the attack, and was not mere bravado.
CG 54	*Antietam*	3	Civil War	Union Army repulsed Gen. Lee's advance along Antietam Creek near Sharpsburg, Maryland, 1862.
CG 55	*Leyte Gulf*	1	World War II	Invasion to reconquer Philippines, and series of naval actions, most notably off Samar and at Surigao Strait, to defend the beachhead against Japanese naval counterattacks, 1944. Largest naval battle in history.
CG 56	*San Jacinto*	3	Texas Independence	American force under Sam Houston defeated Mexican Army under Santa Anna, 1836.
CG 57	*Lake Champlain*	3	Revolution, War of 1812	American inland naval fleets under Arnold (1776) and MacDonough (1814) defeated British attempts to invade New York from Quebec via Lake Champlain.
CG 58	*Philippine Sea*	2	World War II	Naval battle under Adms. Spruance and Mitscher to support the invasion of Saipan, June 1944; fleet air defense battle was the "Marianas Turkey Shoot."
CG 59	*Princeton*	6	Revolution	After victory at Trenton, Gen. Washington counterattacked British reinforcements, causing British to retreat from western New Jersey, 1777.
CG 60	*Normandy*	1	World War II	Gen. Eisenhower's Neptune-Overlord assault on France, D-Day, June 6, 1944, and subsequent breakout from the beachhead through German defensive lines (Operation Cobra).
CG 61	*Monterey*	4	Mexican War	Navy-Marine landing party captured Presidio of Monterey, Calif., 1846. OSD announcement of CG 61's name erroneously cited the unrelated siege of Monterrey, Mexico.
CG 62	*Chancellorsville*	1	Civil War	Confederate tactical triumph under Gens. Robert E. Lee and Stonewall Jackson, routing much larger Union Army near Fredericksburg, Virginia, 1863.
CG 63	*Cowpens*	2	Revolution	American force drove off British unit at a corral near Spartanburg, South Carolina, 1781. First of a series of actions that led the British to concentrate at Yorktown to await reinforcement by sea.

Hull	Name	Ships so Named	Historical Era	Source of Ship's Name
CG 64	*Gettysburg*	3	Civil War	Union Army defeated Confederate attack under Gen. Lee at Gettysburg, Pennsylvania, 1863. Last large Confederate offensive and the largest battle ever fought in the Americas.
CG 65	*Chosin*	1	Korean War	1st Marine Division breakout through Chinese lines to escape from surrounded position at Chosin Reservoir, North Korea, 1950.
CG 66	*Hue City*	1	Vietnam	Successful Marine Corps defense of Hue (pronounced "hway"), South Vietnam, against North Vietnamese Army attack during Tet Offensive, 1968.
CG 67	*Shiloh*	2	Civil War	Union Army under Gen. Grant withstood fierce Confederate attacks around Shiloh Church, near Savannah, Tennessee, 1862. This ultimately secured the Mississippi River for the Union south almost to Vicksburg.
CG 68	*Anzio*	2	World War II	Amphibious assault to outflank German defense lines around Rome, 1944. Assault succeeded but ground forces never broke out from the beachhead.
CG 69	*Vicksburg*	4	Civil War	Union force under Adm. Porter and Gen. Grant captured fortress of Vicksburg, Mississippi, by siege, 1862–63, fatally splitting the Confederacy. Town of Vicksburg refused to celebrate July 4, the day of surrender, as a holiday for the next 80 years.
CG 70	*Lake Erie*	2	War of 1812	Scratch-built fleet under Comm. Oliver Hazard Perry, age 27, defeated British fleet on Lake Erie between Michigan and Ontario, 1813.
CG 71	*Cape St. George*	1	World War II	Five destroyers under Capt. Arleigh Burke attacked five Japanese destroyers ferrying reinforcements to the Solomons, 1943; sank three and stopped reinforcement.
CG 72	*Vella Gulf*	2	World War II	Six destroyers under Cdr. Moosbrugger attacked four Japanese destroyers ferrying reinforcements to the Solomons, 1943; sank three and stopped reinforcement.
CG 73	*Port Royal*	2	Revolution, Civil War	Successful amphibious assault by Union Navy on the Sea Islands, South Carolina, 1861, to seize a support base for the blockade. The tune of a Civil War song about this operation, "Kingdom Coming," is still popular.

* Denotes that a previous warship bore the name but commemorated a person or locale different from that commemorated by the current ship.

Appendix D

Specifications

Spruance (DD 963) Class (fully modernized configuration)

Number	Name	FY	Launched	Commissioned	Fleet	Home port	Configuration
DD 963	*Spruance*	70	10 Nov 73	20 Sep 75	A	Mayport FL	OVFSR1
DD 964	*Paul F. Foster*	70	23 Feb 74	21 Feb 75	P	Long Beach CA	OVFSR1
DD 965	*Kinkaid*	70	25 May 74	10 Jul 76	P	San Diego CA	OVFSR1
DD 966	*Hewitt*	71	24 Aug 74	25 Sep 76	P	Yokosuka JA	PVFSR1
DD 967	*Elliot*	71	19 Dec 74	22 Jan 77	P	San Diego CA	OVFSR1
DD 968	*Arthur W. Radford*	71	1 Mar 75	16 Apr 77	A	Norfolk VA	OVFSR1
DD 969	*Peterson*	71	21 Jun 75	9 Jul 77	A	Norfolk VA	OVFSR2
DD 970	*Caron*	71	24 Jun 75	1 Oct 77	A	Norfolk VA	OVFSR1
DD 971	*David R. Ray*	71	23 Aug 75	19 Nov 77	P	Long Beach CA	PVFSR1
DD 972	*Oldendorf*	71	21 Oct 75	4 Mar 78	P	San Diego CA	OVFSR2
DD 973	*John Young*	72	7 Feb 76	20 May 78	P	San Diego CA	PVFSR2
DD 974	*Comte de Grasse*	72	26 Mar 76	5 Aug 78	A	Norfolk VA	PTFSR2
DD 975	*O'Brien*	72	8 Jul 76	3 Dec 77	P	Yokosuka JA	PVFSR1
DD 976	*Merrill*	72	1 Sep 76	11 Mar 78	P	San Diego CA	OTFSR2
DD 977	*Briscoe*	72	18 Dec 76	3 Jun 78	A	Norfolk VA	PVFSR2
DD 978	*Stump*	72	21 Mar 77	19 Aug 78	A	Norfolk VA	PVFSR2
DD 979	*Conolly*	74	3 Jun 77	14 Oct 78	A	Norfolk VA	PTFSR2
DD 980	*Moosbrugger*	74	23 Jul 77	16 Dec 78	A	Charleston SC	PVFSR1
DD 981	*John Hancock*	74	28 Sep 77	10 Mar 79	A	Mayport FL	PVFSR1
DD 982	*Nicholson*	74	29 Nov 77	12 May 79	A	Charleston SC	PVFSR2
DD 983	*John Rodgers*	74	25 Feb 78	14 Jul 79	A	Charleston SC	OTFSR2
DD 984	*Leftwich*	74	8 Apr 78	25 Aug 79	A	Pearl Harbor HI	PTFSR1
DD 985	*Cushing*	74	17 Jun 78	20 Oct 79	P	Pearl Harbor HI	OVFSR2
DD 986	*Harry W. Hill*	75	10 Aug 78	17 Nov 79	P	San Diego CA	PF
DD 987	*O'Bannon*	75	25 Sep 78	15 Dec 79	A	Charleston SC	PVFSR2
DD 988	*Thorn*	75	14 Nov 78	16 Feb 80	A	Charleston SC	OVFSR1
DD 989	*Deyo*	75	20 Jan 79	22 Mar 80	A	Charleston SC	OTFR1
DD 990	*Ingersoll*	75	10 Mar 79	12 Apr 80	P	Pearl Harbor HI	OTFSR2
DD 991	*Fife*	75	1 May 79	31 May 80	P	Yokosuka JA	OVFSR1
DD 992	*Fletcher*	75	16 Jun 79	12 Jul 80	P	Pearl Harbor HI	OVFSR2
DD 997	*Hayler*	78	27 Mar 82	5 Mar 83	A	Norfolk VA	PVFSR2

All ships were built by Litton Industries Ingalls Shipbuilding Division, Pascagoula, MS.

Displacement: 9,250 tons full load; 7,410 tons light ship (approx.)
Length: 563 ft 4 in overall; 529 ft waterline
Beam: 55 ft
Draft: 22 ft (hull); 32 ft (navigation)
Propulsion: 4 GE LM-2500 gas turbines (80,000 HP, 64 megawatts); arranged in 2 independent COGAG plants each with 2 LM-2500's driving a locked-train double-reduction gear to a 5-bladed controllable-reversible pitch propeller
Electrical: 3 Allison 501K gas turbines, each driving a 2,000kw 60Hz alternator; 3 60Hz–400Hz static frequency converters
Speed: 32.5 knots maximum; 6,000nm range at 20 knots. Fuel: 1,534 tons F-76; 72 tons JP-5 (F-44)
Complement: 24 officers, 330 enlisted (helicopter detachment adds 9 officers and approx 30 enlisted)

Armament:
Surveillance: SSQ-108 Outboard radio direction finding (16 ships)
SPS-40B 2-D air seach radar (SPS-49 in DD 997)
SPS-55 surface search radar
SPS-64 or commercial navigation radar
Naval Tactical Data System, Link 11 data link
Strike: 1 61-cell Mk 41 vertical-launch system for Tomahawk, Standard, and VL-ASROC missiles or 2 4-cell Mk 44 armored box launchers for Tomahawk (DD 974, 976, 979, 983, 984, 989, 990) 8 Harpoon (2 fixed quadruple canister launchers); SWG-1A missile fire control for Harpoon; SWG-2 (AB1) or SWG-3 (VLS) for Tomahawk
Gunnery: 2 5-inch (127mm) 54-cal dual-purpose Mk 45 guns, 18–23 rounds/min
Mk 86 gunfire control: 1 SPQ-9A search radar, 1 SPG-60 height-finding radar, 2 remote optic sights
2 25mm Mk 88 automatic cannon (Bushmaster), up to 200 rounds/min.; aimed by on-mount gunsights
4 .50-inch (12.7mm) machine guns
ASW: 6 12.75-inch (324mm) Mk 32 torpedo tubes in 2 triple mounts, fired through hull doors
Vertical-launch antisubmarine rockets (VL-ASROC) in VLS-equipped ships
Mk 116 underwater fire control: SQQ-89 ASW display system, SQQ-28 sonobuoy processor
SQR-19 tactical towed-array sonar (TACTAS)
SQS-53B active/passive hull-mounted sonar
LAMPS Mk III with 1–2 SH-60B Seahawk helicopters, SRQ-4 data link (Hawk Link)
AAW: 1 8-cell Mk 29 NATO Sea Sparrow launcher; Mk 23 target acquisition system, Mk 91 fire control
2 20mm Mk 15 close-in weapons systems (Vulcan/Phalanx); on-mount Mk 90 radar fire control
ECM/Decoys: SLQ-32(V)3 missile radar warning with active jamming
24 super-rapid-bloom offboard chaff/flare mortars (4 6-barrel Mk 36 launchers)
SLQ-49 floating radar reflector buoys
Nixie SLQ-25 towed antitorpedo decoys

Kidd (DDG 993) Class (New Threat Upgrade configuration)

Number	Name	FY	Launched	Commissioned	Fleet	Home port	Configuration
DDG 993	*Kidd*	79S	11 Aug 79	27 Jul 81	A	Norfolk VA	NKFH
DDG 994	*Callaghan*	79S	1 Dec 79	29 Aug 81	P	San Diego CA	NKFH
DDG-995	*Scott*	79S	1 Mar 80	24 Oct 81	A	Norfolk VA	NKFH
DDG-996	*Chandler*	79S	24 May 80	13 Mar 82	P	San Diego CA	NKFH

All ships were built by Litton Industries Ingalls Shipbuilding Division, Pascagoula, MS.

Displacement: 9,950 tons full load
Length: 563 ft 4 in overall; 529 ft waterline; Beam: 55 ft; Draft: 23 ft (hull); 33 ft (navigation)
Propulsion: 4 GE LM-2500 gas turbines (80,000 HP, 64MW); arranged in 2 independent COGAG plants each with 2 LM-2500's driving a locked-train double-reduction gear to a 5-bladed controllable-reversible pitch propeller
Electrical: 3 Allison 501K gas turbines, each driving a 2,500kw 60Hz alternator; 3 60Hz–400Hz static frequency converters
Speed: 30+ knots maximum; 6,000nm range at 20 knots
Complement: 28 officers, 320 enlisted
Armament:
Surveillance: SPS-48E 3-D air seach radar
SPS-49 2-D air seach radar
SPS-55 surface search radar
SPS-64 navigation radar
Naval Tactical Data System, Link 4A and Link 11 data links
Strike: 8 Harpoon (2 fixed quadruple canister launchers); SWG-1A missile fire control
Gunnery: 2 5-inch (127mm) 54-cal dual-purpose Mk 45 guns, 18–23 rounds/min
Mk 86 gunfire control: 1 SPQ-9A search radar, 1 SPG-60 height-finding radar, 2 remote optic sights
4 .50 inch (12.7mm) machine guns
ASW: 6 12.75-inch (324mm) Mk 32 torpedo tubes in 2 triple mounts, fired through hull doors
Mk 116 underwater fire control system
SQS-53A active/passive hull-mounted sonar
AAW: 2 Mk 26 guided missile launchers; Mod 0 forward (24 Standard SM-2[MR] missiles), Mod 1 aft (44)
1 Mk 74 Tartar-D missile fire control, 2 SPG-51D tracker/illuminator radars; illuminator on SPG-60
4 SYR-1 missile telemetry receivers
2 20mm Mk 15 close-in weapons systems (Vulcan/Phalanx); on-mount Mk 90 radar fire control
ECM/Decoys: SLQ-32(V)2 missile radar warning
24 super-rapid-bloom offboard chaff/flare mortars (4 6-barrel Mk 36 launchers)
Nixie SLQ-25 towed antitorpedo decoys

Number	Name	Baseline	FY	Launched	Commissioned	Fleet	Home port	Configuration
CG 47	*Ticonderoga*	1	78	25 Apr 81	22 Jan 83	A	Norfolk VA	AKFH
CG 48	*Yorktown*	1	80	17 Jan 83	4 Jul 84	A	Norfolk VA	AKFH
CG 49	*Vincennes*	1	81	14 Jan 84	6 Jul 85	P	San Diego CA	AKFR2
CG 50	*Valley Forge*	1	81	23 Jun 84	11 Jan 86	P	San Diego CA	AKFR2
CG 51	*Thomas S. Gates*	1	82	14 Dec 85	22 Aug 87	A	Norfolk VA	AKFR2
CG 52	*Bunker Hill*	2	82	11 Mar 85	20 Sep 86	P	Yokosuka JA	AWFR2
CG 53	*Mobile Bay*	2	82	22 Aug 85	21 Feb 87	P	Yokosuka JA	AWFR2
CG 54	*Antietam*	2	83	14 Feb 86	6 Jun 87	P	Long Beach CA	AWFSR2
CG 55	*Leyte Gulf*	2	83	20 Jun 86	26 Sep 87	A	Mayport FL	AWFSR2
CG 56	*San Jacinto*	2	83	14 Jun 86	23 Jan 88	A	Norfolk VA	AWFSR2
CG 57	*Lake Champlain*	2	84	3 Apr 87	12 Aug 88	P	San Diego CA	AWFSR2
CG 58	*Philippine Sea*	2	84	12 Jul 87	18 Mar 89	A	Mayport FL	AWFSR2
CG 59	*Princeton*	3	84	25 Sep 87	11 Feb 89	P	San Diego CA	AWFSR2
CG 60	*Normandy*	3	85	19 Mar 88	9 Dec 89	A	New York NY	AWFSR2
CG 61	*Monterey*	3	85	23 Oct 88	16 Jun 90	A	Mayport FL	AWFSR2
CG 62	*Chancellorsville*	3	85	15 Jul 88	4 Nov 89	P	San Diego CA	AWFSR2
CG 63	*Cowpens*	3	86	11 Mar 89	9 Mar 91	P	San Diego CA	AWFSR2
CG 64	*Gettysburg*	3	86	22 Jul 89	22 Jun 91	A	Mayport FL	AWFSR2
CG 65	*Chosin*	3	86	1 Sep 89	12 Jan 91	P	Pearl Harbor HI	AWFSR2
CG 66	*Hue City*	3	87	1 Jun 90	14 Sep 91	A	Mayport FL	AWFSR2
CG 67	*Shiloh*	3	87	8 Sep 90	18 Jul 92	P	San Diego CA	AWFSR2
CG 68	*Anzio*	4	87	2 Nov 90	2 May 92	A	Norfolk VA	AWFSR2
CG 69	*Vicksburg*	4	88	2 Aug 91	14 Nov 92	A	Mayport FL	AWFSR2
CG 70	*Lake Erie*	4	88	13 Jul 91	24 Jul 93	P	Pearl Harbor HI	AWFSR2
CG 71	*Cape St. George*	4	88	10 Jan 92	12 Jun 93	A	Norfolk VA	AWFSR2
CG 72	*Vella Gulf*	4	88	13 Jun 92	18 Sep 93	P	Norfolk VA	AWFSR2
CG 73	*Port Royal*	4	88	20 Nov 92	9 Jul 94	P	San Diego CA	AWFSR2

Builders: CG 51, 58, 60, 61, 63, 64, 67, and 70—Bath Iron Works, Bath, ME. All others—Litton Industries Ingalls Shipbuilding Division, Pascagoula, MS.

Displacement: 9,613 tons full load; 7,041 tons light ship (Baseline 1 configuration)
Length: 567 ft overall; 533 ft 8 in waterline
Beam: 55 ft
Draft: 24.5 ft (hull); 32 ft (navigation)
Propulsion: 4 GE LM-2500 gas turbines (100,000 HP, 80MW); arranged in 2 independent COGAG plants each with 2 LM-2500's driving a locked-train double-reduction gear to a 5-bladed controllable-reversible pitch propeller
Electrical: 3 Allison 501K gas turbines, each driving a 2,500kw 60Hz alternator; 3 60Hz–400Hz static frequency converters
Speed: 30+ knots maximum; 6,000nm range at 20 knots
Complement: 29 officers, 350 enlisted
Armament (Baseline 1 configuration):
    Surveillance: SPY-1A Aegis 3-D multifunction air/surface seach/fire control radar
SPS-49 2-D air seach radar
SPS-55 surface search radar
SPS-64 navigation radar
Naval Tactical Data System, Link 4A and Link 11 data links
    Strike: 8 Harpoon (2 fixed quadruple canister launchers); SWG-1A missile fire control
    Gunnery: 2 5-inch (127mm) 54-cal Mk 45 guns, 18–23 rounds/min, 2 25mm Mk 88
Mk 86 gunfire control system with 1 SPQ-9A search radar, 1 remote optic sight
    ASW: 6 12.75-inch (324mm) Mk 32 torpedo tubes in 2 triple mounts, fired through hull doors
Mk 116 underwater fire control
SQS-53A active/passive hull-mounted sonar
    AAW: 2 Mk 26 guided missile launchers for Standard SM-2(MR) missiles, Mod 1forward and aft (44 missiles each)
1 Aegis Mk 7 missile fire control system, 4 SPG-62 illuminator radar sets
4 SYR-1 missile telemetry receivers
2 20mm Mk 15 close-in weapons systems (Vulcan/Phalanx); on-mount Mk 90 radar fire control
    ECM/Decoys: SLQ-32(V)3 missile radar warning with active jamming
24 super-rapid-bloom offboard chaff/flare mortars (4 6-barrel Mk 36 launchers)
Nixie SLQ-25 towed antitorpedo decoys

Armament (Baseline 3–4 configurations; Baseline 2 similar—see text):
  Surveillance: SPY-1B Aegis 3-D multifunction air/surface search/fire control radar
SPS-49 2-D air search radar
SPS-55 surface search radar
SPS-64 navigation radar
Naval Tactical Data System, Link 4A and Link 11 data links
  Strike: 2 61-cell Mk 41 vertical-launch systems for Tomahawk, Standard, and VL-ASROC missiles
8 Harpoon (2 fixed quadruple canister launchers); SWG-1A missile fire control
  Gunnery: 2 5-inch (127mm) 54-cal Mk 45 guns, 18–23 rounds/min; 2 25mm Mk 88
Mk 86 gunfire control: 1 SPQ-9A search radar, 1 remote optic sight
  ASW: 6 12.75-inch (324mm) Mk 32 torpedo tubes in 2 triple mounts, fired through hull doors
Vertical-launch antisubmarine rockets (VL-ASROC) in VLS-equipped ships
Mk 116 underwater fire control, SQQ-89 ASW display system, SQQ-28 sonobuoy processor
SQR-19 tactical towed-array sonar (TACTAS)
SQS-53B active/passive hull-mounted sonar
LAMPS Mk III with 2 SH-60B Seahawk helicopters, SRQ-4 data link (Hawk Link)
  AAW: Standard SM-2(MR) fired from vertical launch systems
1 Aegis Mk 7 missile fire control system, 4 SPG-62 illuminator radar sets
2 20mm Mk 15 close-in weapons systems (Vulcan/Phalanx); on-mount Mk 90 radar fire control
  ECM/Decoys: SLQ-32(V)3 missile radar warning with active jamming
24 super-rapid-bloom offboard chaff/flare mortars (4 6-barrel Mk 36 launchers)
Nixie SLQ-25 towed antitorpedo decoys

Configuration codes:
Sensor: A   Aegis
            N   Tartar-D New Threat Upgrade
            O   Outboard
            P   Patrol capability

Launchers: K   Mk 26
            T   Tomahawk armored box launcher
            V   Vertical launch system
            W   2 vertical launch systems

Gunnery: F   5″ guns for fire support

ASW: H   Helicopter deck and hangar
            R1   RAST for 1 SH-60B Seahawk
            R2   RASTs for 2 SH-60B Seahawks
            S   SQQ-89 ASW system

Notes

Introduction

1. J. W. Devanney, "The DX Competition," U.S. Naval Institute *Proceedings* (hereafter *USNIP*), Aug. 1975.
2. Capt. Robert H. Smith, "A United States Navy for the Future," *USNIP*, Mar. 1971.
3. Robert F. Coulam, *Illusions of Choice: The F-111 and the Problem of Weapons Acquisition Reform* (Princeton: Princeton University Press, 1977), 251.
4. Thomas C. Hone, *Power and Change: The Administrative History of the Office of the Chief of Naval Operations, 1946–86* (Washington, D.C.: Naval Historical Center, 1989), ix.
5. Lawrence J. Korb, *The Fall and Rise of the Pentagon: American Defense Policies in the 1970s* (Westport, Conn.: Greenwood, 1979), chaps. 1–3.
6. Hone, *Power and Change*, 79.
7. Robert McFarlane, letter to *The Economist*, July 6, 1991; Peter Schweizer, *Victory* (New York: Atlantic Monthly Press, 1994).

Chapter 1: Destroyers during the Cold War, 1948–1962

1. Vice Adm. Forrest Sherman, presentation to President Truman, Jan. 14, 1947; rpt. in Michael A. Palmer, *Origins of the Maritime Strategy* (Washington, D.C.: Naval Historical Center, 1988), app. A.
2. Norman Friedman, *U.S. Destroyers: An Illustrated Design History* (Annapolis: Naval Institute Press, 1982), chaps. 9–10.
3. Ibid., chaps. 11–12.
4. David Alan Rosenberg, "U.S. Nuclear War Planning, 1945–1960," in *Strategic Nuclear Targeting*, ed. Desmond Ball and Jeffrey Richelson (Ithaca: Cornell University Press, 1986), 39.
5. Robert J. Donovan, *Tumultuous Years: The Presidency of Harry S Truman, 1949–1953* (New York: Norton, 1982), 105–13.
6. Joel J. Sokolsky, *Seapower in the Nuclear Age: The United States Navy and NATO, 1949–1980* (Annapolis: Naval Institute Press, 1991), 24, 58.
7. Ibid., 58.
8. Naval architect William F. Gibbs, Sept. 1951, quoted in Friedman, *U.S. Destroyers*, 374; *U.S. Destroyers*, 245–49, 270–75.
9. Vice Adm. Raymond Peet, USN (Ret.), interview, Apr. 1993.
10. Cdr. Malcolm Cagle and Cdr. Frank Manson, *The Sea War in Korea* (Annapolis: Naval Institute Press, 1957), chap. 9.
11. Richard Hallion, *The Naval Air War in Korea* (New York: Kensington, 1988), 181–85, 202.
12. Cagle and Manson, *Sea War in Korea*, app. 9.
13. AP dispatch in San Diego *Union-Tribune*, May 24, 1992.
14. Chuck Hansen, *U.S. Nuclear Weapons: The Secret History* (New York: Orion, 1988), 65.
15. Norman Friedman, *U.S. Aircraft Carriers: An Illustrated Design History* (Annapolis: Naval Institute Press, 1983), 255–57.
16. Rosenberg, "U.S. Nuclear War Planning," 43–44, 50–51.
17. Robert W. Love, *History of the U.S. Navy*, vol. 2, *1942–1991* (Harrisburg, Pa.: Stackpole, 1992), 381.
18. Peet interview, Apr. 10, 1993. Capt. William Bainbridge commanded USS *Constitution* when she sank HMS *Java* during the War of 1812. Other than that, he was something of a reprobate.
19. Friedman, *U.S. Destroyers*, 220–21.
20. Paul Nitze, "Running the Navy," *USNIP*, Sept. 1989, 76.
21. Correspondence from Cdr. Douglas Leathem, USNR (Ret.), chief engineer of USS *O'Hare* (DDR 889) during the NTDS tests and the Cuban missile crisis.
22. Rosenberg, "U.S. Nuclear War Planning," 49.
23. Friedman, *U.S. Destroyers*, 168–72, 240–41; Rear Adm. Wayne Meyer, USN, "The Combat Systems of Surface Warships," *USNIP*, May 1977, 122.
24. Capt. Eli M. Vinock, USN, "FRAM Fixes the Fleet," *USNIP*, Aug. 1984.
25. Lt. Edward A. Morgan, USNR, "The DASH Weapons System," *USNIP*, Jan. 1963.
26. Rich Worth, letter to *USNIP*, Dec. 1984, 103–8. A DASH is on display in the Boeing museum near Seattle.
27. Friedman, *U.S. Destroyers*, 287.
28. Rear Adm. Thomas R. Weschler, USN, "The DX/DXG Program," *Naval Engineers Journal*, Dec. 1967.
29. Friedman, *U.S. Destroyers*, 351.
30. Ibid., 350–53.
31. The Navy classified all of these ships as frigates at one time or another. I avoid the term *frigate* because of this confusion.

32. Philip Sims, "We Make Our Ships Ugly by Design," *USNIP*, Nov. 1974.

33. Paul Nitze and Leonard Sullivan, *Securing the Seas* (Boulder, Colo.: Westview, 1979), 44.

34. John Jordan, *Soviet Submarines* (London: Arms & Armour, 1989), 55–70.

35. Ibid., 53–55, 64–66, 77–84; John Jordan, *Soviet Warships* (London: Arms & Armour, 1983), 10–13.

36. Capt. Bryce Inman, USN (Ret.), "From Typhon to Aegis—The Issues and Their Resolution," *Naval Engineers Journal*, May 1988.

37. Peter Tsouras, "Soviet Naval Tradition," in *The Soviet Navy: Strengths and Liabilities*, ed. Bruce W. Watson and Susan M. Watson (Boulder, Colo.: Westview, 1986), 19–23.

38. Charles C. Petersen, "About Face in Soviet Tactics," *USNIP*, Aug. 1983.

39. Richard F. Cross III, "Destroyers, 1971," *USNIP*, May 1971, 249.

40. Michael Beschloss, *The Crisis Years* (New York: Harper Collins, 1991), 494.

41. Leathem correspondence.

42. Norman Polmar and Lt. Cdr. Jurrien Noot, Royal Netherlands Navy, *Submarines of the Russian and Soviet Navies, 1718–1990* (Annapolis: Naval Institute Press, 1990), 171–73.

43. Robert F. Kennedy, *Thirteen Days: A Memoir of the Cuban Missile Crisis* (1967; rpt. New York: Signet, 1969), 69.

44. Leathem correspondence.

45. Adm. Isaac C. Kidd, Jr., USN (Ret.), "The Surface Fleet," in *The U.S. Navy: The View from the Mid-1980s*, ed. James L. George (Boulder, Colo.: Westview, 1985), 82–83. USS *Kidd* (DDG 993) is the namesake of Adm. Kidd's father.

46. Palmer, *Origins of the Maritime Strategy*, 80–82.

47. Sokolsky, *Seapower in the Nuclear Age*, 58–62, 73–75.

Chapter 2: Destroyers on Trial: McNamara, Rickover, and Vietnam

1. Hone, *Power and Change*, 45–46.

2. Cdr. Bruce W. Watson, USN, *Red Navy at Sea: Soviet Naval Operations on the High Seas, 1956–1980* (Boulder, Colo.: Westview, 1982), xix.

3. Quoted in Floyd Kennedy, "David Lamar McDonald," in *The Chiefs of Naval Operations*, ed. Robert W. Love (Annapolis: Naval Institute Press, 1980), 344.

4. Stephen E. Ambrose, *Eisenhower the President* (New York: Simon and Schuster, 1984), 456, 496.

5. James M. Roherty, *Decisions of Robert S. McNamara* (Coral Gables, Fla.: University of Miami Press, 1970).

6. Adm. Anderson directed the quarantine during the Cuban missile crisis as the representative of the Joint Chiefs of Staff. Department of Defense Annual Report, FY 1963, 5.

7. Roherty, *Decisions of Robert S. McNamara*, chap. 4 and app.

8. Hone, *Power and Change*, 65–66.

9. Paul R. Schratz, "Paul Henry Nitze," in *American Secretaries of the Navy*, ed. Paolo Coletta (Annapolis: Naval Institute Press, 1980), 949.

10. Kennedy, *Thirteen Days*, 117–20. President Kennedy was greatly impressed by Barbara Tuchman's *The Guns of August* (New York: Macmillan, 1962). Mrs. Tuchman described how World War I broke out under the pressure of inflexible military plans that then fell apart from shortsighted tactical decisions.

11. Anti-McNamara naval histories include Capt. Paul Ryan, USN (Ret.), *First Line of Defense*, and Love, *History of the U.S. Navy*, vol. 2, *1942–1991*. Typically, both are unreliable. For example, Love attributes cancellation of the obscure Missileer fighter to McNamara and charges him with forcing the Navy to convert old *Essex*-class aircraft carriers to assault ships. In fact, both those decisions were made before Secretary McNamara took office. Even if they were accurate, such criticisms would still be specious because they do not address trade-offs, budgets, problems with weapons performance and support, what different decisions would have been better, or how better decisions could have been reached.

12. Korb, *Fall and Rise of the Pentagon*, 89.

13. Steve Waters, "The DX Contract, 1967–1970," case study prepared for the Harvard Business School (copyright by the President and Fellows of Harvard College, 1974).

14. Naval resentment of Secretary McNamara remains widespread and has distorted official records. For example, articles published during the 1980s quote Navy sources as blaming him for alleged deficiencies in the *Thomaston* (LSD 28) and *Iwo Jima* (LPH 2) classes of amphibious assault ships. Both designs predated his tenure by years.

15. Quoted in Lt. Cdr. Terry Johnson, USN, "Ship Acquisition: The Lost Generation," *USNIP*, Nov. 1972.

16. Meyer, "Combat Systems of Surface Warships."

17. Eugene G. Windchy, *Tonkin Gulf* (New York: Doubleday, 1971); Vice Adm. James Stockdale, USN (Ret.), and Sybil Stockdale, *In Love and War* (Annapolis: Naval Institute Press, 1990), chap. 1 and apps. 1–2; *New York Times* edition of the Pentagon Papers (New York: Bantam, 1971), chap. 5; Steve Edwards, "Stalking the Enemy's Coast," *USNIP*, Feb. 1992.

18. Television interview with one of them, Daniel Ellsberg. It was he who later edited the Pentagon Papers and leaked them to the press. OSD's press announcement on the Tonkin Gulf incident misidentified *Maddox*'s partner as *C. Turner Joy*. Any work so identifying her reveals frivolous research.

19. In June 1971 the *New York Times* and other newspapers began to print the Pentagon Papers, a classified OSD study of American involvement in Vietnam. The Nixon administration obtained an unprecedented prior-restraint court order to suppress publication. The Supreme Court dismissed it without addressing the question of whether prior-restraint orders might be valid in other cases. The Nixon administration prosecuted the former OSD analyst who leaked the study. Illegal evidence-gathering tactics got the case thrown out of court and contributed to the impeachment case against President Nixon in 1974. Congress repealed the Tonkin Gulf Resolution in 1971 and passed its converse, the War Powers Act, over President Nixon's veto in 1973. He and all subsequent presidents have refused to concede that the War Powers Act limits their constitutional authority as commander in chief.

20. Capt. L. D. Caney, USN, commander, Destroyer Squadron 5, letter to *USNIP*, Feb. 1967, 110–12.

21. Capt. Garrette Lockee, USN, "PIRAZ," *USNIP*, Apr. 1969, 143–46.

22. Vice Adm. Malcolm W. Cagle, USN, "Task Force 77 in Action off Vietnam," *Naval Review 1972* (*USNIP*, May 1972).

23. Caney, letter to *USNIP*, Feb. 1967. During the Tonkin Gulf incident *Turner Joy*'s forward Mark 42 gun (Mount 51) was able to fire only a few rounds between malfunctions.

24. Lt. Cdr. David R. Cox, USN, letter to *USNIP*, June 1971, 84.

25. Lt. Cdr. Robert E. Mumford, "Jackstay: New Dimensions in Amphibious Warfare," *Naval Review 1968* (*USNIP*, May 1968).

26. OSD Draft Memorandum for the President, "Major Fleet Escort Forces," Nov. 16, 1966; "lemons": Friedman, *U.S. Destroyers*, 381; "McNamara class": comment to me by an officer assigned to one, 1978.

27. Conversations with multiple commanding officers during the late 1970s. The *Brooke* and *Garcia* classes' reputation as unsatisfactory because of boiler problems apparently led OSD to select them for retirement in 1988. Secretary of the Navy James Webb, badly wounded as a Marine officer in Vietnam, condemned this decision as a return to Vietnam-era policies of neglect of the military and resigned in protest. Another opinion about these ships' retirement was that their early SQS-26 sonars were ineffective against new submarines. However, other ships with the same sonars, such as the *McCloy* (FF 1037) class, remained in service for several more years. Pakistan and Brazil promptly leased most of these ships. Pakistan returned its ships in 1993 after failing to satisfy concern that it was developing a nuclear bomb.

28. Caney, letter to *USNIP*, Feb. 1967.

29. Friedman, *U.S. Destroyers*, 281–83; Worth letter to *USNIP*, Dec. 1984; Norman Polmar, *The Ships and Aircraft of the U.S. Fleet*, 11th ed. (Annapolis: Naval Institute Press, 1978), 111.

30. Cdr. W. J. Hunter, USN, letter to *USNIP*, Jan. 1972.

31. Norman Friedman, *U.S. Naval Weapons* (Annapolis: Naval Institute Press, 1982), 264.

32. Capt. James W. Kehoe, USN, et al., "Seakeeping and Combat Systems Performance—The Operators' Assessment," *Naval Engineers Journal*, May 1983, 260, 265. See also Norman Polmar's introduction to "Ocean Escorts" in the U.S. Navy section of *Jane's Fighting Ships, 1972–1973* (New York: McGraw-Hill, 1972), 472.

33. Friedman, *U.S. Destroyers*, 318–19.

34. Friedman, *U.S. Naval Weapons*, 155–58.

35. Annual Report of the Secretary of Defense, FY 1964, 19.

36. Hone, *Power and Change*, 102–4.

37. Norman Friedman, *The Postwar Naval Revolution* (Annapolis: Naval Institute Press, 1986), 169–71.

38. Richard S. Carleton and Eugene P. Weinert, "Historical Review of the Development and Use of Marine Gas Turbines by the U.S. Navy," Naval Sea Systems Command, 1988.

39. Rear Adm. R. E. Henning, USN (Ret.), interview, Nov. 14, 1990.

40. Carleton and Weinert, "Historical Review," sec. 17.

41. Ibid.; James Willis, interview, Jan. 14, 1991.

42. The -OG suffix (as in CODOG: combined diesel or gas turbine) indicated that only one power source at a time could drive the propeller shaft, while -AG (as in COGAG: combined gas turbine and gas turbine) indicated that both could deliver power to the shaft simultaneously. The *Spruance* class has a COGAG plant.

43. Carleton and Weinert, "Historical Review."

44. Friedman, *U.S. Destroyers*, 362–68.

45. Vice Adm. John T. Hayward, USN (Ret.), letter to *USNIP*, Apr. 1976.

46. U.S. Naval Institute Oral History Collection, Vice Adm. Thomas R. Weschler, USN. A senior official from the AsstSecDef(I&L) office first suggested this plan to Adm. Sonenshein in 1963. Correspondence from Rear Adm. Sonenshein, Jan. 28, 1991.

47. Carleton and Weinert, "Historical Review." *Admiral Wm. M. Callaghan* honors a different officer than does USS *Callaghan* (DDG 994).

48. Adm. Elmo Zumwalt, USN (Ret.), *On Watch* (New York: Quadrangle, 1976), 73.

49. Friedman, *U.S. Destroyers*, 222–24, 318–19.

50. Adm. Horacio Rivero, USN (Ret.), "As I Recall . . . ," *USNIP*, July 1979.

51. Ibid.; Friedman, *U.S. Destroyers*, 318–20.

52. Weschler Oral History; Rivero, "As I Recall."

53. Department of Defense FY 1967 budget, Feb. 23, 1966.

54. Russell Murray II, Principal Deputy Assistant Secretary of Defense for Systems Analysis and later a member of the CNO Executive Board, interview, Oct. 31, 1990.

55. Friedman, *U.S. Destroyers*, 319–20, 340.

56. Peet interview, Feb. 1, 1993.

57. Capt. Zeke Foreman, USN (Ret.), interview, Sept. 11, 1990. Capt. Foreman was on the NavMat ASW project office staff.

58. Carleton and Weinert, "Historical Review," sec. 47.

Chapter 3: The Navy Adopts Total Package Procurement

1. Deborah Shapley, *Promise and Power: The Life and Times of Robert McNamara* (Boston: Little, Brown, 1993), 206, 230.

2. Peet interview, Apr. 1993.

3. Murray interview, Oct. 31, 1990. The phrase "whiz kids" is quoted as a cliché of the era, not as Murray's words. Murray was an executive at Grumman before joining OASD(SA).

4. Lt. Cdr. Charles J. DiBona, "Can We Modernize U.S. Shipbuilding?" *USNIP*, Jan. 1966.

5. Ibid.

6. Robert Johnson, "The Changing Nature of the U.S. Navy Ship Design Process," *Naval Engineers Journal*, Apr. 1980, 90ff.

7. *U.S. Defense Program for FY-1968*, dated Jan. 23, 1967.

8. Murray interview, Oct. 31, 1990.

9. Ibid; Nitze and Sullivan, *Securing the Seas*, 6.

10. Lt. Cdr. John Carpenter, USN, and Lt. Cdr. Peter Finne, USN, "Navy Ship Procurement," *Naval Engineers Journal*, Dec. 1972, tab. 1.

11. Frederick Taylor, "Navy Aims to Pressure Yards into Streamlining as It Modernizes Fleet," *Wall Street Journal*, Mar. 25, 1967.

12. Rear Adm. Nathan Sonenshein, USN, "Results of FDL

Contract Definition," *Naval Engineers Journal,* Oct. 1967, 822.

13. Carpenter and Finne, "Navy Ship Procurement."
14. Henning interview, Nov. 14, 1990.
15. Waters, "DX Contract"; Weschler, "DX/DXG Program."
16. J. P. Banko, *Spruance DD 963 Class Destroyers Lessons Learned,* Naval Sea Systems Command, Feb. 1980, 5.
17. Peet interview, Apr. 10, 1993.
18. Rear Adm. Nathan Sonenshein, USN (Ret.), interview, Nov. 10–11, 1990.
19. Richard Stubbing, *The Defense Game* (New York: Harper & Row, 1986), 179–82.
20. Assistant Secretary of the Navy (Installations and Logistics) Graeme Bannerman, "Multi-Year Ship Procurement and Other Ship Acquisition Concepts," *Naval Engineers Journal,* Dec. 1967.
21. Sonenshein, "Results of FDL Contract Definition," 820; Hone, *Power and Change,* 79.
22. Banko, *Spruance Lessons Learned,* 9–13.
23. Richard F. Cross III, manager of program development for the General Dynamics DD 963 project, interview, Sept. 9, 1990.
24. Correspondence from Rear Adm. Sonenshein, Aug. 1, 1992.
25. Johnson, "Changing Nature of the U.S. Navy Ship Design Process," 98.
26. Sonenshein interview, Nov. 10–11, 1990; Bannerman, "Multi-Year Ship Procurement," 927–930.
27. Sonenshein interview, Nov. 10–11, 1990.
28. Annual Report of the Secretary of the Navy, FY 1966, 322.
29. Lt. Cdr. Harold J. Sutphen, USN, "The Failure of the Fast Deployment Logistics Ship," *Naval War College Review,* Oct. 1970, 69–86; Sonenshein interview, Nov. 10–11, 1990.
30. Leopold, "Should the Navy Design Its Own Ships?"
31. Johnson, "Changing Nature of the U.S. Navy Ship Design Process."
32. Willis C. Barnes, "1950–1972: Korea and Vietnam," in *Naval Engineering and American Sea Power,* ed. Rear Adm. R. W. King, USN (Ret.) (Baltimore: Nautical & Aviation Publishing, 1989), 309–13.
33. Johnson, "Changing Nature of the U.S. Navy Ship Design Process."
34. Ibid.; Reuven Leopold, "Should the Navy Design Its Own Ships?" *USNIP,* May 1975, 152–53.
35. Murray interview, Oct. 31, 1990.
36. Sonenshein, "Results of FDL Contract Definition."
37. Vice Adm. J. B. Colwell, USN, deputy CNO for fleet operations and readiness, quoted in Sutphen, "Failure of the FDL Ship."
38. Henning interview, Nov. 14, 1990.
39. Sutphen, "Failure of the FDL Ship."
40. Murray interview, Oct. 31, 1990.
41. Andrew Gibson and Cdr. Jacob Shuford, USN, "DE-SERT SHIELD and Strategic Sealift," *Naval War College Review,* spring 1991.
42. Tsouras, "Soviet Naval Tradition," 19–20.
43. Michael MccGwire, "Gorshkov's Navy," *USNIP,* Aug. 1989.
44. Ibid.
45. Cecil Brownlow and Barry Miller, "USSR Improves Cruise Missile Capabilities," *Aviation Week and Space Technology,* Nov. 8, 1971.
46. Charles C. Petersen, "Aircraft Carrier Development in Soviet Naval Theory," *Naval War College Review,* Jan.–Feb. 1984.
47. Correspondence from Norman Friedman, Nov. 1993; "Soviet/Russian Navy Air-to-Surface Weapons," in *World Naval Weapons Systems, 1994 Update* (Annapolis: Naval Institute Press, 1994).
48. Petersen, "About Face in Soviet Tactics," 58–63.
49. Watson, *Red Navy at Sea,* 1–19.
50. Nitze and Sullivan, *Securing the Seas,* 68–72.
51. Sokolsky, *Seapower in the Nuclear Age,* 92–102.
52. Ibid.
53. Schratz, "Paul Henry Nitze."
54. Desmond Ball, "The Development of the SIOP, 1960–1983," in *Strategic Nuclear Targeting,* ed. Ball and Richelson, 70.
55. Joel Sokolsky, "The U.S. Navy and Nuclear ASW Weapons," *USNIP,* Dec. 1984, 155, who cites Navy documents as sources for the quoted passages. A more extensive discussion is in his *Seapower in the Nuclear Age,* 66–70.
56. Nitze and Sullivan, *Securing the Seas,* 12.
57. See app. A.
58. U.S. Defense Program 1969–73 and Defense Budget FY 1969, dated Jan. 22, 1968. This budget requested the first five DD 963s.
59. Roherty, *Decisions of Robert S. McNamara,* 161–63.
60. Ball, "Development of the SIOP," tab. 3.1 and pp. 63, 70.
61. Murray interview, Oct. 31, 1990. After the retirement of the CVS fleet, in 1975 the CA/CVAN designation was shortened to CV/CVN, since the attack carriers now operated ASW aircraft and ASW operation centers.

Chapter 4: The DX/DXG Project

1. Capt. E. R. Zumwalt, "A Course for Destroyers," *USNIP,* Nov. 1962.
2. Murray interview, Oct. 31, 1990.
3. OSD Draft Memorandum for the President, "Major Fleet Escort Ship Forces," Nov. 16, 1966 (rpt. in app. A); Murray interview, Oct. 31, 1990.
4. Murray interview, Oct. 31, 1990.
5. Peet interview, Apr. 16, 1993.
6. "Implementing a DX/DXG Program," Sept. 23, 1966, cited in Friedman, *U.S. Destroyers,* 372–73.
7. OSD declassified this document for use in the present book.
8. Sonenshein interview, Nov. 10–11, 1990.
9. See app. A.
10. Representative Les Aspin, "The Litton Ship Fiasco," *The Nation,* Dec. 11, 1972.
11. See app. A.
12. Banko, *Spruance Lessons Learned.*
13. Friedman, *U.S. Destroyers,* 320.
14. ASD (Program Analysis and Evaluation) memorandum, Feb. 4, 1975, quoted in Thomas C. Hone, "The Program Manager as Entrepreneur: AEGIS and RADM Wayne Meyer," in *Decision Analysis* 3, no. 3 (1987) (Oxford: Brassey's Defense Publishers, 1987), 211.
15. Banko, *Spruance Lessons Learned,* 16.

16. Ibid., 21–23.

17. Peet interview, Apr. 16, 1993. While DX/DXG program coordinator, Ray Peet's official rank was captain, but he had been selected for promotion to rear admiral.

18. DX/DXG Concept Formulation plan, Feb. 1, 1967, as quoted in Friedman, *U.S. Destroyers*, 373; "Destroyers Join the 'Jet Set,' " *Armed Forces Management*, Dec. 1969, an article based on an interview with Adm. Henning; Peet interview, Feb. 1, 1993.

19. Peet interview, Feb. 1 and Apr. 16, 1993.

20. U.S. Naval Institute Oral History Collection, Vice Adm. Raymond E. Peet, USN (Ret.).

21. Zumwalt, *On Watch*, 73.

22. Vice Adm. John T. Hayward, USN (Ret.), letter to *USNIP*, Aug. 1976.

23. Weschler Oral History.

24. Correspondence from Capt. L. D. Caney, Sept. 12, 1992.

25. Weschler Oral History.

26. Vice Adm. Thomas Weschler, interview, Sept. 5, 1992.

27. Bannerman, "Multi-Year Ship Procurement," 927–30.

28. Sonenshein interview, Nov. 10–11, 1990.

29. Weschler Oral History.

30. Ibid.

31. Roherty, *Decisions of Robert S. McNamara*, 172.

32. Sonenshein interview, Nov. 10–11, 1990.

33. Correspondence from Capt. Wayne Hughes, who served on the Major Fleet Escort Study project, Feb. 21, 1994; Friedman, *U.S. Destroyers*, 373–74.

34. Zumwalt, *On Watch*, 101–4; Friedman, *U.S. Destroyers*, 373–74.

35. Secretary McNamara recommended approval in a Draft Memorandum for the President, reproduced in part in app. A. It is dated "Revised Jan. 5, 1968"; the original version was written in mid-1967 (Zumwalt, *On Watch*, 101–3).

36. See app. A.

37. U.S. Defense Program 1969–73 and Defense Budget FY 1969, Jan. 22, 1968.

38. Annual Report of the Secretary of the Navy, FY 1970, 99.

39. Capt. John R. Baylis, USN (Ret.), interview, Sept. 19, 1990.

40. Henning interview, Nov. 14, 1990; Baylis interview, Sept. 19, 1990.

41. "CC-280" and "ASWCCS," in *World Naval Weapons Systems, 1989/90* (Annapolis: Naval Institute Press, 1989).

42. Antony Preston, "Canada," in *Conway's All the World's Fighting Ships, 1947–1982* (Annapolis: Naval Institute Press, 1983).

43. Infrared target: Baylis interview, Sept. 19, 1990.

44. Ibid.

45. Weschler interview, Sept. 5, 1992; Banko, *Spruance Lessons Learned*, 10.

46. Carleton and Weinert, "Historical Review."

47. Weschler Oral History.

48. Willis interview, Jan. 14, 1991.

49. Reuven Leopold, "Gas Turbines in the U.S. Navy: Analysis of an Innovation and Its Future Prospects as Viewed by a Ship Designer," *Naval Engineers Journal*, Apr. 1975; Carleton and Weinert, "Historical Review," sec. 40–41.

50. Rear Adm. Nathan Sonenshein, USN, "What's New in Navy Petroleum," *Naval Engineers Journal*, Aug. 1969.

51. Weschler interview, Sept. 5, 1992.

52. Dean Rains et al., "Design Appraisal—DD 963," *Naval Engineers Journal*, Oct. 1976.

53. Baylis interview, Sept. 19, 1990; Friedman, *U.S. Destroyers*, 342.

54. Correspondence from Vice Adm. Weschler, Dec. 1992; Banko, *Spruance Lessons Learned*, 10.

55. "Destroyer," in *Encyclopaedia Britannica*, 1954 ed.

56. Weschler correspondence, Dec. 1992.

57. Weschler Oral History.

58. Johnson, "Ship Acquisition: The Lost Generation."

59. Roherty, *Decisions of Robert S. McNamara*, 166–69.

60. Banko, *Spruance Lessons Learned*, 21–23, 54.

61. Ibid., 15.

62. Weschler, "DX/DXG Program," 931.

63. Michael Gettler, "Navy Planning RFP's in Fall for New Class of Destroyers," *Technology Week*, May 15, 1967.

64. Waters, "DX Contract."

65. Baylis interview, Sept. 19, 1990.

66. Friedman, *U.S. Destroyers*, 373.

67. Gerald Boatwright and John Couch, "The Gas Turbine Propulsion System for the U.S. Navy *Spruance* (DD 963) Class Destroyers," paper presented to American Society of Mechanical Engineers, Mar. 1971.

68. Weschler correspondence, Dec. 7, 1992.

69. Sonenshein interview, Nov. 10–11, 1990.

70. Banko, *Spruance Lessons Learned*, 6.

71. Matthew Forrest, comments published in "Discussion Regarding the DX/DXG Program," *Naval Engineers Journal*, Dec. 1967.

72. Rear Adm. Mark Woods, USN (Ret.), interview, Mar. 25, 1991.

73. Murray interview, Oct. 31, 1990.

74. Woods interview, Mar. 25, 1991.

75. Foreman interview, Sept. 11, 1990.

76. Friedman, *U.S. Destroyers*, 342.

77. Weschler Oral History.

78. Friedman, *U.S. Destroyers*, 375.

79. Correspondence from Vice Adm. Weschler, Dec. 7, 1992.

80. MAT-09X (Weschler) Memo 28–70, 1 May 1970, "Concept Formulation and Contract Definition." Chap. 12 covers this.

81. *Jane's Fighting Ships, 1972–1973*, 248; Joan Potter, "A History of India," unpublished paper, Challenger Junior High School, San Diego, 1991.

82. Banko, *Spruance Lessons Learned*, 38–39.

83. Correspondence from Adm. Nathan Sonenshein, Aug. 29, 1991.

84. Waters, "DX Contract."

85. Watson, *Red Navy at Sea*, 11.

86. Henry Trewhitt, *McNamara* (New York: Harper & Row, 1971), 242–45.

87. Peter Ognibene, "Getting and Spending: Litton Industries and the Seven Rules of Big-Time Defense Contracting," *Harper's Magazine*, May 1973.

88. James S. Ennes, *Assault on the Liberty* (New York: Ivy Books, 1979), 87–92. Lt. Ennes, wounded in the attack, shows the military's bitterness about Secretary McNamara in attributing the recall of the F-4s to McNamara

himself. However, his basis for the charge is weak, and former CNO Adm. Thomas Moorer's foreword to the book says, "So far, no one knows" who ordered the recall.

89. Polmar and Noot, *Submarines of the Russian and Soviet Navies*, 164; Love, *History of the U.S. Navy*, vol. 2, *1942–1991*, 615–16.

90. Polmar and Noot, *Submarines of the Russian and Soviet Navies*, 192–93.

91. Abraham Rabinovich, *The Boats of Cherbourg* (New York: Seaver Books/Henry Holt, 1988), 180–82.

Chapter 5: The DD 963 Design Competition

1. Devanney, "DX Competition," 21–22.
2. Robert Browning, "A Memoir of Contract Definition," *Naval Engineers Journal*, Nov. 1987, 75–76.
3. Banko, *Spruance Lessons Learned*, 18.
4. Waters, "DX Contract."
5. Banko, *Spruance Lessons Learned*, 71.
6. Annual Report of the Secretary of the Navy, FY 1970, 97.
7. Waters, "DX Contract."
8. Ibid. Only five LHAs were funded and built. The TPP termination charge was half the cost of one LHA.
9. Weschler Oral History.
10. Waters, "DX Contract"; Woods interview, Mar. 25, 1991; Weschler Oral History.
11. Friedman, *U.S. Destroyers*, 357, 365–66.
12. Cross interview, Sept. 9, 1990.
13. Reuven Leopold, interview, Sept. 11, 1990.
14. Ibid.
15. Reuven Leopold and Wolfgang Reuter, "Three Winning Designs—FDL, LHA, DD-963: Method and Selected Features," *Transactions* of the Society of Naval Architects and Marine Engineers, 1971, 297.
16. Ibid.
17. Joseph Conrad, *The Mirror of the Sea* (Garden City, N.Y.: Doubleday, ca. 1906), 28.
18. Leopold and Reuter, "Three Winning Designs."
19. During the 1950s the *Farragut* (DLG 6/DDG 37) class was initially designed to mount a quintuple antiship torpedo bank amidships. See Friedman, *U.S. Destroyers*, 294–97.
20. Leopold interview, Sept. 11, 1990.
21. John Charles Roach and Herbert Meier, "Visual Effectiveness in Modern Warship Design," *Naval Engineers Journal*, Dec. 1979, and their very similar "Warships Should Look Warlike," *USNIP*, June 1979.
22. Palmer, *Origins of the Maritime Strategy*, 21–22; Vice Adm. Stansfield Turner, USN, "Missions of the U.S. Navy," *Naval War College Review*, Mar.–Apr. 1974.
23. Leopold and Reuter, "Three Winning Designs."
24. Leopold, "Gas Turbines in the U.S. Navy," fig. 5.
25. Devanney, "DX Competition," n. 9. Reuven Leopold agreed with him on this point.
26. David M. Krepchin of M. Rosenblatt & Son, Mar. 1990.
27. Devanney, "DX Competition." Devanney attributed this error to himself.
28. Boatwright and Couch, "Gas Turbine Propulsion System."
29. Ibid.; Leopold interview, Sept. 11, 1990.
30. Cdr. Ken Smith, USN, force engineer on the staff of

Cdr., Naval Surface Force, U.S. Pacific Fleet, interview, Sept. 1990.

31. Devanney, "DX Competition."
32. Capt. James Kehoe, USN, "Destroyer Seakeeping: Ours and Theirs," in *USNIP*, Nov. 1973.
33. Rains et al., "Design Appraisal—DD 963," 45–46.
34. Malcolm Dick, president emeritus, Gibbs & Cox, interview, July 28, 1992.
35. Correspondence from H. E. Buttelmann, president, Gibbs & Cox, Nov. 23, 1992.
36. Henning interview, Nov. 14, 1990.
37. Waters, "DX Contract."
38. Cross interview, Sept. 9, 1990.
39. Foreman interview, Sept. 11, 1990.
40. Information from design summary in Dick Cross's collection.
41. Weschler Oral History.
42. Thomas Buermann, president emeritus of Gibbs & Cox, interview, July 28, 1992.
43. Banko, *Spruance Lessons Learned*, 18.
44. Johnson, "Changing Nature of the U.S. Navy Ship Design Process," 99.
45. Willis interview, Jan. 14, 1991.
46. Friedman, *U.S. Destroyers*, 375.
47. Willis interview, Jan. 14, 1991.
48. Capt. Clark M. Graham, USN, interview, Nov. 6, 1990.

Chapter 6: Design Selection and Completion

1. Weschler Oral History.
2. Mosher quoted in Baltimore *Sun*, May 11, 1969, and elsewhere. On Adm. Rickover's opposition to gas turbines at congressional hearings, see Norman Polmar and Thomas Allen, *Rickover* (New York: Simon and Schuster, 1982), 228–36. Rickover later named two nuclear submarines for the ranking members of the House Armed Services Committee of this period: USS *William H. Bates* (SSN 680) and USS *L. Mendel Rivers* (SSN 686).
3. Correspondence from Bennett to Rivers, June 11, 1969; copy provided to me by Dick Cross.
4. Waters, "DX Contract."
5. Weschler Oral History.
6. Foreman interview, Sept. 11, 1990.
7. Banko, *Spruance Lessons Learned*, 18.
8. Waters, "DX Contract."
9. Cross interview, Sept. 9, 1990.
10. Waters, "DX Contract"; Sonenshein interview, Nov. 10–11, 1990.
11. Baylis interview, Sept. 19, 1990.
12. See, for example, Coulam, *Illusions of Choice*, 251.
13. Sonenshein interview, Nov. 10–11, 1990.
14. Correspondence from H. E. Buttelmann, Nov. 23, 1992.
15. Buermann interview, July 28, 1992.
16. U.S. Naval Institute Oral History Collection, Capt. L. Colbus, USN (Ret.).
17. Waters, "DX Contract."
18. Sonenshein interview, Nov. 10–11, 1990.
19. Carleton and Weinert, "Historical Review."
20. Ibid.
21. Information from Adm. Sonenshein.
22. Waters, "DX Contract"; Colbus Oral History.
23. Foreman interview, Sept. 11, 1990.

24. General Accounting Office Report, Aug. 26, 1970 (Comptroller General of the United States report serial B-170269).
25. Waters, "DX Contract."
26. Secretary of the Navy John Chaffee, quoted in *Wall Street Journal*, June 9, 1970.
27. GAO Report, Aug. 26, 1970.
28. Stubbing, *Defense Game*, 198.
29. GAO Report, Aug. 26, 1970.
30. Weschler Oral History.
31. GAO Report, Aug. 26, 1970.
32. All of Senator Smith's letters and the Navy's responses were published in *Congressional Record—Senate*, June 29, 1970.
33. "The Struggle over a Destroyer Contract," *Business Week*, June 27, 1970. Adm. Sonenshein estimated the total increase as over $600 million; quoted at press conference, June 23, 1970, in *Navy Times*, July 8, 1970.
34. "Struggle over a Destroyer Contract."
35. See, for example, Stubbing, *Defense Game*.
36. Johnson, "Changing Nature of the U.S. Navy Ship Design Process," states that the buildup and disestablishment of the competition teams and the shifting of naval architects among them nearly destroyed several ship design firms.
37. Sonenshein letter, Aug. 29, 1991.
38. Waters, "DX Contract"; Stubbing, *Defense Game*, 204.
39. "Navy's Plan to Build 30 Destroyers at 1 Yard May Be Revamped," *Wall Street Journal*, June 10, 1970.
40. Zumwalt, *On Watch*, 73. He was promoted to full admiral and became Chief of Naval Operations on July 1, 1970.
41. Banko, *Spruance Lessons Learned*, 18.
42. Destroyer (DD 963/DDG) project manager memorandum serial 922–PMS389, "DD 963 Engineering Change Proposal for main propulsion turbines," Dec. 23, 1970.
43. Dean Rains and Ronald d'Arcy, "Considerations in the DD 963 Propulsion System Design," *Naval Engineers Journal*, Aug. 1972, 66–67.
44. Jack W. Abbott, "Integration of Modern Machinery Systems," *Naval Engineers Journal*, June 1973, 80–82.
45. Capt. John Collins, USN, "USS *Spruance* (DD 963) Class, Designed for Change—The Payoff Is Now," *Naval Engineers Journal*, Apr. 1975.
46. Sonenshein interview, Nov. 10–11, 1990.
47. Inman, "From Typhon to Aegis."
48. Watson, *Red Navy at Sea*, 8–18, 28–29; Nitze and Sullivan, *Securing the Seas*, 66–73.
49. Capt. H. E. Reichert, "LAMPS in the Gulf of Tonkin," *USNIP*, Mar. 1973.
50. Rear Adm. Richard J. Grich and Cdr. Robert E. Bruninga, "Electromagnetic Environment Engineering," *Naval Engineers Journal*, May 1987, 203.
51. Adm. Isaac C. Kidd, Jr., USN, "The View from the Bridge of the Sixth Fleet Flagship," *USNIP*, Feb. 1972.
52. Zumwalt, *On Watch*, 436.
53. Henry Kissinger, *Years of Upheaval* (Boston: Little, Brown, 1982), 606.
54. Quoted in *USNIP*, Oct. 1975.
55. Zumwalt, *On Watch*, 72.
56. Ibid., 73–74, 102, 150–51.
57. Ibid., 74–76.

1. Weschler Oral History.
2. Sonenshein interview, Nov. 10–11, 1990.
3. "Litton Shapes up a Ship Contract," *Business Week*, Mar. 24, 1973.
4. *Annual Defense Department Report FY 1973*, Feb. 8, 1972, 101.
5. James D. Hessman, "Navy, Litton Look ahead to Long-Term Partnership but All Is Not Roses down among the Magnolia Blossoms," *Armed Forces Journal*, Mar. 15, 1971.
6. Banko, *Spruance Lessons Learned*, 13, 16, 34, 40, 49–50.
7. Naval architects refer to "double-line" systems as those that drawings must show scaled in correct dimensions (double lines show the diameter of a pipe section). Single lines indicate systems that are so small and cheap to install that the building crew can place them wherever is convenient. Submarine designers use double lines for all systems because of the internal tightness of such ships.
8. Sonenshein interview, Nov. 10–11, 1990.
9. Joseph F. Yurso, "Decline of the Seventies," in *Naval Engineering and American Sea Power*, ed. King, 344–45.
10. "The U.S. Navy's *Spruance* Class Destroyers," *Warship International*, no. 3 (1974), 252–64.
11. Banko, *Spruance Lessons Learned*, 49.
12. Ken Smith interview, Sept. 1990.
13. Carleton and Weinert, "Historical Review," 53.
14. Sonenshein interview, Nov. 10–11, 1990.
15. Friedman, *U.S. Destroyers*, 322; Willis interview, Jan. 14, 1991; Leopold interview, Sept. 11, 1990.
16. Hone, *Power and Change*, 79. The quoted passage is from Friedman, *U.S. Destroyers*, 371.
17. Woods interview, Mar. 25, 1991.
18. Banko, *Spruance Lessons Learned*, 76–77.
19. Ibid., 77–78.
20. Willis interview, Jan. 14, 1991; Banko, *Spruance Lessons Learned*, 80–81.
21. My personal experience.
22. Capt. Robert H. Smith, USN, "A United States Navy for the Future," *USNIP*, Mar. 1971.
23. Capt. Robert H. Smith, USN, letter to *USNIP*, Mar. 1973.
24. Adm. Rickover telephoned his acclamation both to the Naval Institute (three times in one day) and to Capt. Smith. See Secretary's Notes, *USNIP*, June 1971.
25. Capt. W. J. Ruhe, USN (Ret.), "A Future for the Destroyer?" *USNIP*, Aug. 1971.
26. All quotations from letters to *USNIP*, June 1971 through July 1972.
27. Banko, *Spruance Lessons Learned*, 81–82.
28. Kidd, "View from the Bridge of the Sixth Fleet Flagship," 23.
29. Charles C. Petersen, "Soviet Tactics for Warfare at Sea" (Alexandria: Center for Naval Analyses, 1982); rpt. slightly modified as "About Face in Soviet Tactics." Altimeter-fusing is my surmise based on unclassified publications.
30. Sims, "Ugly by Design," 113ff.; Reuven Leopold, "U.S. Naval Ship Design: Platforms *vs.* Payloads," *USNIP*, Aug. 1975, 30ff.

31. L. Edgar Prina, "Pentagon Studies Russian Threat to New Destroyer," San Diego *Union*, June 17, 1971.

32. Ognibene, "Getting and Spending."

33. Sonenshein interview, Nov. 10–11, 1990.

34. Banko, *Spruance Lessons Learned*, 21.

35. "Defense Contracts Escalate Again," *Business Week*, Apr. 22, 1972; Stubbing, *Defense Game*, 199–201. Stubbing says Litton's first claim was in Mar. 1973, but the *Business Week* report makes it clear that the first claim was in 1972.

36. Zumwalt, *On Watch*, 150–51.

37. Quoted in "Navy Seeks 50 Small Escorts for Modernization of Fleet," *Wall Street Journal*, Sept. 1, 1971; rpt. in *USNIP*, Dec. 1971.

38. "Behind the Write-offs at Litton Industries," *Business Week*, Sept. 23, 1972.

39. "The Litton Comeback That Didn't Come Off," *Business Week*, June 3, 1972.

40. Banko, *Spruance Lessons Learned*, 45–46.

41. Korb, *Fall and Rise of the Pentagon*, 169–73.

42. "A Fox in the Chicken Coop?" *New Republic*, Jan. 20, 1973.

43. Both titles are from issues of the *New Republic*.

44. See Representative Les Aspin (D-Wisc.), "The Litton Ship Fiasco," *The Nation*, Dec. 11, 1972, and "Litton: Incompetent as Ever," *The Nation*, June 28, 1975.

45. Criticism of Litton, Ash, and the DD 963 class appears throughout the *New Republic* issues of Dec. 11, 1972; Jan. 6–13, 20, and 27, 1973; and Feb. 3 and 17, 1973; and in *Harper's Magazine*, May 1973, which cites Capt. Smith's *USNIP* essay as evidence that the *Spruance* class would be ineffective and unneeded.

46. Reuven Leopold and E. K. Straubinger, interview, Sept. 11, 1990.

47. Devanney, "DX Competition."

48. Boatwright and Couch, "Gas Turbine Propulsion System."

49. Abbott, "Integration of Modern Machinery Systems."

50. Devanney, "DX Competition."

51. Boatwright and Couch, "Gas Turbine Propulsion System."

52. Devanney, "DX Competition." The bias refers to the frictional drag error described in chap. 5.

53. J. W. Devanney, letter to *USNIP*, Feb. 1976.

54. Interview with a classmate.

55. Leopold interview, Sept. 11, 1990; *Naval Engineers Journal*, July 1987, 73.

56. Coulam, *Illusions of Choice*.

57. Letter to *USNIP*, Aug. 1977.

58. Cdr. Strafford Morss, USNR (Ret.), "*Ticonderoga*: Another *Hood?*" *USNIP*, Aug. 1982.

59. Adm. W. J. Lisanby, letter to *USNIP*, Sept. 1978.

60. Naval Sea Systems Command memorandum, "Controllable Pitch Propeller History," 13 Mar. 1991.

61. House Armed Services Committee testimony, Mar. 1975.

62. Capt. C. A. Bartholomew, USN, *Mud, Muscle and Miracles: Marine Salvage in the U.S. Navy* (Washington, D.C.: Naval Historical Center and Naval Sea Systems Command, 1990), 353–55; Rear Adm. George Wagner, USN, interview, Oct. 24, 1992.

63. Banko, *Spruance Lessons Learned*, 56–58.

64. Stubbing, *Defense Game*, 202.

65. Quoted in ibid., 203.

66. Friedman, *U.S. Destroyers*, 377. A complicating factor in evaluating TPP contract costs is that in 1972 Congress put a five–year life on appropriations. Thus the Navy needed new appropriations to pay for contract work performed after 1975 even when the original (FY 1970) contract had covered that work.

67. Stubbing, *Defense Game*, 203–4.

Chapter 8: Fleet Introduction, Engineering, and Supply

1. Capt. Carl Kowalski, SC, USN, on the SupShip Pascagoula staff at the time, interview, Mar. 1991.

2. Banko, *Spruance Lessons Learned*, 58.

3. Norman Polmar, *The Ships and Aircraft of the U.S. Fleet*, 12th ed. (Annapolis: Naval institute Press, 1981), lists *Comte de Grasse* in the Pacific Fleet because that was her intended assignment.

4. USS *Comte de Grasse* (DD 974) Command History File.

5. Banko, *Spruance Lessons Learned*, 11.

6. Ibid., 10.

7. Sims, "Ugly by Design," 113–18.

8. David A. Blank et al., *Introduction to Naval Engineering* (Annapolis: Naval Institute Press, 1985), 244–63.

9. Ken Smith interview, Sept. 1990.

10. Rains and d'Arcy, "Considerations in the DD 963 Propulsion System Design."

11. This message crossed my desk.

12. Blank, *Introduction to Naval Engineering*, 262.

13. Norman Polmar, *The Ships and Aircraft of the U.S. Fleet*, 14th ed. (Annapolis: Naval Institute Press, 1987), 150; John Jordan, "*Spruance* Class," *Defence* (UK), Apr. 1979.

14. *Jane's All the World's Weapons Systems, 1971–1972* (London: Sampson Low Marston, 1971).

15. *Combat Fleets of the World, 1990/91* (Annapolis: Naval Institute Press, 1990), 671.

16. "U.S. Navy Unveils *Sea Shadow* Stealth Vessel," *Aviation Week and Space Technology*, Apr. 26, 1993.

17. Cdr. John Preisel, "The Evolution of Machinery Control Systems Aboard US Navy Gas Turbine Ships," *Naval Engineers Journal*, May 1989, 102–6.

18. Leopold interview, Sept. 1990.

19. Lt. (jg) Jim Stavridis, "Handling a *Spruance*-Class Destroyer," *USNIP*, Oct. 1979.

20. Rains et al., "Design Appraisal—DD 963."

21. A. Erich Baitis and Louis Schmidt, "Ship Roll Stabilization in the U.S. Navy," *Naval Engineers Journal*, May 1989.

22. Robert E. McKeown and David Robinson, "Submarines," in *Soviet Navy: Strengths and Liabilities*, ed. Watson and Watson, 57–71.

23. *USNIP*, Jan. 1985, 115.

24. Adm. James Watkins, USN, "Increased Force Levels and Warfighting Readiness through Better Ship Design," *Naval Engineers Journal*, June 1981.

25. Dennis A. Libby, letter to *Naval Engineers Journal*, June 1982.

26. Statement at DD 963 supply conference in San Diego, 1980.

27. Leopold interview, Sept. 11, 1990, confirming Devanney.

28. Watkins, "Increased Force Levels." He became CNO and later Secretary of Energy.

29. James G. Mellis et al., "Is Automation the Magic Potion for Manning Problems?" *Naval Engineers Journal*, Apr. 1982.

30. Ibid.; Arvin Plato, letter to *USNIP*, Dec. 1975, 71.

31. Boatwright and Couch, "Gas Turbine Propulsion System," 4.

32. Rains et al., "Design Appraisal—DD 963," 45.

33. Norman Friedman, *Modern Warship Design and Development* (New York: Mayflower, 1979), 173–75.

34. Leopold and Reuter, "Three Winning Designs," 4.

Chapter 9: Gunnery, Electronics, and Antimissile Weapons

1. Jordan, "*Spruance* Class."

2. Hip Pocket was a package of ASCM defenses carried by some destroyers toward the end of the Vietnam conflict. See Friedman, *U.S. Naval Weapons*, 161.

3. Ken Smith interview, Sept. 1990; USS *Kinkaid* (DD 965) Command History File, 1981.

4. "Tactical Data Systems" section introduction, in *World Naval Weapons Systems, 1991/92*.

5. Capt. Erick Swenson et al., "NTDS—A Page in Naval History," *Naval Engineers Journal*, May 1988.

6. Ronald Levine, "Supercomputers," *Scientific American*, Jan. 1982.

7. Lt. Cdr. T. J. McKearney, "The Offensive Surface Ship," *USNIP*, Dec. 1983, 66.

8. Naval Education and Training Support Command, *Surface Ship Operations* (Washington, D.C.: U.S. Government Printing Office, 1978), 213.

9. "Harpoon," in *World Naval Weapons Systems, 1991/92*; Cdr. S. J. Froggett, "Tomahawk Has Arrived," *USNIP*, Dec. 1985.

10. As late as 1943 World War II ASW doctrine for destroyers, even for the heavily armed *Fletcher* class, prescribed ramming surfaced submarines. Postwar Navy planners may have assumed that such desperate tactics suggested that U-boat pressure hulls had withstood 5-inch/38 gunfire even at point-blank range. In 1947 OpNav demanded higher-energy 5-inch/54 guns for postwar destroyers, beginning with the *Mitscher* (DD 927/DL 2) class, specifically to sink surfaced submarines. The wartime preference for ramming in fact had been to guarantee that the submarine could not successfully submerge, so that when she "went sinker" she was a confirmed kill, not a new underwater threat. OpNav dropped the ASW rationale for 5-inch/54 guns but still required them for shore bombardment and surface action. Friedman, *U.S. Destroyers*, 182, 242–43; Peet Oral History.

11. Cdr. P. A. Kissling and Lt. R. J. Hoffman, *Amphibious Warfare Indoctrination* (Coronado: Naval Amphibious School, 1994).

12. Friedman, *U.S. Naval Weapons*, 69.

13. "5/54 Mk 42" and "5/54 Mk 45," in *World Naval Weapons Systems, 1991/92*.

14. *Surface Warfare*, Sept. 1975, 31.

15. "TAS Mk 23," "Mk 91 FCS," and "Sparrow and Sea Sparrow," in *World Naval Weapons Systems, 1991/92*; *USNIP*, Nov. 1985, 133–37.

16. Norman Polmar, "Shooting at the Arrow," *USNIP*, Mar. 1988.

17. Rear Adm. Richard Grich and Cdr. Robert Bruninga, "Electromagnetic Environment Engineering," *Naval Engineers Journal*, May 1987.

18. Joseph Weis, "Identification Friend of Foe . . . (Maybe)," *USNIP*, Dec. 1989; "IFF," in *World Naval Weapons Systems, 1991/92*, xxv.

19. "Furuno Radars" and "Pathfinder 1500," in *World Naval Weapons Systems, 1991/92*.

20. Lee Jackson, Naval Command Control and Ocean Surveillance Center, interview, Nov. 3, 1992.

21. *Signal*, Sept. 1975.

22. "Satellite Information Exchange System," in *World Naval Weapons Systems, 1991/92*; *Shipboard Electronics Material Officer* (NavEdTra 10478-A1), 5.22–5.30.

23. William R. Bigas, "Get ESM out of the Closet," *USNIP*, Sept. 1991.

24. Ibid.

25. Ibid.

26. Lt. Cdr. Larry DiRita, "Exocets, Air Traffic, and the Air Tasking Order," *USNIP*, Aug. 1992.

27. "WLR-1/WLR-11," in *World Naval Weapons Systems, 1991/92*.

28. Bigas, "Get ESM out of the Closet."

29. Correspondence from Norman Friedman, Nov. 1993.

30. "SLQ-32/Sidekick," in *World Naval Weapons Systems, 1991/92*; Bigas, "Get ESM out of the Closet."

31. *USNIP*, Jan. 1983, 101–3.

32. James W. Rawles, "Countering Anti-Ship Missiles," *Defense Electronics*, Sept. 1990.

33. David Brown, *The Royal Navy and the Falklands War* (Annapolis: Naval Institute Press, 1987), 228.

34. Norman Friedman, "Stealth Applied," *USNIP*, July 1993; unclassified information from Cdr. Gregory Dies, USNR, 1992.

35. Weschler interview, Sept. 5, 1992.

36. Conversation with Mike Coumatos, 1979.

37. "Kaman Seasprite," in *Jane's All the World's Aircraft, 1974–1975* (London: Macdonald, 1974), 362–63.

38. "ALQ-142" and "APS-124," in *World Naval Weapons Systems, 1991/92*.

39. Rear Adm. David Bill, "Surface Combat Systems in Littoral Warfare," *Surface Warfare*, Mar.–Apr. 1993.

40. Lt. Cdr. Vincent Rossitto, "A Concept of Operations for the Rapid Anti-Ship Missile Integrated Defense System (RAIDS) Prototype," *Naval Engineers Journal*, May 1990; "RAIDS," in *World Naval Weapons Systems, 1991/92*.

41. "RAM (RIM-116A)," in *World Naval Weapons Systems, 1991/92*; *Surface Warfare*, Jan.–Feb. 1993.

Chapter 10: Strike Weapons

1. "Harpoon and SLAM," in *World Naval Weapons Systems, 1991/92*.

2. Wagner interview, Oct. 24, 1992.

3. Zumwalt, *On Watch*, 81.

4. "Mark/B61" and "W-80," in Hansen, *U.S. Nuclear Weapons*; Norman Polmar, "The U.S. Navy," *USNIP*, Feb. 1991, 105.

5. Cdr. Miles Libbey, USNR, "Tomahawk," *USNIP*, May 1984, 155.

6. Carrier Group 7 staff, "Tomahawk Equals True Value," *USNIP*, Sept. 1990.

7. Froggett, "Tomahawk Has Arrived," 123–26.

8. *Aviation Week and Space Technology*, July 25, 1993.

9. *Aviation Week and Space Technology*, Jan. 17, 1994.

10. "Tomahawk," in Friedman, *U.S. Naval Weapons*.

11. Rear Adm. Walter Locke, USN (Ret.), "Tomahawk Tactics: The Midway Connection," *USNIP*, June 1992.

12. Fleet Adm. Chester Nimitz and E. B. Potter, *The Great Sea War* (Englewood Cliffs, N.J.: Prentice-Hall, 1960), 232–35. E. B. Potter is not a relative.

13. Locke, "Tomahawk Tactics."

14. "OSIS," in *World Naval Weapons Systems, 1991/92*.

15. Locke, "Tomahawk Tactics."

16. Froggett, "Tomahawk Has Arrived"; "JOTS," "TEPEE/TWCS," and "GLOBIXS and TADIXS," in *World Naval Weapons Systems, 1991/92*.

17. *Aviation Week and Space Technology*, Jan. 17, 1994.

18. USS *Hewitt* (DD966) Command History File, 1984.

19. Lt. Cdr. Rodney P. Rempt, "Vertical Missile Launchers: Part I," *USNIP*, Oct. 1977.

20. John F. Lehman, *Command of the Seas* (New York: Macmillan, 1988) 169; Scott Truver, "Tactical Vertical Launch Systems," *Naval Forces* 11, no. 3 (1990).

21. *Surface Warfare*, July–Aug. 1992, 13.

22. Norman Friedman, *Naval Radar* (Annapolis: Naval Institute Press, 1981), 127–28.

23. James Bamford, *The Puzzle Palace* (Boston: Houghton Mifflin, 1982), 162.

24. Ibid., 162.

25. "OSIS," in *World Naval Weapons Systems, 1991/92*.

26. Unclassified information provided to House Armed Services Committee for FY 1976 budget, Mar. 1975.

27. Ibid.

28. Ibid.

29. Ball, "Development of the SIOP," 72–78.

30. "SSQ-108," in *World Naval Weapons Systems, 1991/92*.

31. "SRD-19," in ibid.

32. "SLR-16," in ibid.

33. "SLR-16," "SRD-19," and "SSQ-108," in ibid.

34. Milan Vego, "Soviet Shipboard Tactical Communications," *Signal*, Nov. 1988.

35. Froggett, "Tomahawk Has Arrived."

36. USS *Caron* (DD 970) Command History File, 1981.

37. "Satellite Information Exchange System: TACINTEL," in *World Naval Weapons Systems, 1991/92*; "USQ-64," in ibid., addendum.

38. "JOTS II (USQ-112A)/VIDS" and "GLOBIXS and TADIXS," in *World Naval Weapons Systems, 1991/92*, addendum.

39. Rossitto, "Concept of Operations for RAIDS Prototype," fig. 1.

40. Schweizer, *Victory*, 230–31; Stuart Slade, "Naval Electronic Warfare," in Robert Gardiner, ed., *Navies in the Nuclear Age* (London: Brassey's, 1993), 195.

41. John Jordan, *Soviet Warships, 1945 to the Present* (London: Arms & Armour, 1992), 176–77.

42. The destroyer group commander, Rear Adm. V. P. Yeremin, was in a position to know. His flagship was the *Sovremennyy*-class destroyer *Bezuprechny*, calling at Portsmouth, U.K., in 1990. "Soviet C² Deficiencies Highlighted," *International Defense Review*, Aug. 1990.

43. Correspondence from Norman Friedman, Nov. 3, 1993. Still another theory is that they were a response to the Aegis-armed strike cruiser plan.

44. *USNIP*, Feb. 1984, 119. USS *Barry* (DD 933), the museum ship at Washington Navy Yard, was one of the ASW-modernized DD 931–class destroyers.

45. Sokolsky, *Seapower in the Nuclear Age*, 185.

46. Quoted in David Alan Rosenberg, "It Is Hardly Possible to Imagine Anything Worse," *Naval War College Review*, summer 1988.

47. Love, *History of the U.S. Navy*, vol. 2, *1942–1991*, 722; interview with Lt. Col. Robert McFarlane, USMC (Ret.), Oct. 18, 1994.

48. Quoted in Eric Arnett, *Limiting Sea-Launched Cruise Missiles* (Washington, D.C.: American Association for the Advancement of Science, 1989).

49. Cdr. S. J. Froggett, USN, "Tomahawk's Role," *USNIP*, Feb. 1987.

50. Capt. Linton Brooks, " 'New' as in Nuclear Land Attack Tomahawk," *USNIP*, Apr. 1985.

51. Ibid.; "Standard Missile-2," in *World Naval Weapons Systems, 1991/92*.

52. Capt. Linton Brooks, USN, for the National Security Council, quoted in Arnett, *Limiting Sea-Launched Cruise Missiles*.

53. Michael R. Beschloss and Strobe Talbott, *At the Highest Levels* (Boston: Little, Brown, 1993), 162–63.

54. Norman Friedman, *Desert Victory* (Annapolis: Naval Institute Press, 1991), 205.

55. *Surface Warfare*, Sept.–Oct. 1992.

56. Friedman, *Desert Victory*, 185.

57. *Surface Warfare*, Sept.–Oct. 1992; U.S. Department of Defense, *Conduct of the Persian Gulf War* (Washington, D.C.: U.S. Department of Defense, 1992), 116.

58. See, for example, *Aviation Week and Space Technology*, Apr. 27, 1992, 18–20. Wagner interview, Oct. 24, 1992.

59. I witnessed this incident.

60. Surface Navy Assn. littoral warfare symposium, San Diego, Sept. 28, 1994.

61. *Aviation Week and Space Technology*, Jan. 17, 1994.

62. Lt. Cdr. Kevin Baxter, "Tomahawk," *Surface Warfare*, May/June 1994.

Chapter 11: Antisubmarine Warfare

1. Friedman, *U.S. Destroyers*, 369–77.

2. McKeown and Robinson, "Submarines."

3. Lt. Cdr. Wayne Brown, "The Navy's Poor Relation," *USNIP*, Oct. 1987, 149.

4. NavEdTra, *Surface Ship Operations*, 228.

5. William D. O'Neill, "Passive Sonar Has the Edge," *USNIP*, Apr. 1978, 105–7; "Antisubmarine Warfare," in *World Naval Weapons Systems, 1989/90*.

6. NavEdTra, *Surface Ship Operations*, 224–27; "SSQ-36 (XBT)," in *World Naval Weapons Systems, 1989/90*.

7. Jordan, *Soviet Submarines*, 104–6; USS *Hewitt* (DD 966) Command History File, 1985.

8. *Jane's Fighting Ships, 1972–1973*, 462.

9. "SQS-26/53," in *World Naval Weapons Systems, 1991/92*.

10. Lt. G. S. Capen, "We Need to Rethink Littoral ASW," *Surface Warfare*, Sept.–Oct. 1993.

11. "SQS-35," in *World Naval Weapons Systems, 1991/92*; Polmar, *Ships and Aircraft of the U.S. Fleet*, 11th ed., 95.

12. "SQR-14/15," in *World Naval Weapons Systems, 1991/*

92; *Surface Warfare*, Jan.–Feb. 1987, 13; Polmar, *Ships and Aircraft of the U.S. Fleet*, 14th ed., 522–23.

13. "ASROC (RUR-5A)" and "Vertical Launch ASROC (VLA)," in ibid.

14. NavEdTra, *Surface Ship Operations*, 255.

15. Norman Friedman, "SUBROC, ASROC, and Terrier Retire," *USNIP*, July 1989.

16. No nuclear capability: *USNIP*, July 1983, 126.

17. Hansen, *U.S. Nuclear Weapons*, 86–88, 164–66, 206–9; Sokolsky, "U.S. Navy and Nuclear ASW Weapons."

18. *Spruance* (DD 963) and *Moosbrugger* (DD 980) Command History Files.

19. Capt. James W. Kehoe et al., "Seakeeping and Combat Systems—The Operators' Assessment," *Naval Engineers Journal*, May 1983, 256–66.

20. Lt. Dennis Stokowski, USN, "The FFG-7s in War and Peace," *USNIP*, Apr. 1984.

21. "SQS-26/53," in *World Naval Weapons Systems, 1991/92*.

22. "SQQ-89" and "Mk 116," in ibid.

23. Cdr. Gregory Dies, interview, July 30, 1992.

24. A mobile inshore undersea warfare unit is a road-mobile naval command that operates a TSQ-108 surveillance van, in effect a coastal combat information center. The van includes secure radio circuits, an X-band radar, and an SQR-17 sonar processor to monitor moored or drifting sonobuoys that the unit deploys by Zodiac boats.

25. John J. Engelhardt, "The Soviets on Submerged Unsinkability, 1959–84," *Submarine Review*, Apr. 1988.

26. Dan Manningham, "LAMPS III," *USNIP*, Mar. 1978, 159–61.

27. Cdr. George Galdorisi, "LAMPS-III: Carrier Battle Group Synergist," *USNIP*, Aug. 1986, 98–99; *Surface Warfare*, Jan.–Feb. 1987, 8–10.

28. Lt. Cdr. R. E. Hammond, USN, and Lt. Pat Tierney, USN, "The LAMPShip Team," *USNIP*, Mar. 1978, 154–58.

29. Edward N. Comstock et al., "Seakeeping Performance Comparison of Air Capable Ships," *Naval Engineers Journal*, Apr. 1982.

30. Rear Adm. Raymond Winkel and Dan Manningham, "LAMPS: The Ship System with Wings," *USNIP*, Mar. 1980, 114–17.

31. Lt. Cdr. J. D. Beaulieu, "Launching and Landing LAMPS—Safely," *USNIP*, Apr. 1986.

32. Comstock et al., "Seakeeping Performance."

33. Stokowski, "FFG-7s in War and Peace."

34. Capt. William J. Frigge, "Winning Battle Group ASW," *USNIP*, Oct. 1987.

35. Capt. Wayne Hughes, *Fleet Tactics: Theory and Practice* (Annapolis: Naval Institute Press, 1986), 166–69.

36. "BQQ-2," in *World Naval Weapons Systems, 1989/90*; Rear Adm. Fred Gustavson, ComSubGru 5, unclassified presentation to Surface Navy Assn., San Diego, Aug. 11, 1994.

37. "SUS," in ibid.

38. Capen, "We Need to Rethink Littoral ASW."

39. Information from Surface Navy Association Shallow Water ASW Symposium, San Diego, Mar. 22, 1994.

40. Robert L. Scheina, "Where Were Those Argentine Subs?" *USNIP*, Mar. 1984; Brown, *Royal Navy and Falklands War*, 156–57.

41. Capt. James H. Patton, USN (Ret.), letter to *USNIP*, July 1988, 71.

42. Lt. Cdr. David Frieden, *Principles of Naval Weapons Systems* (Annapolis: Naval Institute Press, 1985), 349–50.

43. "Nixie (SLQ-25)/SLQ-36," in *World Naval Weapons Systems, 1991/92*.

44. Richard A. Holden, "Surface Ship Combat System Upgrade Engineering," *Naval Engineers Journal*, May 1988, fig. 8.

45. "Antisubmarine Warfare," in *World Naval Weapons Systems, 1991/92*; Lt. John Hutsebaut, "Still No Cure for the Common Torpedo," *USNIP*, Oct. 1985.

46. *Surface Warfare*, Jan.–Feb. 1987, 16.

47. "Soviet/Russian Navy ASW Weapons Systems," in *World Naval Weapons Systems, 1994 Update*, 128.

48. "Sonars and Underwater Fire Control Systems," in *World Naval Weapons Systems, 1994 Update*.

49. Norman Friedman, interview, Mar. 1992.

50. Capt. William D. O'Neil, "Winning the ASW Technology Race," *USNIP*, Oct. 1988.

Chapter 12: The *Kidd* (DDG 993) Class

1. Weschler interview, Sept. 5, 1992.

2. MAT 09X/ms (Weschler) Memo 28–70, 1 May 1970, "Concept Formulation and Contract Definition."

3. Kissinger, *Years of Upheaval*, 667–69.

4. See chap. 3.

5. Michael A. Palmer, *Guardians of the Gulf* (New York: Free Press/Macmillan, 1992), 90.

6. William Forbis, *Fall of the Peacock Throne* (New York: Harper & Row, 1980), 278–79.

7. See, for example, ibid., 281.

8. R. N. Frye, *The Cambridge History of Iran*, vol. 4 (New York: Cambridge University Press, 1975), 215–16. Frye transliterates the name as Anushirvan.

9. I owe this insight to Yeoman Second Class Musime Shine, USNR.

10. Palmer, *Guardians of the Gulf*, 91–92.

11. The most recent was in 1992.

12. J. S. Keating, "Mission to Mecca: The Cruise of the *Murphy*," *USNIP*, Jan. 1976. The article included a photograph of the sheep pen.

13. Palmer, *Guardians of the Gulf*, 27–28.

14. Secretary of Defense Dick Cheney, televised statement, Aug. 1990.

15. Thomas K. Korb, "USS *Kidd*," *Sea Classics*, Jan. 1988.

16. Ibid.; on the 1914 incident, see Tuchman, *Guns of August*, chap. 10.

17. *Combat Fleets of the World*, 1982–83 and all subsequent editions.

18. My experience. For a published example, see Lt. Cdr. John G. Morgan, "The *Kidd* DDG: The Non-Nuclear Standard," *USNIP*, July 1981, 99–101.

19. Congressional Budget Office, *Rationale and Plans for the 600-Ship Navy*, Sept. 1985, tab. 1.

20. Erwin Straubinger, JJH, Inc., interview, Sept. 1990.

21. At least one Mk 26 magazine has single-key access in 1982, indicating no capability to carry nuclear ASROC.

22. Edward J. Walsh, "An Alternative to Aegis," *Sea Power*, Feb. 1989.

23. Ibid.

24. Roy J. Biondi and Bradford E. Kruger, "The UNIMAST Concept—A Major Departure in Shipboard Radar An-

tenna Installation Philosophy," *Naval Engineers Journal*, Apr. 1981, 83.

25. Ibid.
26. "SLQ-34," in *World Naval Weapons Systems, 1991/92*, 531.
27. David Miller and Chris Miller, *Modern Naval Combat* (London: Salamander Books, 1986), 50.
28. Norman Friedman, "U.S. Naval Weapons and Combat Systems Development in 1985," *USNIP*, May 1986, 82–83.
29. USS *Kidd* (DDG 993) Command History File, 1989.
30. This discussion is based on a Litton-Ingalls Shipbuilding brochure, "The DD 963 Variants Program," about 1977.
31. James L. George, "Maintaining a Western Carrier Capability," *USNIP*, Oct. 1977.
32. *Aviation Week and Space Technology*, Apr. 21, 1980, 65–75.
33. Harold Flak, "Superiority at Sea—An Affordable System for the 1990s," *Naval Engineers Journal*, Apr. 1982; and comments on his paper in *Naval Engineers Journal*, June 1982.
34. Quoted in Polmar, *Ships and Aircraft of the U.S. Fleet*, 12th ed., 104.
35. Litton-Ingalls Shipbuilding, "DD 963 Variants Program."
36. Philip Covich and Michael Hammes, "Repeat Ship Designs: Facts and Myths," *Naval Engineers Journal*, May 1983, 106.
37. Correspondence from Leonard Schwartz, 1992; *USNIP*, Jan. 1981, 120n.
38. Norman Polmar, "Cruisers and Destroyers: Losing Out," *USNIP*, Apr. 1983, 99.

Chapter 13: The Shield of the Fleet: The Aegis Cruisers

1. FY 1964 Defense Budget.
2. Friedman, *U.S. Destroyers*, 224–25.
3. Friedman, *Naval Radar*, 180.
4. Inman, "From Typhon to Aegis," 62–65; and Friedman, *U.S. Naval Weapons*, 154–55 and 177–78.
5. Inman, "From Typhon to Aegis."
6. Rear Adm. Wayne Meyer, USN (Ret.), interview in *USNIP*, Oct. 1988.
7. Inman, "From Typhon to Aegis."
8. Brown, *Royal Navy and Falklands War*, 154, 220–23.
9. Capt. Bryce Inman, "The Two Cent Decision," *Naval Engineers Journal*, May 1990, 24.
10. Inman, "From Typhon to Aegis."
11. Ibid.
12. Sonenshein interview, Nov. 10–11, 1990.
13. Inman, "From Typhon to Aegis"; and his reply to comments on the same article in *Naval Engineers Journal*, July 1988, 126.
14. Quoted in Hone, "Program Manager as Entrepreneur."
15. Ibid.
16. Zumwalt, *On Watch*, 73–75.
17. Cited in *Jane's Fighting Ships, 1972–1973*.
18. Friedman, *U.S. Destroyers*, 321.
19. Hone, "Program Manager as Entrepreneur."
20. Cdr. Todd Blades, "DDG-47: Aegis on Its Way to Sea," *USNIP*, Jan. 1979, says that Zumwalt authorized the DG/Aegis design in Jan. 1972.
21. Friedman, *U.S. Destroyers*, 321–22.
22. Leopold interview, Sept. 1990.
23. Leopold, "Gas Turbines in the U.S. Navy."
24. Friedman, *U.S. Destroyers*, 342–44.
25. Zumwalt, *On Watch*, 121; Polmar and Allen, *Rickover*, 393–99.
26. According to a well-placed participant, "Because of a variety of factors, *mostly political*, the CGN program was curtailed and eventually limited to four ships, all equipped with Tartar-D" (my emphasis). Robert Beers, letter to *USNIP*, Aug. 1985.
27. Francis Duncan, *Rickover and the Nuclear Navy* (Annapolis: Naval Institute Press, 1990), 165–68.
28. Zumwalt, *On Watch*, 121; Polmar and Allen, *Rickover*, 393–99.
29. Friedman, *U.S. Destroyers*, 343, and his *U.S. Cruisers* (Annapolis: Naval Institute Press, 1974), 420–21.
30. Unclassified Navy testimony to House Armed Services Committee, 1977, in University of California, San Diego, library accession no. H201-14.3, pp. 578–84.
31. Friedman, *U.S. Cruisers*, 419–22.
32. Leopold interview, Sept. 1990.
33. Hone, "Program Manager as Entrepreneur."
34. Beers, letter to *USNIP*, Aug. 1985.
35. Friedman, *U.S. Destroyers*, 345.
36. Hone, "Program Manager as Entrepreneur."
37. Capt. Alva M. Bowen, USN (Ret.), and Ronald O'Rourke, "DDG-51 and the Future Surface Navy," *USNIP*, May 1985, 182.
38. This section is based on Robert C. Staimen, "Aegis Cruiser Weight Reduction and Control," *Naval Engineers Journal*, May 1987, and comments on his article, *Naval Engineers Journal*, July 1987.
39. A. D. Baker, letter in response to Rear Adm. Wayne Meyer and Capt. Bart Dalla Mura, "Aegis," *USNIP*, July 1977.
40. Lt. Cdr. James Stavridis, "Handling a *Ticonderoga*," *USNIP*, Jan. 1987.
41. Ibid; Senior Chief Boatswain's Mate K. L. Cook, interview, Jan. 31, 1993.
42. Stavridis, "Handling a *Ticonderoga*."
43. *New York Times*, Aug. 17, 1982.
44. Hone, "Program Manager as Entrepreneur."
45. *New York Times*, Aug. 17, 1982.
46. Frederick H. Hartmann, *Naval Renaissance: The U.S. Navy in the 1980s* (Annapolis: Naval Institute Press, 1990), 136.
47. *New York Times*, Aug. 17, 1982.
48. Hartmann, *Naval Renaissance*, 136–41.
49. Quoted in *Naval Engineers Journal*, Sept. 1984, 13.
50. *Navy Times*, Jan. 15, 1990.
51. Ibid.
52. Eli Brookner, "Phased-Array Radars," *Scientific American*, Feb. 1985, 94–102.
53. "SPY-1/FARS" and "Aegis," in *World Naval Weapons Systems, 1991/92*.
54. *Navy Times*, Aug. 25, 1986, 35–36.
55. Meyer and Dalla Mura, "Aegis."
56. *Navy Times*, Aug. 25, 1986, 35–36.
57. *USNIP*, Oct. 1983, 26.
58. Capt. Joseph McClane and Cdr. James McClane, "The *Ticonderoga* Story: Aegis Works," *USNIP*, May 1985.
59. Rear Adm. R. G. Guilbault, letter to *USNIP*, Jan. 1986, 100; correspondence from Rear Adm. Donald Dyer, USN, Apr. 25, 1994.

60. Rear Adm. Richard Berry, quoted in *Naval Engineers Journal*, Sept. 1984, 17.

61. Lehman, *Command of the Seas*, 370.

62. Vice Adm. Joseph Metcalf, "Revolution at Sea," *USNIP*, Jan. 1988.

63. Rodney Rempt, "Killing Scuds from the Sea," and Cdr. John Carey, "Fielding a Theater Ballistic Missile Defense," both in *USNIP*, June 1993; *Aviation Week and Space Technology*, Mar. 28, 1994.

64. Holden, "Surface Ship Combat System Upgrade Engineering," fig. 8.

Chapter 14: Operations

1. Hammond and Tierney, "LAMPShip Team."

2. USS *Elliot* (DD 967) Command History File, 1979; my experience as this ship's supply officer during 1978–80.

3. *Ranger* was too large for a floating dry dock, so the Subic Bay ship repair facility built a cofferdam around her damaged bow and repaired it with temporary plating stiffened by concrete. This gave her stem its curious shape as seen in photographs in *The Ships and Aircraft of the U.S. Fleet*, 12th and 13th eds. (Annapolis: Naval Institute Press, 1981, 1984), 60 and 93, respectively. *Ranger* was permanently repaired at Yokosuka.

4. Floyd D. Kennedy, "From SLOC Protection to a National Maritime Strategy: The U.S. Navy under Carter and Reagan, 1977–1984," in *In Peace and War*, 2nd ed., ed. Kenneth J. Hagan (Westport, Conn.: Greenwood, 1984), 350.

5. Watson, *Red Navy at Sea*, 63–68 and 162–64.

6. *Aviation Week and Space Technology*, May 28 and June 4, 1979.

7. Conversation with me.

8. *Playboy*, Apr. 1979, or a month on either side of it. It identified her as Denise, from Baltimore; she called herself Dee when I knew her. She had posed for the magazine before she wed an Annapolis graduate.

9. Information from Cdr. Gregory Dies, USNR, Aug. 1992.

10. USS *Arthur W. Radford* (DD 968) Command History File, 1979; Dies interview.

11. William J. Aceves, "Diplomacy at Sea: U.S. Freedom of Navigation Operations in the Black Sea," *Naval War College Review*, spring 1993.

12. USS *Caron* (DD 970) Command History File, 1979 and 1980.

13. Jeffrey Richelson, *The U.S. Intelligence Community*, 2nd ed. (Cambridge: Harper & Row, 1989), 188–89.

14. Ibid.

15. Friedman, *Desert Victory*, 65.

16. Col. W. Hays Parks, USMC, "Crossing the Line," *USNIP*, Nov. 1986, 42.

17. Ibid.

18. USS *Caron* (DD 970) Command History File, 1981. Published accounts do not list *Caron* among ships present, but she was there.

19. Col. W. Hays Parks, USMC, "Righting the Rules of Engagement," *USNIP*, May 1989.

20. USS *Moosbrugger* (DD 980) Command History File, 1979.

21. USS *Caron* (DD 970) Command History File, 1982; USS *Deyo* (DD 989) Command History File, 1982.

22. Newspaper and news magazine clippings in ibid.

23. Christopher C. Wright, "U.S. Naval Operations in 1982," *USNIP*, May 1983, 225.

24. USS *Caron* (DD 970) Command History File, 1984, 1987; *Navy Times*, Nov. 26, 1984.

25. USS *Kidd* (DDG 993) Command History File, 1983; G. Mueller and F. Adler, *Outlaws of the Ocean* (New York: Hearst Marine Books, 1985), 90–91; QM1 Robert Laahs, USCG, " 'Joint' Operation," *USNIP*, June 1984.

26. USS *Stump* (DD 978) Command History File, 1989.

27. Most published accounts refer to the flight as KAL-007.

28. Murray Sayle, "Closing the File on Flight 007," *New Yorker*, Dec. 13, 1993.

29. Ibid.; USS *Elliot* (DD 967) Command History File, 1983.

30. *Elliot* Cruise Book for 1983 in Navy Department Library; Bartholomew, *Mud, Muscle and Miracles*, 417–18.

31. For example, R. W. Johnson, *Shootdown* (New York: Viking, 1986).

32. *Aviation Week and Space Technology*, June 21, 1993.

33. Frances Wright Norton, "Caribbean Naval Activity," in *Soviet Navy: Strengths and Liabilities*, ed. Watson and Watson, 211; Maj. Mark Adkin, *Urgent Fury* (Lexington, Mass.: Lexington Books, 1989), 22, 110. USS *Caron* (DD 970) Command History File, 1983; C. C. Wright, "U.S. Naval Operations in 1983," *USNIP*, May 1984. Adkin is often unreliable: several of his episodes are invented (see n34, n38); he makes misleading errors in times and sequences of actual events; his denial of any American skill or leadership is absurd; and his consequent attribution of the U.S. victory to endless good luck is ridiculous.

34. USS *Caron* (DD 970) Command History File, 1983; correspondence from Rear Adm. Donald A. Dyer, USN, Mar. 21, 1994; Kevin Dockery, *SEALs in Action* (New York: Avon, 1991), 263–67; information from officer briefed by the SEAL platoon commander. In a Jan. 1984 public-affairs note *Caron* reported rescuing 12 commandos at 0430, but *Moosbrugger* logged a report of 16 and it may have been 18 (*Caron* reported 41 total rescues at Grenada). I suggest *Caron*'s writer copied 12 from that night's rescue; he also erred in saying lookouts spotted the raft, which had sunk. Adkin and others err in dating the pre-dawn rescue Oct. 24 and claiming it somehow occurred twice.

35. Rear Adm. Donald A. Dyer, USN, interview, Feb. 28, 1994; he still has this note. Daniel P. Bolger, *Americans at War* (Novato: Presidio, 1988), 300. JP-5 transfer amount is estimated from 62% used of 72 tons capacity by Oct. 27 while supporting an SH-2F and SH-3H.

36. Dyer interview, Feb. 28, 1994.

37. USS *Caron* (DD 970) Command History File, 1983; Silver Star citation for SEAL commander in Orr Kelly, *Brave Men, Dark Waters* (Novato: Presidio, 1992), 209.

38. Adkin, *Urgent Fury*, 181–91, is fiction as to the Beausejour cover fire and "hesitation and delay" in refueling helicopters and rescuing the wounded.

39. Dyer interview, Feb. 28, 1994.

40. Capt. Richard M. Butler, "Operational Sea Stories from URGENT FURY," Naval War College, 1986.

41. Bolger, *Americans at War*, 335–36.

42. Butler, "Operational Sea Stories from URGENT FURY."

43. Dyer interview, Feb. 28, 1994; USS *Caron* (DD 970) Command History File, 1983.

44. Dyer interview, Feb. 28, 1994.

45. USS *John Rodgers* (DD 983) Command History File, 1983.

46. Benis M. Frank, *U.S. Marines in Lebanon, 1982–1984* (Washington, D.C.: USMC History and Museums Division, 1987), chronology and chap. 6.

47. Ibid.

48. Wright, "U.S. Naval Operations in 1983."

49. Information from Lt. Cdr. Eric Massa, USN, aboard *New Jersey* in 1984; Lehman, *Command of the Seas,* 334.

50. USS *Caron* (DD 970) Command History File, 1984; Dyer correspondence, Apr. 25, 1994.

51. News briefing by Secretary of Defense Weinberger, Oct. 11, 1985; Lehman, *Command of the Seas,* 364–66.

52. Scott Truver, "Maritime Terrorism, 1985," *USNIP,* May 1986; USS *Yorktown* (CG 48) Command History File, 1985; USS *Caron* (DD 970) Command History File, 1985.

53. Bolger, *Americans at War,* 383–86, 437 n. 3.

54. *New York Times,* Mar. 28, 1986, 12.

55. Ibid.; authoritative Navy source.

56. Adm. V. Chernavin, Soviet Navy, in *Izvestia,* Mar. 22, 1986, quoted in UPI dispatch from Moscow.

57. Aceves, "Diplomacy at Sea," cites *Washington Post,* June 8, 1984. The *Post*'s reporter, Rick Atkinson, later spread the Desert Storm carbon-wire Tomahawk rumor.

58. Quoted in ibid., n. 26.

59. Ibid., citing a CIA message to Department of State.

60. See, for example, *Time,* Mar. 31, 1986.

61. USS *Yorktown* (CG 48) Command History File, 1988; USS *Caron* (DD 970) Command History File, 1988; *Navy Times,* Feb. 22, 1988.

62. Capt. William L. Schachte, OSD deputy assistant judge advocate general (international law), "The Black Sea Challenge," *USNIP,* June 1988.

63. Lt. Cdr. Ronald Neubauer, "The Right of Innocent Passage for Warships in the Territorial Sea: A Response to the Soviet Union," *Naval War College Review,* spring 1988; and letters in *Naval War College Review,* autumn 1988 and spring 1989.

64. USS *Peterson* (DD 969) Command History File, 1979.

65. SECDEF Washington DC message 1421152Z MAY 86.

66. USS *David R. Ray* (DD 971) Command History File, 1986.

67. Palmer, *Guardians of the Gulf,* 123.

68. John Bulloch and Harvey Morris, *The Gulf War* (London: Methuen, 1989), 233–35.

69. Tamara Moser Melia, *"Damn the Torpedoes": A Short History of U.S. Naval Mine Countermeasures, 1777–1991* (Washington, D.C.: Naval Historical Center, 1991), 121; McFarlane interview, Oct. 18, 1994.

70. Quoted in "SS *Bridgeton:* The First Convoy," *USNIP,* May 1988, 52.

71. USS *Kidd* msg 261150Z JUL 87, in USS *Kidd* (DDG 993) Command History File, 1987.

72. Ronald O'Rourke, "The Tanker War," *USNIP,* May 1988, tab. 7; *Navy Times,* Dec. 21, 1987.

73. Norman Cigar, "The Soviet Navy in the Persian Gulf: Naval Diplomacy in a Combat Zone," *Naval War College Review,* spring 1989, 76.

74. USS *Kidd* (DDG 993) Command History File, 1987.

75. Palmer, *Guardians of the Gulf,* 133.

76. Ronald O'Rourke, "The Gulf War," *USNIP,* May 1989.

77. USS *Kidd* (DDG 993) Command History File, 1987; O'Rourke, "Gulf War." The Navy Office of Information in Los Angeles tried to help a Hollywood firm produce a movie about the capture of *Iran Ajr.* After-action reports were classified, so the film was not made.

78. Adm. William J. Crowe, USN, chairman, Joint Chiefs of Staff, endorsement (Aug. 18, 1988) on Rear Adm. W. M. Fogarty, "Formal Investigation into the Circumstances Surrounding the Downing of Iran Air Flight 655 on July 3, 1988," July 28, 1988. Hereafter the basic letter is cited as Fogarty Investigation.

79. Frank Elliott, "The Navy in 1987," *USNIP,* May 1988, 146–47; USS *John Young* (DD 973) Command History File, 1987; USS *Kidd* (DDG 993) Command History File, 1987.

80. USS *Bunker Hill* (CG 52) Command History File, 1987.

81. *Pravda,* Dec. 6, 1987, quoted in Capt. William J. Manthorpe, "The Soviet Navy, 1987," *USNIP,* May 1988, 233, and in Cigar, "Soviet Navy in the Persian Gulf," 75 and 87, n. 107.

82. USS *Chandler* (DD 996) Command History File, 1988.

83. Department of Defense press briefing, reported in San Diego *Tribune,* Feb. 16, 1988; Michael Palmer, Naval Historical Center, interview, July 26, 1990.

84. USS *Merrill* (DD 976) Command History File, 1988.

85. Col. William M. Rakow, USMC, "Marines in the Gulf—1988," *Marine Corps Gazette,* Dec. 1988.

86. Capt. J. B. Perkins, USN, "The Surface View: Operation Praying Mantis," in *USNIP,* May 1989; Rakow, "Marines in the Gulf"; USS *Merrill* (DD 976) Command History File, 1988.

87. Perkins, "Surface View: Operation Praying Mantis."

88. Ibid.

89. Capt. Bud Langston, USN, and Lt. Cdr. Don Bringle, USN, "The Air View: Operation Praying Mantis," *USNIP,* May 1989.

90. Perkins, "Surface View: Operation Praying Mantis."

91. Lt. Col. David Evans, USMC (Ret.), "*Vincennes:* A Case Study," *USNIP,* Aug. 1993.

92. Palmer, *Guardians of the Gulf,* 146.

93. Crowe endorsement on Fogarty Investigation.

94. Fogarty Investigation.

95. Ibid.

96. *Newsweek,* July 13, 1992. Parts of this article, such as its claim that radio messages supposedly from the freighter were a Navy deception to lure out the Iranian boats, were fiction.

97. Fogarty Investigation.

98. Evans, "*Vincennes:* A Case Study."

99. Ibid.

100. Fogarty Investigation.

101. Capt. Will Rogers, USN (ret.), interview, Jan. 10, 1994.

102. Evans, "*Vincennes:* A Case Study"; Cdr. David Carlson, USN, letter to *USNIP,* Sept. 1989.

103. Capt. Will Rogers, "Counterbattery," *USNIP,* Aug. 1993; Gen. George B. Crist, USMC, commander in chief, U.S. Central Command, endorsement on Fogarty Investigation, Aug. 5, 1988. Gen. Crist's analysis is the only reliable section of the official investigation.

104. Crist endorsement on Fogarty Investigation.

105. Fogarty Investigation.

106. Ibid.

107. Reports that it was a Mecca-bound pilgrimage flight are incorrect.

108. USS *Vincennes* (CG 49) Command History File, 1988. The ship submitted this accurate report only after Capt. Rogers was transferred off.

109. USS *Spruance* (DD 963) Command History File, 1988.

110. Palmer, *Guardians of the Gulf*, 148.

111. See, for example, Los Angeles *Times* stories during the following days.

112. Rogers interview, Jan. 10, 1994.

113. HQ, USMC SO/LIC (Col. J. G. Magee, USMC) letter 1000 of Oct. 19, 1990, a letter prompted by a chance meeting between me and Gen. Al Gray, commandant of the Marine Corps, in the check-out line at the Pentagon bookstore. After Gen. Gray appeared in a cameo on the TV series *Major Dad*, my daughter told friends, "My dad knows a TV star."

114. Lt. Col. T. W. Parker, USMC, "Operation SHARP EDGE," *USNIP*, May 1991.

115. Information presented at Programming, Planning and Budgeting System course in the Pentagon, Aug. 1985.

116. *Conduct of the Persian Gulf War*, 206.

117. USS *Leftwich* (DD 984) Command History File, 1990; *Surface Warfare*, July–Aug. 1991, 14–15; *Conduct of the Persian Gulf War*, 190–96.

118. USS *Fife* (DD 991) Command History File, 1991; Friedman, *Desert Victory*, 72–73; *Conduct of the Persian Gulf War*, 59, 62; *Fife* and *Oldendorf* cruise books.

119. *Conduct of the Persian Gulf War*, 118–19.

120. USS *Paul F. Foster* (DD 964) Command History File, 1991; RM3 Pickens, aboard *Paul F. Foster* at the time, interview, Apr. 1993.

121. Lt. John van Patten, "*Princeton* Completes Persian Gulf Tour," press release from the ship, Apr. 24, 1991.

122. *Conduct of the Persian Gulf War*, 206.

123. *Surface Warfare*, July–Aug. 1991, 10; Pickens interview, Apr. 1993.

124. USS *Valley Forge* (CG 50) Command History File, 1992.

125. *Aviation Week and Space Technology*, Jan. 25, 1993.

126. Ibid., July 5, 1993.

Conclusion

1. President George Bush, *National Security Strategy of the United States* (Washington, D.C.: The White House, 1991).

2. Hughes, *Fleet Tactics*, 203.

3. Ibid., 196–99.

4. Tracy Kidder, *The Soul of a New Machine* (Boston: Little, Brown, 1981).

5. Hartmann, *Naval Renaissance*, 136ff.

6. See comments by Adm. Stansfield Turner in Paul Ryan, *The Iranian Rescue Mission: Why It Failed* (Annapolis: Naval Institute Press, 1985), 135.

7. Quoted in Evans, "*Vincennes*: A Case Study."

Glossary

This does not list all acronyms and weapon designations that appear in this book, but it does list those most frequently used.

2-D, 3-D radar	2-D = range and bearing, 3-D = range, bearing, altitude
AAW	Antiaircraft warfare
ABL	Armored box launcher for TASM and TLAM
Aegis	Computer-based combat control system on *Ticonderoga* class
ASCM	Antiship cruise missile
ASROC	Antisubmarine rocket
ASW	Antisubmarine warfare
CDS	Command and decision system or combat direction system
CG, CGN	Guided missile cruiser
CIC	Combat information center
CIWS	Close-in weapon system (Phalanx)
CNO	Chief of Naval Operations
COGAG	Combined gas-and-gas, a propulsion plant design in which two gas turbines can simultaneously power one propeller shaft
ComInt	Communications intelligence
CRP	Controllable-reversible pitch (applies to propellers)
DD, DDG	Destroyer hull sequence number prefix. DDG denotes guided missile destroyer.
DFM	Diesel fuel—marine (also called NDF and F-76)
DX/DXG	Code designating *Spruance*-class design requirements specification
ECM	Electronic countermeasures (radar jamming, etc.)
ESM	Electronic support measures (passive surveillance systems to intercept radio and radar emissions). Similar to intelligence collection; the distinction is that ESM focuses on providing immediate combat data.
FF/FFG	Frigate (originally DE/DEG, destroyer escort)
FY	Fiscal year, ending September 30 (June 30 until 1976)
GFE	Government-furnished equipment
IFF	Identification friend or foe; part of many military radars

JP-5	Grade of jet propulsion fuel suitable for shipboard storage
LAMPS	Light airborne multipurpose system (see Seasprite and Seahawk)
LHA	*Tarawa*-class amphibious assault ship
Link 11	NTDS data link
Mark 26	Twin-arm guided missile launcher on *Kidd*- and early *Ticonderoga*-class ships
Mark 32	Triple 12.75" torpedo tube mounting for ASW torpedoes
Mark 45	5"/54 single-gun mount
Mark 74	Tartar-D (Digital Tartar) missile fire control system
Mark 86	Computer-based gunfire control system for 5" guns
NavOrd	Naval Ordnance Systems Command (part of NavSea since 1973)
NavSea	Naval Sea Systems Command, headquarters for design of warships, radar, sonar, and ship-launched weapons
NavSEC	Naval Ship Engineering Center (now part of NavSea)
NavShips	Naval Ship Systems Command (part of NavSea since 1973)
NTDS	Naval tactical data system, real-time software to track contacts and to designate specific weapons to attack targets
OASD(SA)	Office of the Assistant Secretary of Defense for Systems Analysis (now Director, Program Analysis and Evaluation)
OpNav	Office of the Chief of Naval Operations
OSD	Office of the Secretary of Defense
Outboard	Communications intelligence collection system on most *Spruance*-class destroyers for detection and classification of radio transmitters and for analysis of transmissions
SAR	Search and rescue
SH-2F Seasprite	Early destroyer-based helicopter for LAMPS Mark I
SH-60B Seahawk	Current destroyer-based helicopter for LAMPS Mark III
SIOP	Single integrated operational plan, the set of

weapon-target assignments that the president would order the military to execute at the outbreak of nuclear war

SLQ-32 — Shipboard electronic warfare and countermeasures system. SLQ-32(V)2 is a radar-warning system for chaff launchers, while SLQ-32(V)3 adds active jamming.

SPG-n — Shipboard radar for gunfire control or missile guidance: SPG-51D, SPG-60, SPG-62

SPQ-9A — Gunfire control radar to engage surface and shore targets (part of Mark 86 gunfire control system for 5" guns)

SPS-n — Shipboard mechanically scanning search radar: SPS-40B, SPS-48E, SPS-49, SPS-55, SPS-64

SPY-1 — Aegis multifunction electronically scanning radar for search and missile fire control

SQQ-89 — Computer-based ASW display system, used primarily to coordinate SQR-19 long-range sonar and SH-60B helicopter

SQR-19 — Very long range passive towed-array sonar

SQS-53 — Shipboard hull-mounted active/passive sonar

Standard — Medium-range antiaircraft missile used by *Kidd* and *Ticonderoga* classes: SM-1(MR), SM-2(MR)

Tartar-D — Digital Tartar, the fire control system for Standard missiles on the *Kidd* class

TASM — Tomahawk antiship missile

TLAM — Tomahawk land-attack missile. TLAM-C is the most common version, carrying a high-explosive warhead

TPP — Total package procurement

UYK-n — Military digital computer: UYK-7, UYK-43, etc.

VL-ASROC, VLA — Vertical-launch antisubmarine rocket

VLS — Vertical launch system for guided missiles

Index

torpedo defense, 105, 155–59, 162–63, 192
torpedoes, 155–59
    Mk 32 tubes, 155–57, *157*
Total Package Procurement (TPP), 4, 27–29, 75, 83, 88,
    239–40. *See also* FDL ship TPP; Office of the
    Secretary of Defense; *Spruance* class; *Tarawa* class
Trenton (LPD 14), 224–26, 229, 231, 233
Trost, Adm. Carlisle, 221, 229
Truman, Harry S., 6–7
Truxtun (CGN 35), 8
Turner Joy (DD 951), 15
Type 21 frigate *Arrow* (F 173) (U.K.), 162
Type 22 frigate (U.K.), 137, 232
Type 42 destroyer (U.K.), 120, 162, 180–81
Type 81 frigate (Tribal class) (U.K.), 19
Typhon guided missiles, 11, 18–19, 21, 179–81

Udaloy class (USSR), 139, *140*
underway replenishment, 48, 86, 104, 106–7
    ships, AOE 6 Supply class, 160
U.S. Air Force, 6–8, 20, 25, 33, 179
    and DD 963s, 209–10, 215, 219, 230
    devises TPP, 28–30
U.S. Army, 6, 29–31, 231
    and DD 963s, 209–13, 226, 231–33
U.S. Coast Guard, 206, 232
    and *Hamilton*-class cutters, 27, 41, 46, 72
U.S. Marine Corps, 29–31, 141
    and DD 963s, 213–15, 225, 230
Urgent Fury, Operation. *See* Grenada
USSR, 32, 77, 86, 139, 142, 180, 218, 226. *See also*
    individual ships; aircraft, USSR; antiship cruise
    missiles, USSR; submarines, USSR
    collapse of, 4, 139, 142, 198, 230
    and first salvo tactics, 10–11, 31–32, 79, 104, 145
    military expansion of, 24, 31–33, 38, 75, 77, 198
    naval operations of, before 1975, 12, 51, 77–79
    reaction of, to *Spruance* class, 139–40
    strategy of, 5–7, 13, 31–34, 141, 147, 160
    surface ships of, 75, 77–79, 83–84, 87–88; and DD 963
    operations, 200–202, 204, 216–19, 221

U.S. Navy challenges claims of, 204, 217–19
UYK-7/UYK-43 computers, 113, 156, 171, 181, 193

Valley Forge (CG 50), 189, 221, 232, 235
vertical launch system (VLS), 136, 194, 197, 234, 237
    installation of, 112, 118, 129, 162
    mentioned, 230, 232
    and Tomahawk land-attack missiles, 131, 134, 189–90
Vessey, Gen. John, 208
Vietnam Heroico, 210–11
Vietnam War, 15–16, 25, 33, 77–78, 214
    influence of, on DD 963, 16–18, 33–34, 126, 181
    political effects of, 2, 5, 31, 24, 79
Vincennes (CG 49), 16, 124, 189, 226–29, *228*, 240
Virginia (CGN 38), 214
Virginia class, 50, 164, 169, 185–87
Vladivostok, 200
VL-ASROC. *See* ASROC
VLS. *See* vertical launch system
V/STOL ship designs, 175–78, 186–87

Wainwright (CG 28), 225
Watergate scandal, 16, 78, 89, 186
Watkins, Adm. James, 105, 192
Weschler, Vice Adm. Thomas, 17, 39, 54, 238–39
    and DXG, 49–50, 164
    as DX program manager, 38–40, 44–47, 50, 69, 75, 164
    originates LAMPS, 44–45, 126
Withington, Rear Adm. Frederick, 180–81
Woods, Rear Adm. Mark, 76–77, 183
Worden (DLG/CG 18), 78, 120
WSC-3 SatCom, 95, 111, 122, 138, 171

Yom Kippur War, 78–79, 90, 204
Yorktown (CG 48), 189, 197, 215–18

Zumwalt, Adm. Elmo, 14–15, 77–78, 89–91, 97, 131, 147
    and DG/Aegis, 184–85
    and Major Fleet Escort Study, 27, 35, 40–42
    mentioned, 176, 186
    and *Oliver Hazard Perry* class, 50, 85, 88–89, 164–65
    and *Spruance* class, 38, 75, 79, 88, 94, 238

About the Author

Michael Potter has firsthand knowledge of *Spruance*-class destroyers: he helped to prepare one for her maiden deployment in 1979 and served on the first *Spruance*-class destroyer operation in the Indian Ocean and Arabian Sea. Before that, he participated in the evacuations of Cambodia and South Vietnam and was on the staff of Commander Naval Air Force, U.S. Pacific Fleet.

As an officer in the Naval Reserve, he has had assignments in the Pentagon, the Naval War College, Japan, Alaska, and elsewhere and has served as the commanding or executive officer of three units. In civilian life he is a program manager in the computer industry, specializing in defense systems.

Captain Potter earned a B.A. in economics from the University of Michigan and an M.S. in systems management from the University of Southern California. He lives in San Diego, California, with his wife, Jan, and two children. This is his first book.

The **Naval Institute Press** is the book-publishing arm of the U.S. Naval Institute, a private, nonprofit society for sea service professionals and others who share an interest in naval and maritime affairs. Established in 1873 at the U.S. Naval Academy in Annapolis, Maryland, where its offices remain, today the Naval Institute has more than 100,000 members worldwide.

Members of the Naval Institute receive the influential monthly magazine *Proceedings* and discounts on fine nautical prints and on ship and aircraft photos. They also have access to the transcripts of the Institute's Oral History Program and get discounted admission to any of the Institute-sponsored seminars offered around the country.

The Naval Institute also publishes *Naval History* magazine. This colorful bimonthly is filled with entertaining and thought-provoking articles, first-person reminiscences, and dramatic art and photography. Members receive a discount on *Naval History* subscriptions.

The Naval Institute's book-publishing program, begun in 1898 with basic guides to naval practices, has broadened its scope in recent years to include books of more general interest. Now the Naval Institute Press publishes more than seventy titles each year, ranging from how-to books on boating and navigation to battle histories, biographies, ship and aircraft guides, and novels. Institute members receive discounts on the Press's nearly 400 books in print.

For a free catalog describing Naval Institute Press books currently available, and for further information about subscribing to *Naval History* magazine or about joining the U.S. Naval Institute, please write to:

Membership & Communications Department
U.S. Naval Institute
118 Maryland Avenue
Annapolis, Maryland 21402-5035
Or call, toll-free, (800) 233-USNI.